PRAISE FOR THE MANGA GUIDE SERIES

"A fun and fairly painless lesson on what many consider to be a less-than-thrilling subject."
—SCHOOL LIBRARY JOURNAL

"This is really what a good math text should be like. . . . It presents statistics as something fun, and something enlightening."
—GOOD MATH, BAD MATH

"The most enjoyable tech book I've ever read."
—LINUX PRO MAGAZINE

"If you want to introduce a subject that kids wouldn't normally be very interested in, give it an amusing storyline and wrap it in cartoons."
—MAKE

"This is a solid book and I wish there were more like it in the IT world."
—SLASHDOT

"Great for anyone wanting an introduction or a refresher on statistics."
—PARENTING SQUAD

"A light, impressively non-oppressive read, especially considering the technical nature of its subject."
—AIN'T IT COOL NEWS

WOW!

"Makes accessible a very intimidating subject, letting the reader have fun while still delivering the goods."
—GEEKDAD, WIRED.COM

"Definitely way better than trying to comprehend a bland statistics book."
—DR. DOBB'S CODETALK

"I would have killed for these books when studying for my school exams 20 years ago."
—TIM MAUGHAN

"An awfully fun, highly educational read."
—FRAZZLEDDAD

THE MANGA GUIDE™ TO CALCULUS

THE MANGA GUIDE™ TO
CALCULUS

HIROYUKI KOJIMA
SHIN TOGAMI
BECOM CO., LTD.

THE MANGA GUIDE TO CALCULUS. Copyright © 2009 by Hiroyuki Kojima and Becom Co., Ltd

The Manga Guide to Calculus is a translation of the Japanese original, *Manga de Wakaru Bibun Sekibun*, published by Ohmsha, Ltd. of Tokyo, Japan, © 2005 by Hiroyuki Kojima and Becom Co., Ltd.

This English edition is co-published by No Starch Press and Ohmsha, Ltd.

13 12 11 10 09   1 2 3 4 5 6 7 8 9

ISBN-10: 1-59327-194-8
ISBN-13: 978-1-59327-194-7

SUSTAINABLE FORESTRY INITIATIVE
Label applies to the text stock
Certified Fiber Sourcing
www.sfiprogram.org

Publisher: William Pollock
Author: Hiroyuki Kojima
Illustrator: Shin Togami
Producer: Becom Co., Ltd.
Production Editor: Megan Dunchak
Developmental Editor: Tyler Ortman
Technical Reviewers: Whitney Ortman-Link and Erika Ward
Compositor: Riley Hoffman
Proofreader: Cristina Chan
Indexer: Sarah Schott

For information on book distributors or translations, please contact No Starch Press, Inc. directly:

No Starch Press, Inc.
555 De Haro Street, Suite 250, San Francisco, CA 94107
phone: 415.863.9900; fax: 415.863.9950; info@nostarch.com; http://www.nostarch.com/

Library of Congress Cataloging-in-Publication Data

Kojima, Hiroyuki, 1958-
  [Manga de wakaru bibun sekibun. English]
  The manga guide to calculus / Hiroyuki Kojima, Shin Togami, and Becom Co., Ltd.
       p. cm.
  Includes index.
  ISBN-13: 978-1-59327-194-7
  ISBN-10: 1-59327-194-8
  1. Calculus--Comic books, strips, etc. I. Togami, Shin. II. Becom Co. III. Title.
  QA300.K57513 2009
  515--dc22
                                        2008050189

CONTENTS

PREFACE

There are some things that only manga can do.

You have just picked up and opened this book. You must be one of the following types of people.

The first type is someone who just loves manga and thinks, "Calculus illustrated with manga? Awesome!" If you are this type of person, you should immediately take this book to the cashier—you won't regret it. This is a very enjoyable manga title. It's no surprise—Shin Togami, a popular manga artist, drew the manga, and Becom Ltd., a real manga production company, wrote the scenario.

"But, manga that teaches about math has never been very enjoyable," you may argue. That's true. In fact, when an editor at Ohmsha asked me to write this book, I nearly turned down the opportunity. Many of the so-called "manga for education" books are quite disappointing. They may have lots of illustrations and large pictures, but they aren't really manga. But after seeing a sample from Ohmsha (it was *The Manga Guide to Statistics*), I totally changed my mind. Unlike many such manga guides, the sample was enjoyable enough to actually read. The editor told me that my book would be like this, too—so I accepted his offer. In fact, I have often thought that I might be able to teach mathematics better by using manga, so I saw this as a good opportunity to put the idea into practice. I guarantee you that the bigger manga freak you are, the more you will enjoy this book. So, what are you waiting for? Take it up to the cashier and buy it already!

Now, the second type of person is someone who picked up this book thinking, "Although I am terrible at and/or allergic to calculus, manga may help me understand it." If you are this type of person, then this is also the book for you. It is equipped with various rehabilitation methods for those who have been hurt by calculus in the past. Not only does it explain calculus using manga, but the way it explains calculus is fundamentally different from the method used in conventional textbooks. First, the book repeatedly presents the notion of what calculus really does. You will never understand this through the teaching methods that stick to *limits* (or ε-δ logic). Unless you have a clear image of what calculus really does and why it is useful in the world, you will never really understand or use it freely. You will simply fall into a miserable state of memorizing formulas and rules. This book explains all the formulas based on the concept of the *first-order approximation*, helping you to visualize the meaning of formulas and understand them easily. Because of this unique teaching method, you can quickly and easily proceed from differentiation to integration. Furthermore, I have adopted an original method, which is not described in ordinary textbooks, of explaining the differentiation and integration of trigonometric and exponential functions—usually, this is all Greek to many people even after repeated explanations. This book also goes further in depth than existing manga books on calculus do, explaining even Taylor expansions and partial differentiation. Finally, I have invited three regular customers of calculus—physics, statistics, and economics—to be part of this book and presented many examples to show that calculus is truly practical. With all of these devices, you will come to view calculus not as a hardship, but as a useful tool.

I would like to emphasize again: All of this has been made possible because of manga. Why can you gain more information by reading a manga book than by reading a novel? It is because manga is visual data presented as animation. Calculus is a branch of mathematics that describes dynamic phenomena—thus, calculus is a perfect concept to teach with manga. Now, turn the pages and enjoy a beautiful integration of manga and mathematics.

HIROYUKI KOJIMA
NOVEMBER 2005

NOTE: *For ease of understanding, some figures are not drawn to scale.*

PROLOGUE: WHAT IS A FUNCTION?

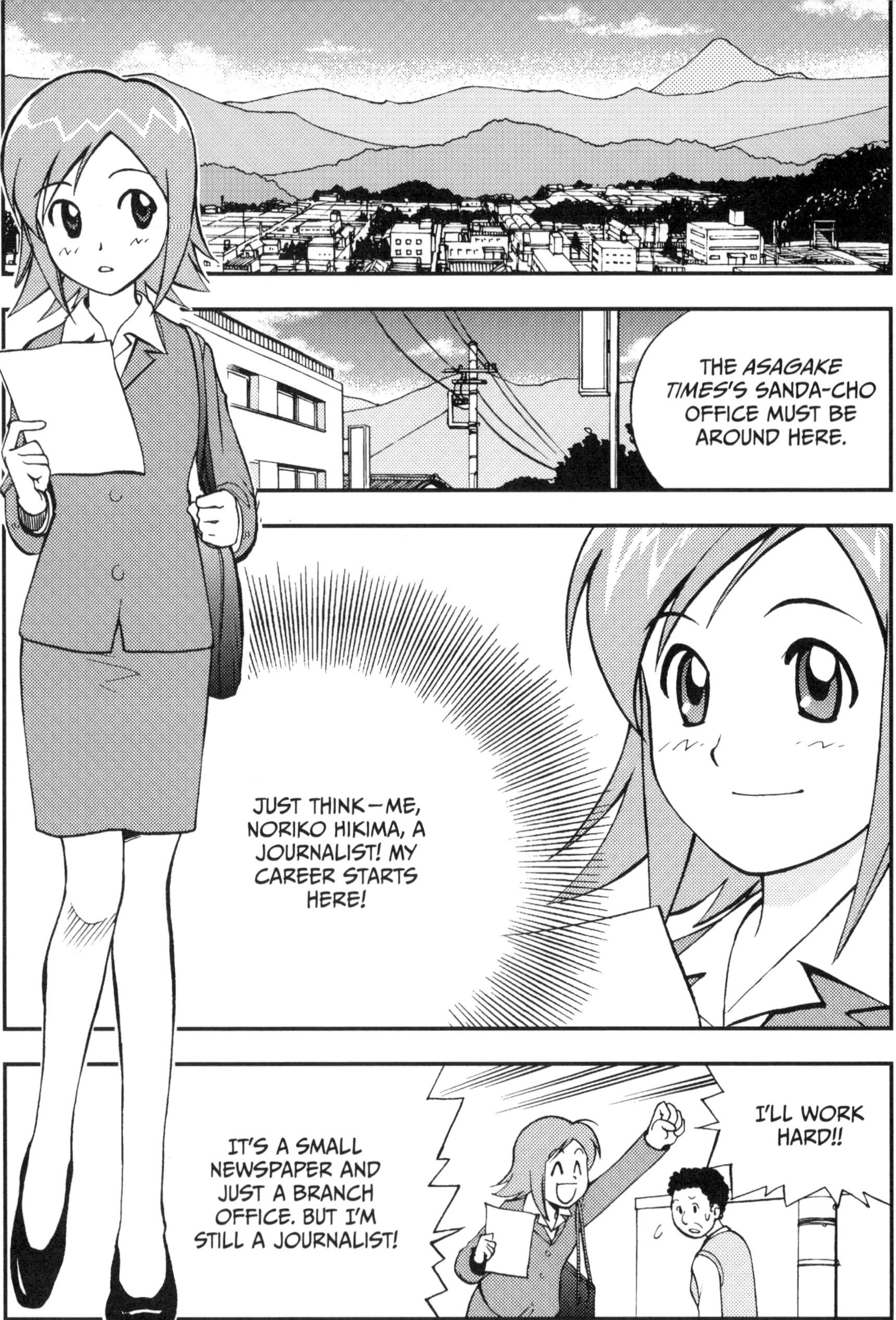
THE *ASAGAKE TIMES*'S SANDA-CHO OFFICE MUST BE AROUND HERE.
JUST THINK—ME, NORIKO HIKIMA, A JOURNALIST! MY CAREER STARTS HERE!
IT'S A SMALL NEWSPAPER AND JUST A BRANCH OFFICE. BUT I'M STILL A JOURNALIST!
I'LL WORK HARD!!

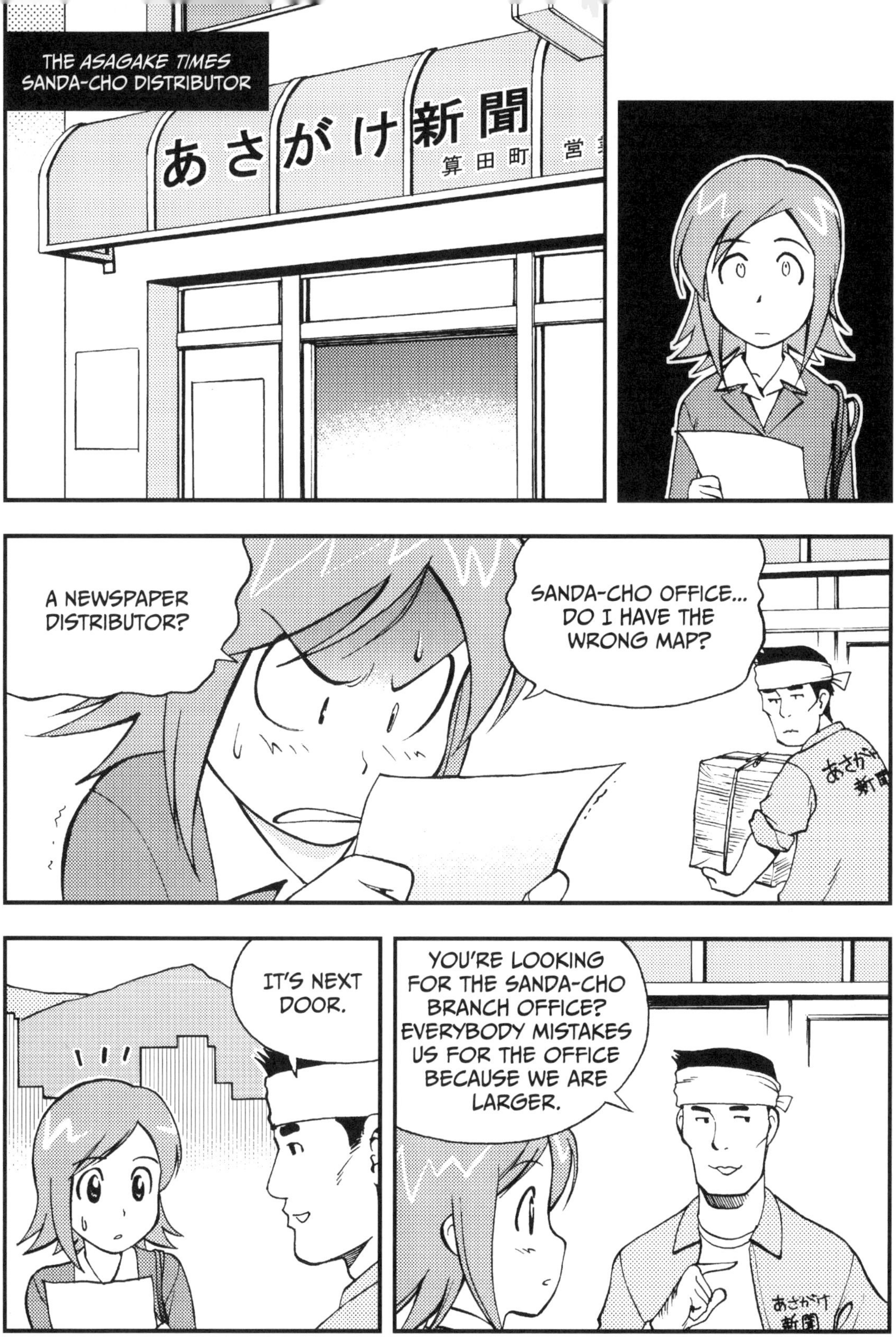
THE ASAGAKE TIMES SANDA-CHO DISTRIBUTOR
あさがけ新聞
算田町 営
A NEWSPAPER DISTRIBUTOR?
SANDA-CHO OFFICE... DO I HAVE THE WRONG MAP?
IT'S NEXT DOOR.
YOU'RE LOOKING FOR THE SANDA-CHO BRANCH OFFICE? EVERYBODY MISTAKES US FOR THE OFFICE BECAUSE WE ARE LARGER.
あさがけ新聞

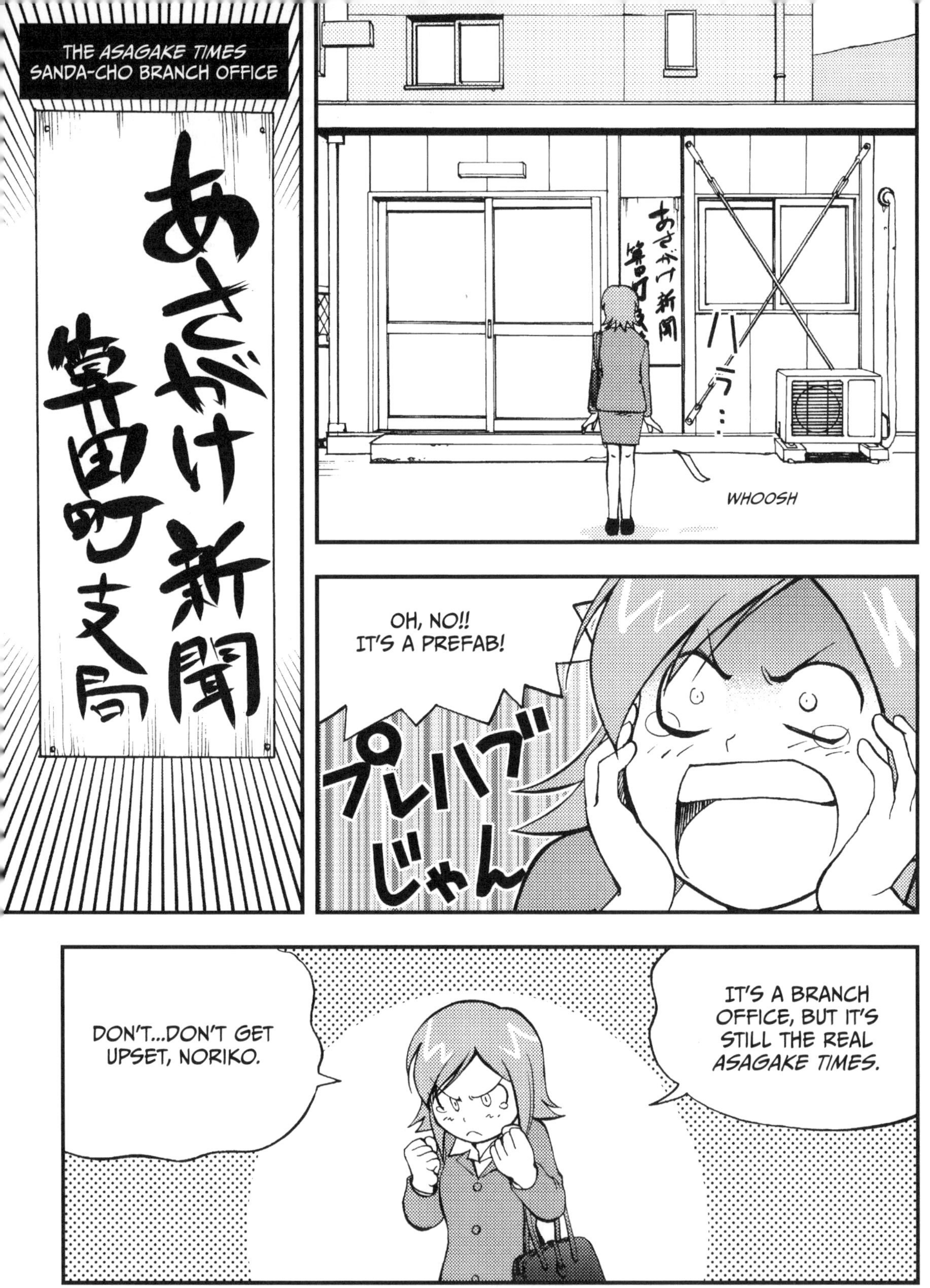
THE ASAGAKE TIMES
SANDA-CHO BRANCH OFFICE
あさがけ新聞
算田町支局
あさがけ新聞
ハラ…
WHOOSH
OH, NO!! IT'S A PREFAB!
プレハブじゃん
DON'T...DON'T GET UPSET, NORIKO.
IT'S A BRANCH OFFICE, BUT IT'S STILL THE REAL ASAGAKE TIMES.

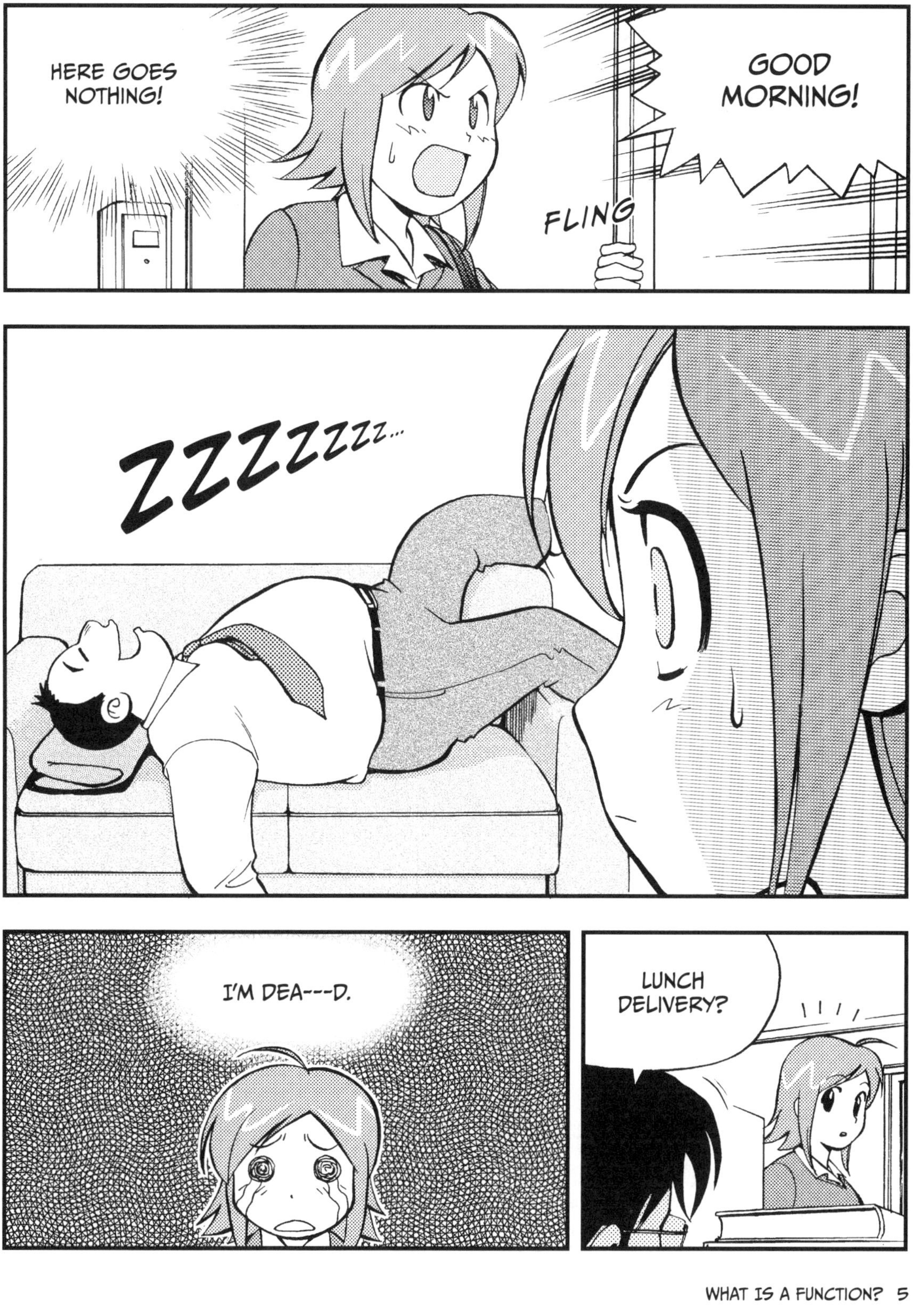
HERE GOES NOTHING!
GOOD MORNING!
FLING
ZZZZzzz...
I'M DEA---D.
LUNCH DELIVERY?

WILL YOU LEAVE IT, PLEASE?

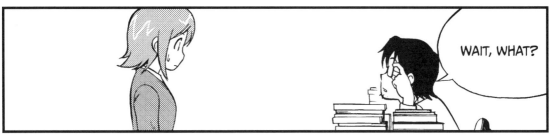

WAIT, WHAT?

OH, YOU HAVE BEEN ASSIGNED HERE TODAY.

I'M NORIKO HIKIMA.

LONG TRIP, WASN'T IT? I'M KAKERU SEKI, THE HEAD OF THIS OFFICE.

THE BIG GUY THERE IS FUTOSHI MASUI, MY ONLY SOLDIER.

JUST TWO OF THEM...

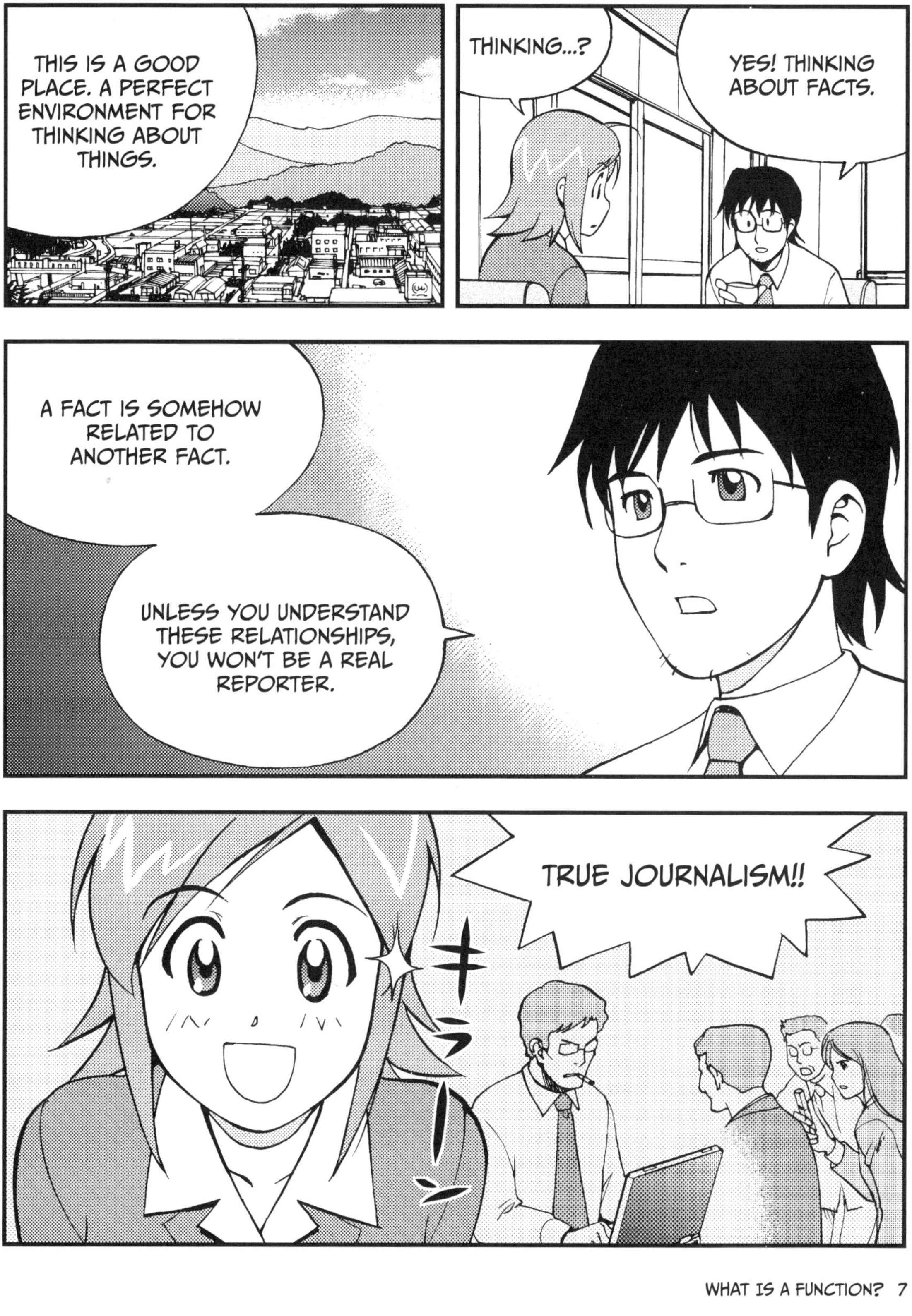
THIS IS A GOOD PLACE. A PERFECT ENVIRONMENT FOR THINKING ABOUT THINGS.
THINKING...?
YES! THINKING ABOUT FACTS.
A FACT IS SOMEHOW RELATED TO ANOTHER FACT.
UNLESS YOU UNDERSTAND THESE RELATIONSHIPS, YOU WON'T BE A REAL REPORTER.
TRUE JOURNALISM!!

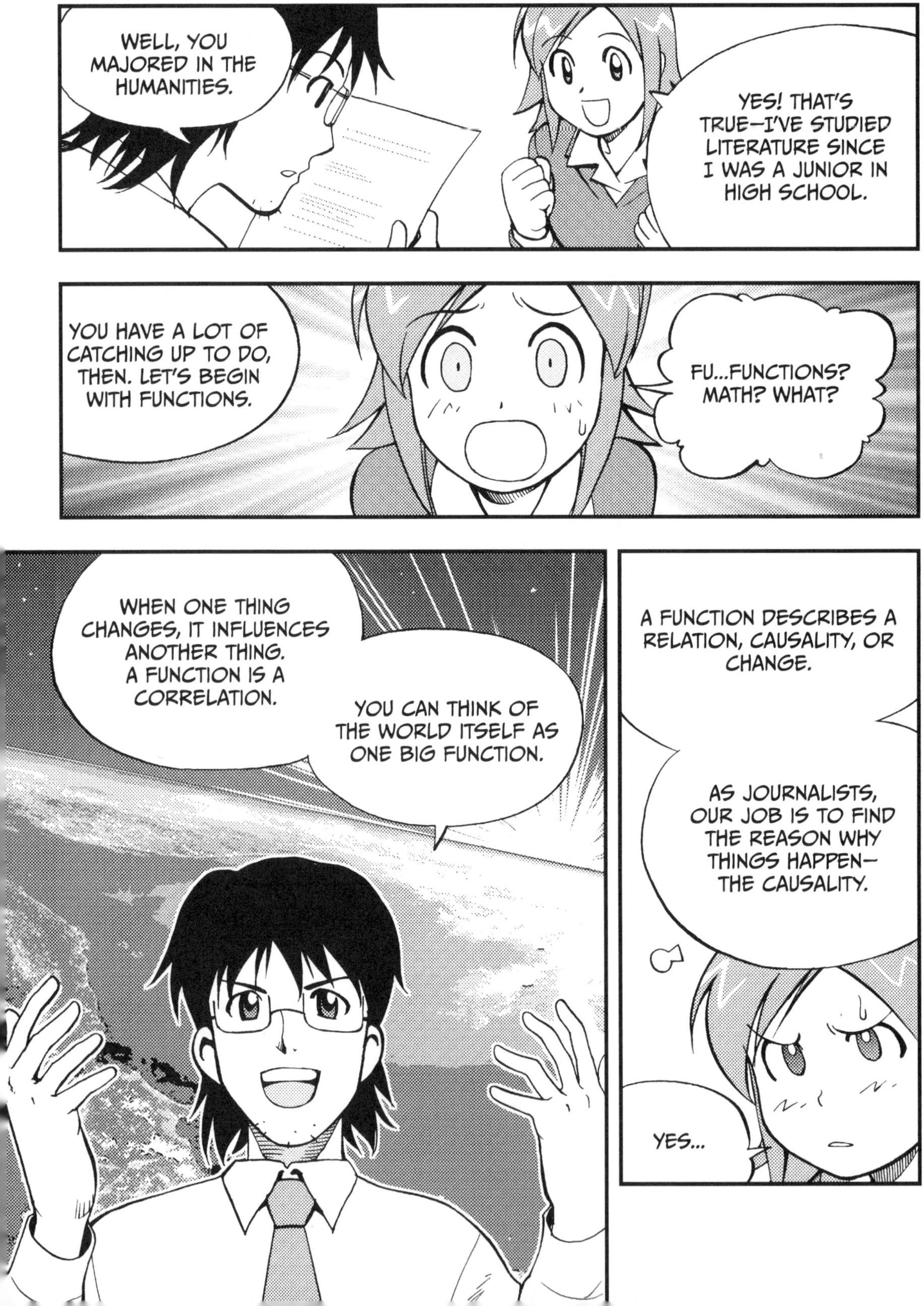
WELL, YOU MAJORED IN THE HUMANITIES.
YES! THAT'S TRUE—I'VE STUDIED LITERATURE SINCE I WAS A JUNIOR IN HIGH SCHOOL.
YOU HAVE A LOT OF CATCHING UP TO DO, THEN. LET'S BEGIN WITH FUNCTIONS.
FU...FUNCTIONS? MATH? WHAT?
WHEN ONE THING CHANGES, IT INFLUENCES ANOTHER THING. A FUNCTION IS A CORRELATION.
YOU CAN THINK OF THE WORLD ITSELF AS ONE BIG FUNCTION.
A FUNCTION DESCRIBES A RELATION, CAUSALITY, OR CHANGE.
AS JOURNALISTS, OUR JOB IS TO FIND THE REASON WHY THINGS HAPPEN—THE CAUSALITY.
YES...

DID YOU KNOW A FUNCTION IS OFTEN EXPRESSED AS $y = f(x)$?

NOPE!!

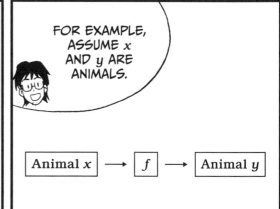

FOR EXAMPLE, ASSUME x AND y ARE ANIMALS.

$$\boxed{\text{Animal } x} \longrightarrow \boxed{f} \longrightarrow \boxed{\text{Animal } y}$$

ASSUME x IS A FROG. IF YOU PUT THE FROG INTO BOX f AND CONVERT IT, TADPOLE y COMES OUT OF THE BOX.

BUT, UH... WHAT IS f?

THE f STANDS FOR *FUNCTION*, NATURALLY.

f IS USED TO SHOW THAT THE VARIABLE y HAS A PARTICULAR RELATIONSHIP TO x.

AND WE CAN ACTUALLY USE ANY LETTER INSTEAD OF f.

function

IN THIS CASE, f EXPRESSES THE RELATIONSHIP OR RULE BETWEEN "A PARENT" AND "AN OFFSPRING."

A PARENT

AN OFFSPRING

AND THIS RELATIONSHIP IS TRUE OF ALMOST ANY ANIMAL. IF x IS A BIRD, y IS A CHICK.

OKAY! NOW LOOK AT THIS.

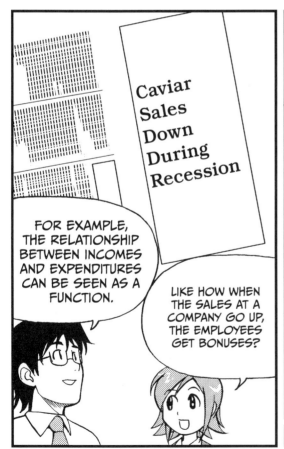

Caviar Sales Down During Recession

FOR EXAMPLE, THE RELATIONSHIP BETWEEN INCOMES AND EXPENDITURES CAN BE SEEN AS A FUNCTION.

LIKE HOW WHEN THE SALES AT A COMPANY GO UP, THE EMPLOYEES GET BONUSES?

X-43 Scram Jet Reaches Mach 9.6 — New World Record

THE SPEED OF SOUND AND THE TEMPERATURE CAN ALSO BE EXPRESSED AS A FUNCTION. WHEN THE TEMPERATURE GOES UP BY 1°C, THE SPEED OF SOUND GOES UP BY 0.6 METERS/SECOND.

AND THE TEMPERATURE IN THE MOUNTAINS GOES DOWN BY ABOUT 0.5°C EACH TIME YOU GO UP 100 METERS, DOESN'T IT?

YOO-HOO!

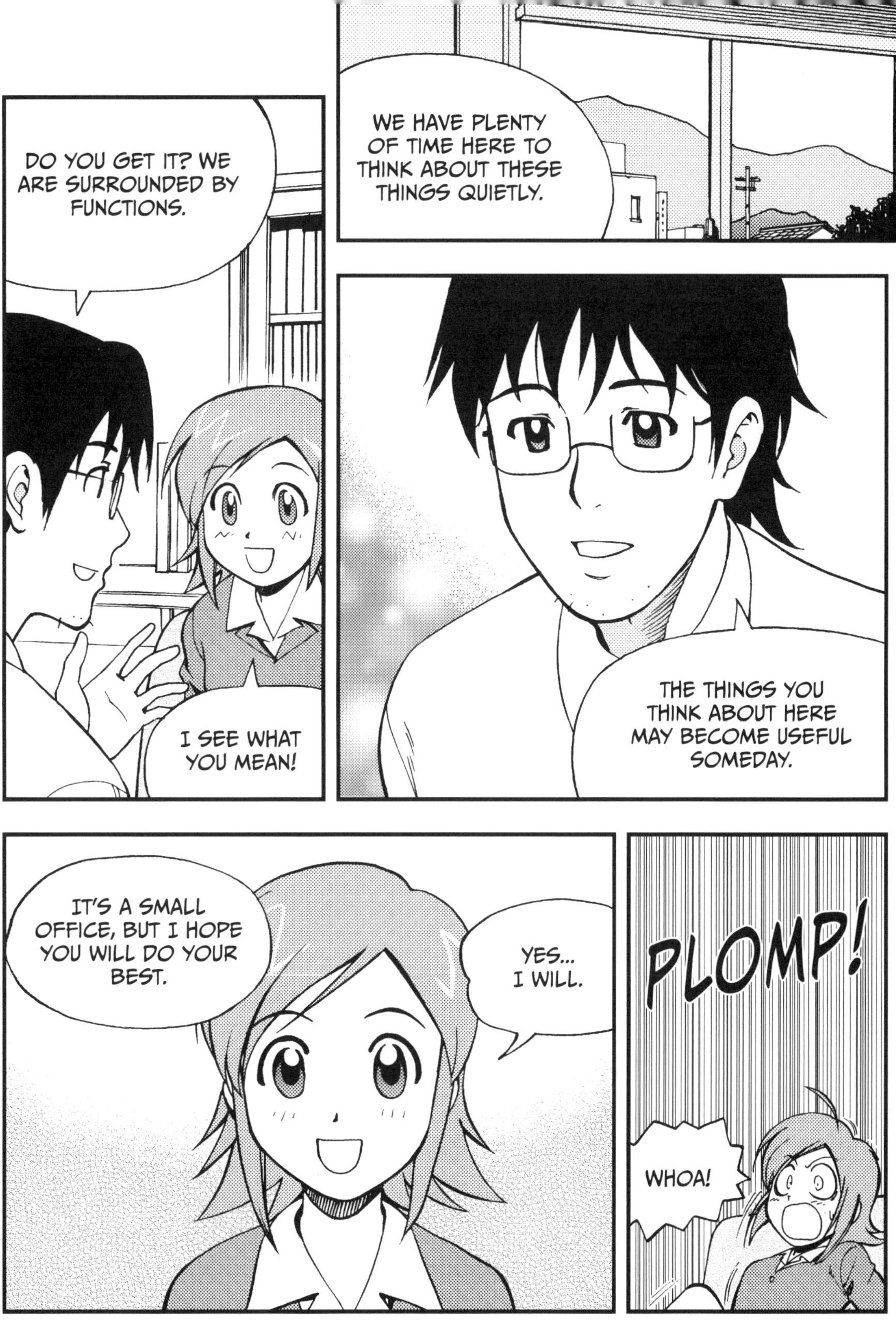
DO YOU GET IT? WE ARE SURROUNDED BY FUNCTIONS.
I SEE WHAT YOU MEAN!
WE HAVE PLENTY OF TIME HERE TO THINK ABOUT THESE THINGS QUIETLY.
THE THINGS YOU THINK ABOUT HERE MAY BECOME USEFUL SOMEDAY.
IT'S A SMALL OFFICE, BUT I HOPE YOU WILL DO YOUR BEST.
YES... I WILL.
PLOMP!
WHOA!

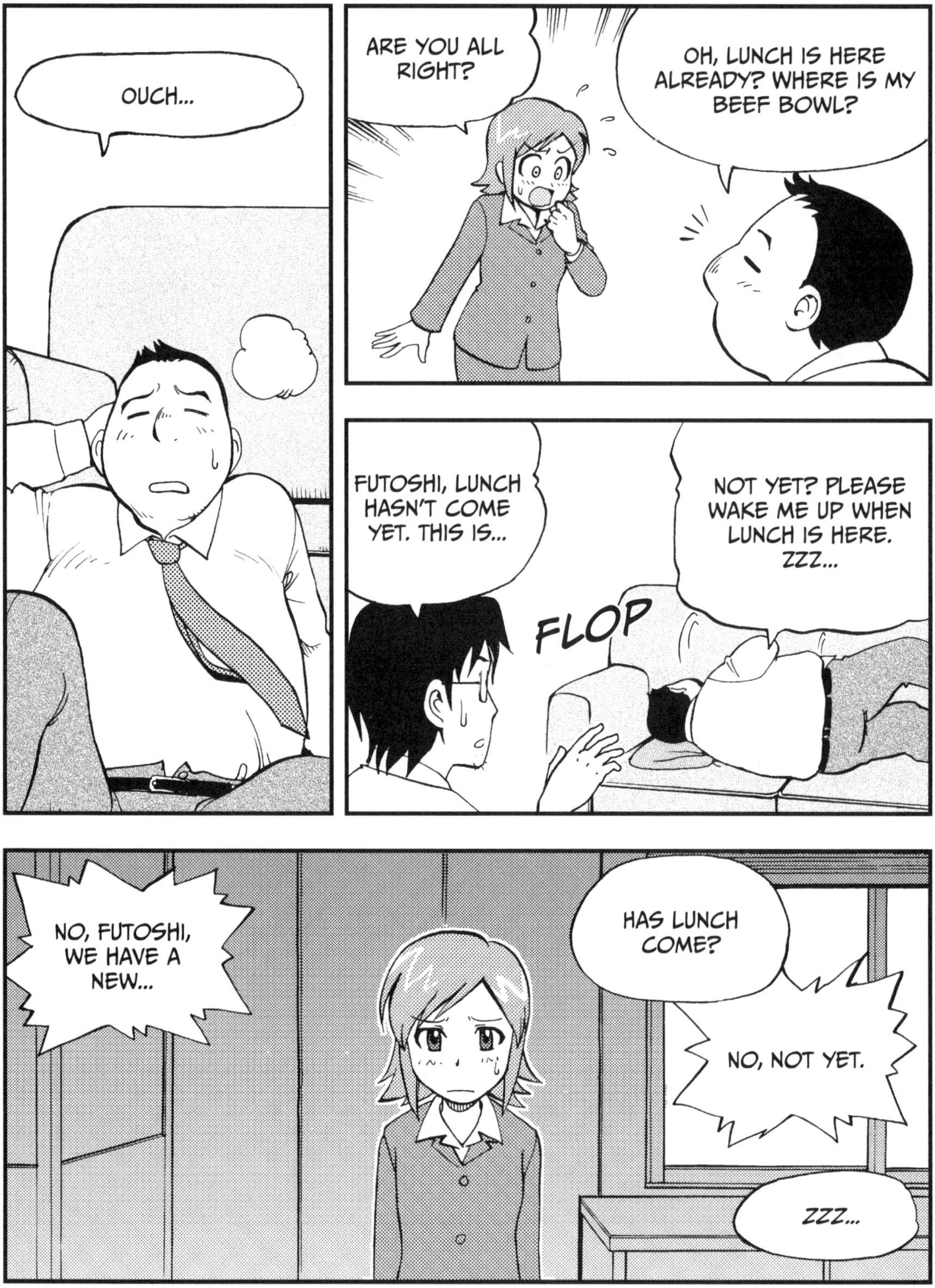
OUCH...
ARE YOU ALL RIGHT?
OH, LUNCH IS HERE ALREADY? WHERE IS MY BEEF BOWL?
FUTOSHI, LUNCH HASN'T COME YET. THIS IS...
NOT YET? PLEASE WAKE ME UP WHEN LUNCH IS HERE. ZZZ...
FLOP
NO, FUTOSHI, WE HAVE A NEW...
HAS LUNCH COME?
NO, NOT YET.
ZZZ...

TABLE 1: CHARACTERISTICS OF FUNCTIONS

SUBJECT	CALCULATION	GRAPH
Causality	The frequency of a cricket's chirp is determined by temperature. We can express the relationship between y chirps per minute of a cricket at temperature $x°$C approximately as $y = g(x) = 7x - 30$ $x = 27° \quad 7 \times 27 - 30$ The result is 159 chirps a minute.	When we graph these functions, the result is a straight line. That's why we call them linear functions. 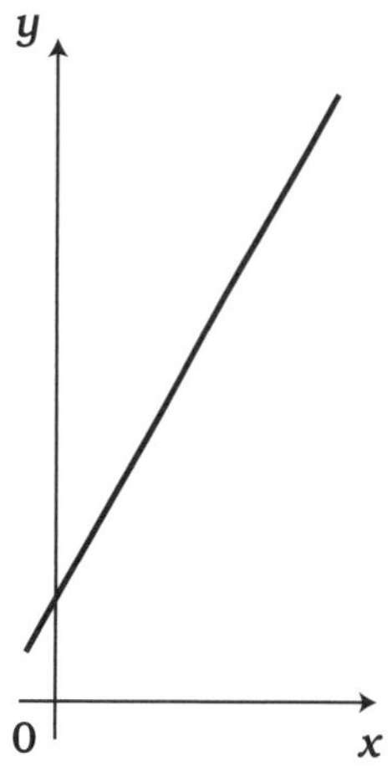
Changes	The speed of sound y in meters per second (m/s) in the air at $x°$C is expressed as $y = v(x) = 0.6x + 331$ At 15°C, $y = v(15) = 0.6 \times 15 + 331 = 340$ m/s At –5°C, $y = v(-5) = 0.6 \times (-5) + 331 = 328$ m/s	
Unit Conversion	Converting x degrees Fahrenheit (°F) into y degrees Celsius (°C) $y = f(x) = \dfrac{5}{9}(x - 32)$ So now we know 50°F is equivalent to $\dfrac{5}{9}(50 - 32) = 10°\,$C	
	Computers store numbers using a binary system (1s and 0s). A binary number with x bits (or binary digits) has the potential to store y numbers. $y = b(x) = 2^x$ (This is described in more detail on page 131.)	The graph is an exponential function. 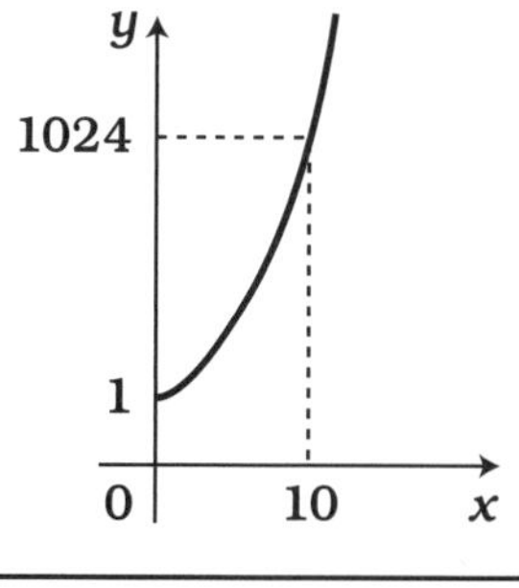

THE GRAPHS OF SOME FUNCTIONS CANNOT BE EXPRESSED BY STRAIGHT LINES OR CURVES WITH A REGULAR SHAPE.

The stock price P of company A in month x in 2009 is
$$y = P(x)$$

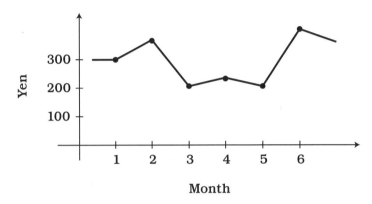

$P(x)$ cannot be expressed by a known function, but it is still a function.

If you could find a way to predict $P(7)$, the stock price in July, you could make a big profit.

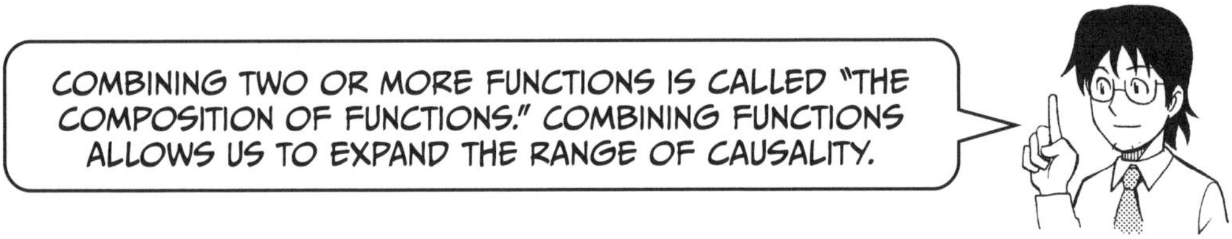

COMBINING TWO OR MORE FUNCTIONS IS CALLED "THE COMPOSITION OF FUNCTIONS." COMBINING FUNCTIONS ALLOWS US TO EXPAND THE RANGE OF CAUSALITY.

A composite function of f and g

$$x \longrightarrow \boxed{f} \longrightarrow f(x) \longrightarrow \boxed{g} \longrightarrow g(f(x))$$

EXERCISE

1. Find an equation that expresses the frequency of z chirps/minute of a cricket at $x°$F.

1
LET'S DIFFERENTIATE A FUNCTION!

APPROXIMATING WITH FUNCTIONS

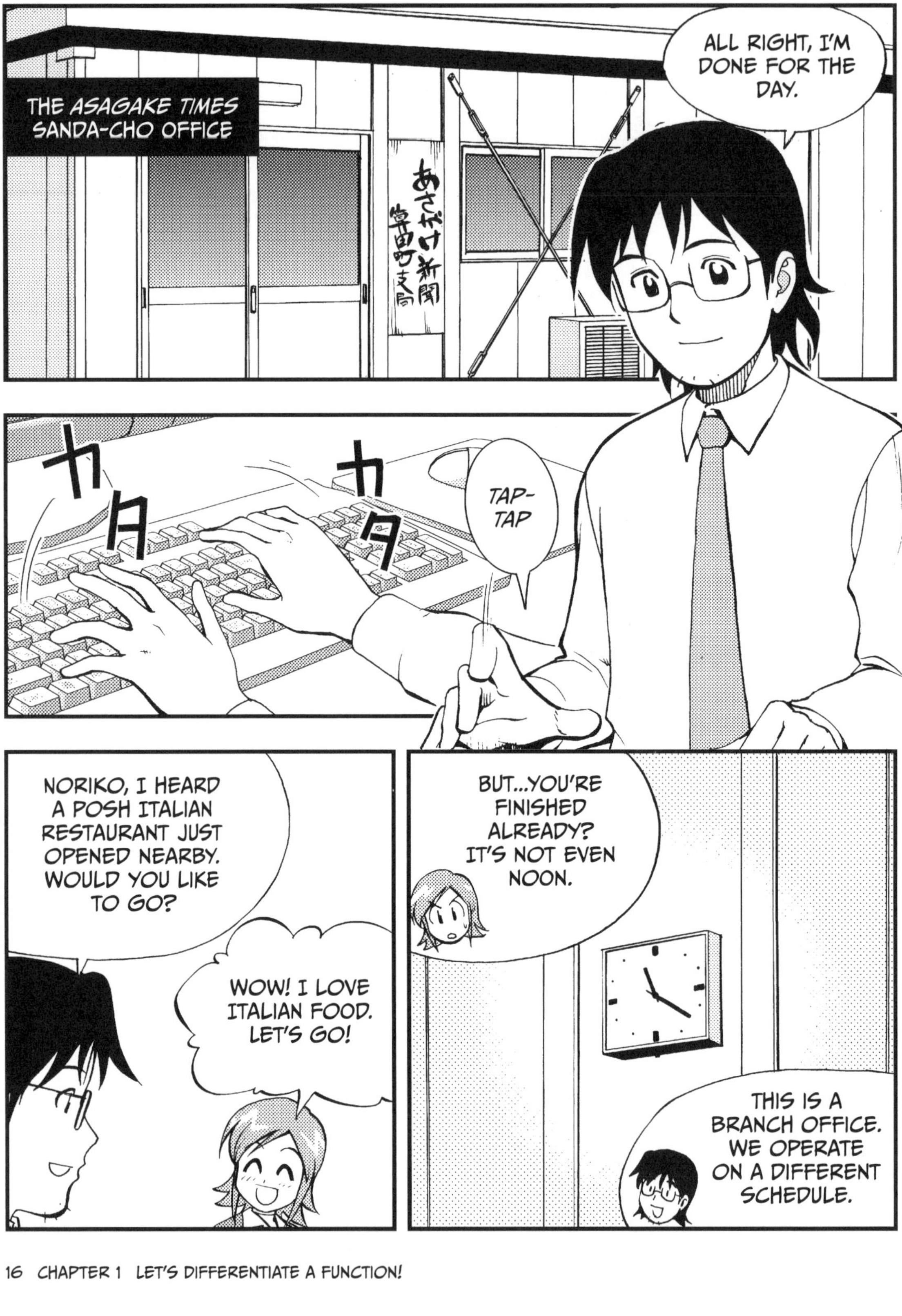
THE *ASAGAKE TIMES* SANDA-CHO OFFICE
ALL RIGHT, I'M DONE FOR THE DAY.
カタ
カタ
TAP-TAP
NORIKO, I HEARD A POSH ITALIAN RESTAURANT JUST OPENED NEARBY. WOULD YOU LIKE TO GO?
WOW! I LOVE ITALIAN FOOD. LET'S GO!
BUT...YOU'RE FINISHED ALREADY? IT'S NOT EVEN NOON.
THIS IS A BRANCH OFFICE. WE OPERATE ON A DIFFERENT SCHEDULE.

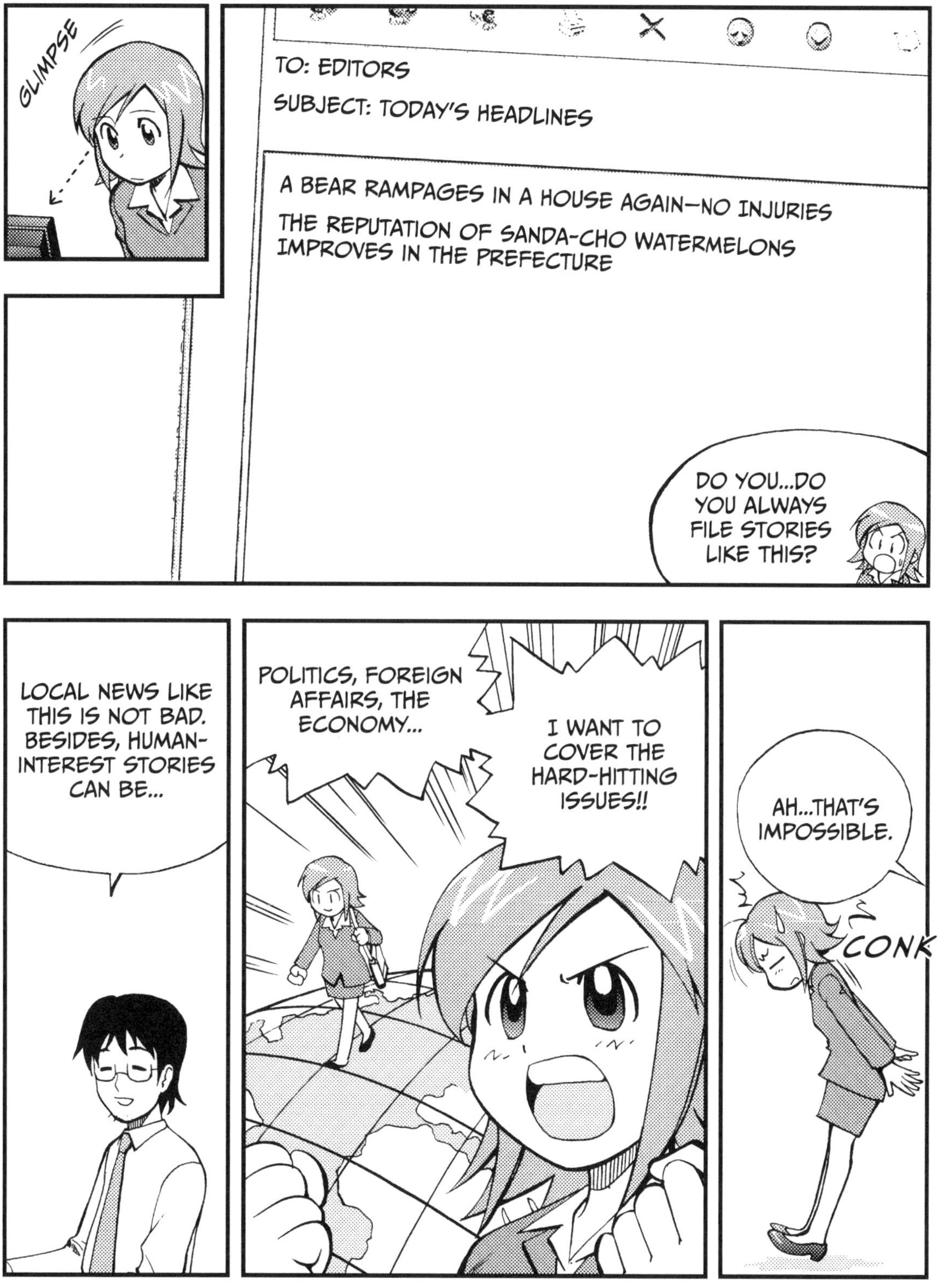
GLIMPSE
TO: EDITORS
SUBJECT: TODAY'S HEADLINES
A BEAR RAMPAGES IN A HOUSE AGAIN—NO INJURIES
THE REPUTATION OF SANDA-CHO WATERMELONS IMPROVES IN THE PREFECTURE
DO YOU...DO YOU ALWAYS FILE STORIES LIKE THIS?
LOCAL NEWS LIKE THIS IS NOT BAD. BESIDES, HUMAN-INTEREST STORIES CAN BE...
POLITICS, FOREIGN AFFAIRS, THE ECONOMY...
I WANT TO COVER THE HARD-HITTING ISSUES!!
AH...THAT'S IMPOSSIBLE.
CONK

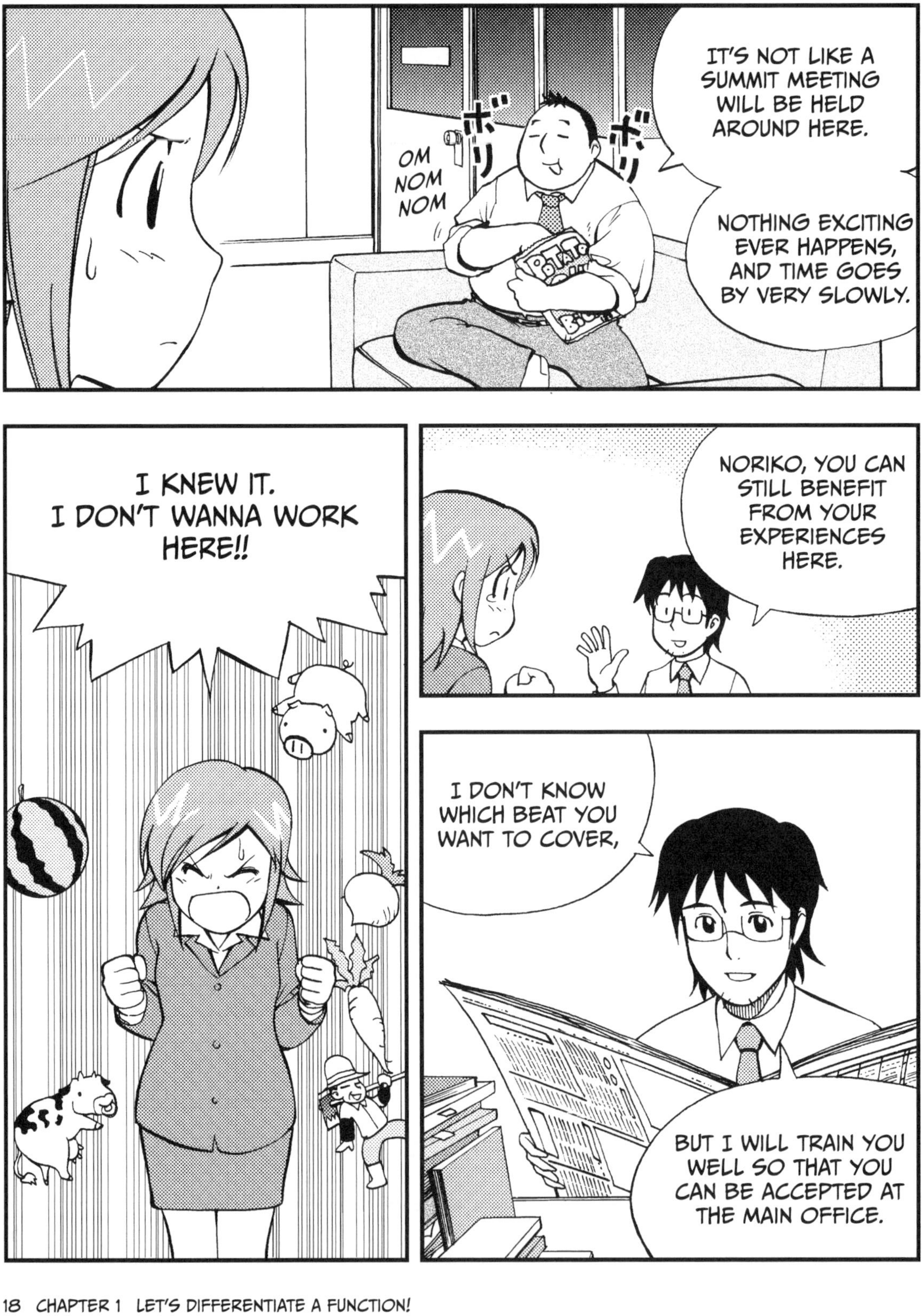

ボリ
OM NOM NOM
ボリ
IT'S NOT LIKE A SUMMIT MEETING WILL BE HELD AROUND HERE.
NOTHING EXCITING EVER HAPPENS, AND TIME GOES BY VERY SLOWLY.
I KNEW IT. I DON'T WANNA WORK HERE!!
NORIKO, YOU CAN STILL BENEFIT FROM YOUR EXPERIENCES HERE.
I DON'T KNOW WHICH BEAT YOU WANT TO COVER,
BUT I WILL TRAIN YOU WELL SO THAT YOU CAN BE ACCEPTED AT THE MAIN OFFICE.

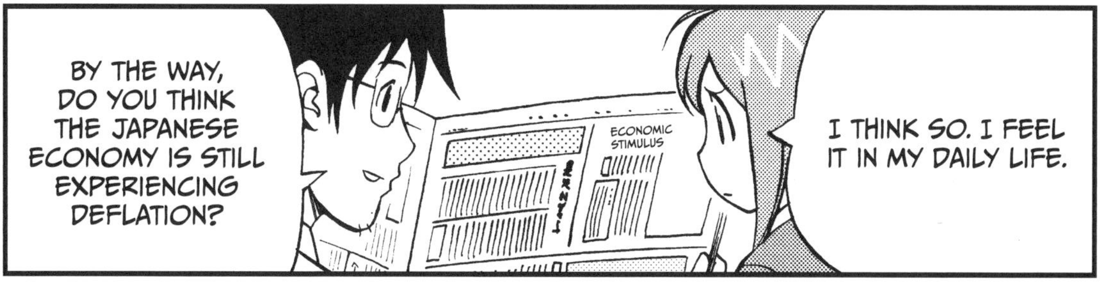
BY THE WAY, DO YOU THINK THE JAPANESE ECONOMY IS STILL EXPERIENCING DEFLATION?
ECONOMIC STIMULUS
I THINK SO. I FEEL IT IN MY DAILY LIFE.

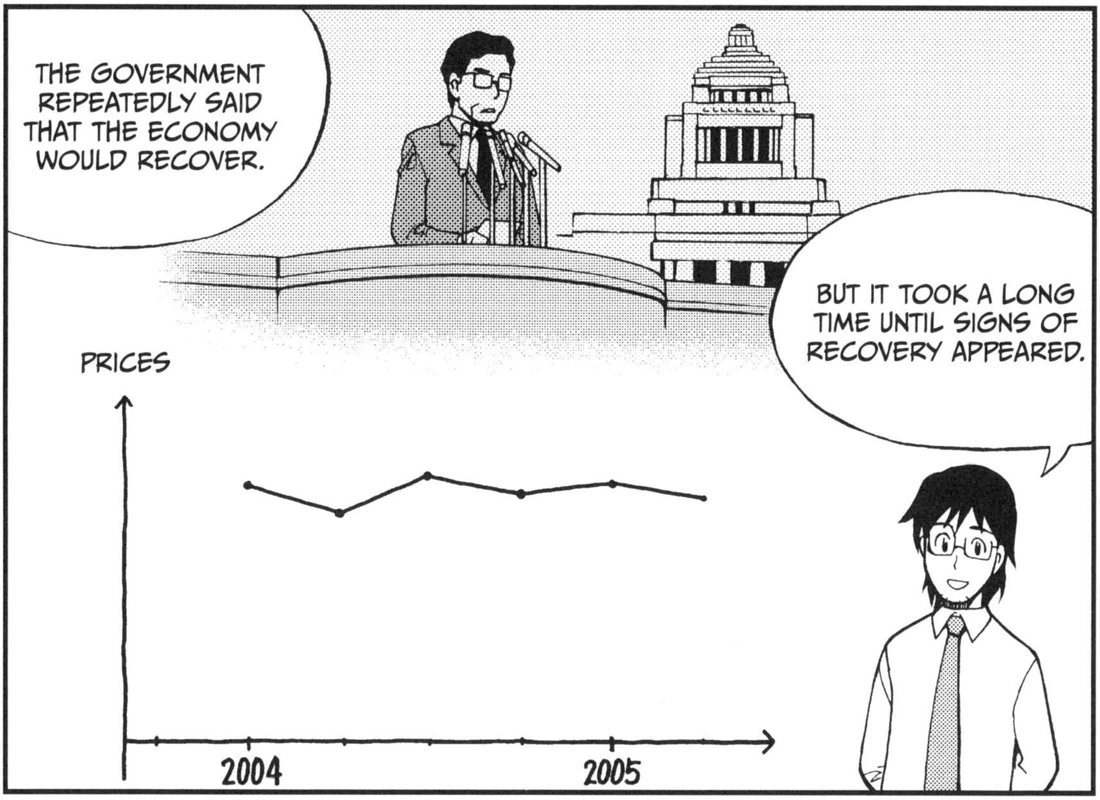
THE GOVERNMENT REPEATEDLY SAID THAT THE ECONOMY WOULD RECOVER.
PRICES
2004
2005
BUT IT TOOK A LONG TIME UNTIL SIGNS OF RECOVERY APPEARED.

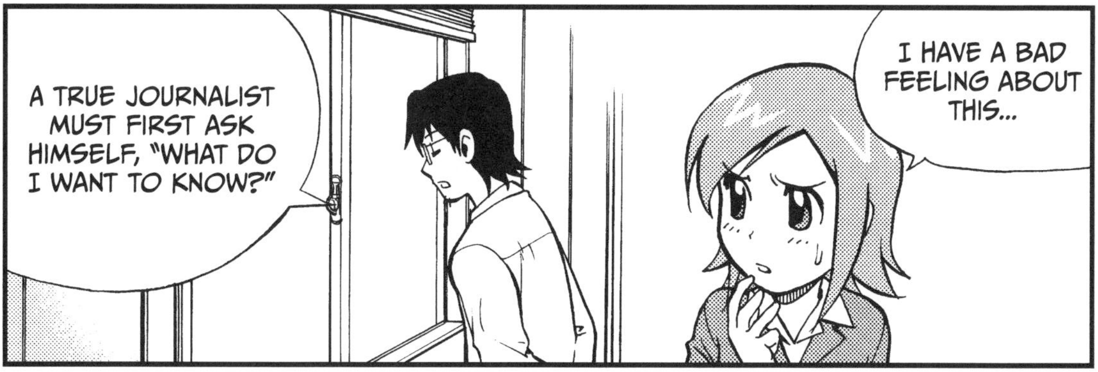
A TRUE JOURNALIST MUST FIRST ASK HIMSELF, "WHAT DO I WANT TO KNOW?"
I HAVE A BAD FEELING ABOUT THIS...

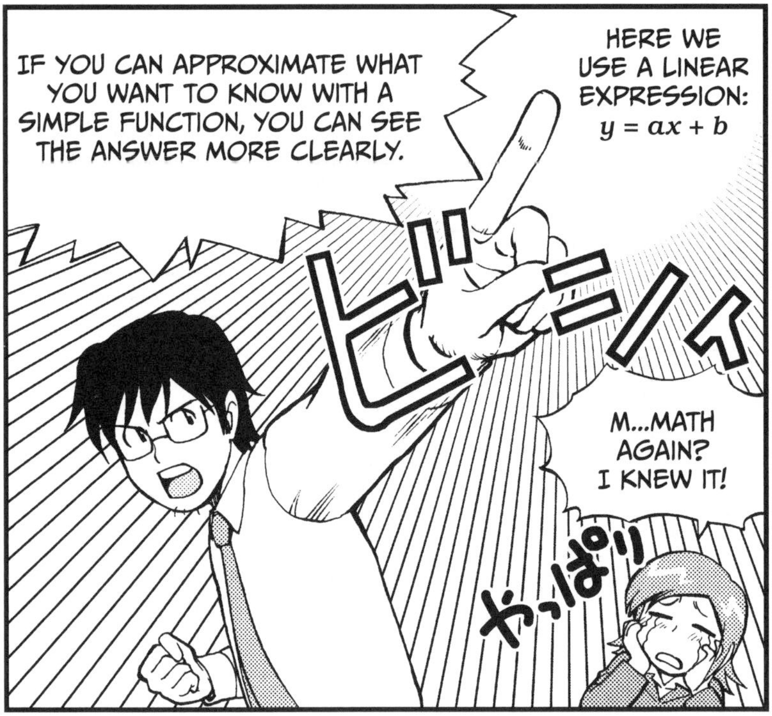
IF YOU CAN APPROXIMATE WHAT YOU WANT TO KNOW WITH A SIMPLE FUNCTION, YOU CAN SEE THE ANSWER MORE CLEARLY.
HERE WE USE A LINEAR EXPRESSION: $y = ax + b$
ビシィ
M...MATH AGAIN? I KNEW IT!
やっぱり

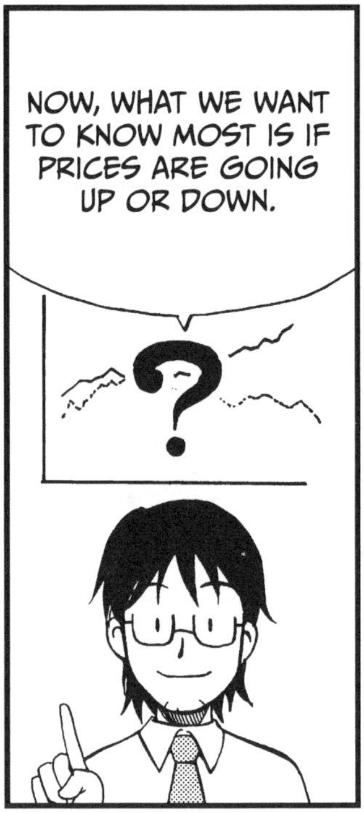
NOW, WHAT WE WANT TO KNOW MOST IS IF PRICES ARE GOING UP OR DOWN.
?

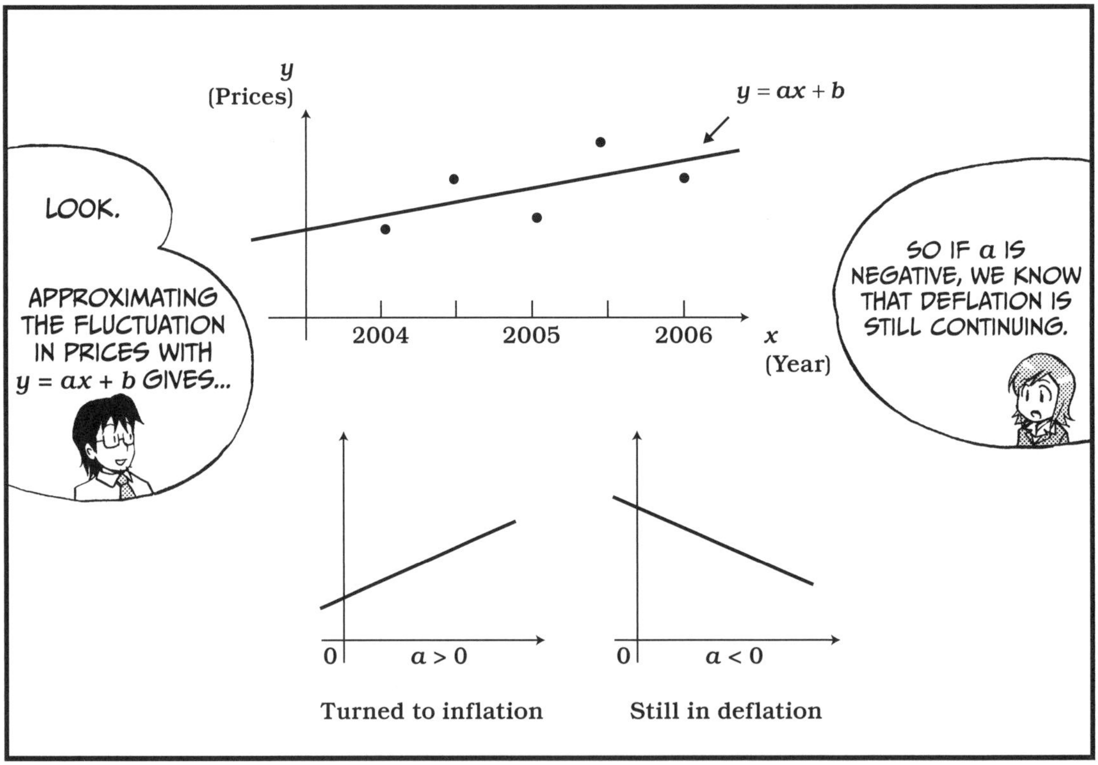
LOOK.
APPROXIMATING THE FLUCTUATION IN PRICES WITH $y = ax + b$ GIVES...
y (Prices)
$y = ax + b$
2004
2005
2006
x (Year)
SO IF a IS NEGATIVE, WE KNOW THAT DEFLATION IS STILL CONTINUING.
0
$a > 0$
Turned to inflation
0
$a < 0$
Still in deflation

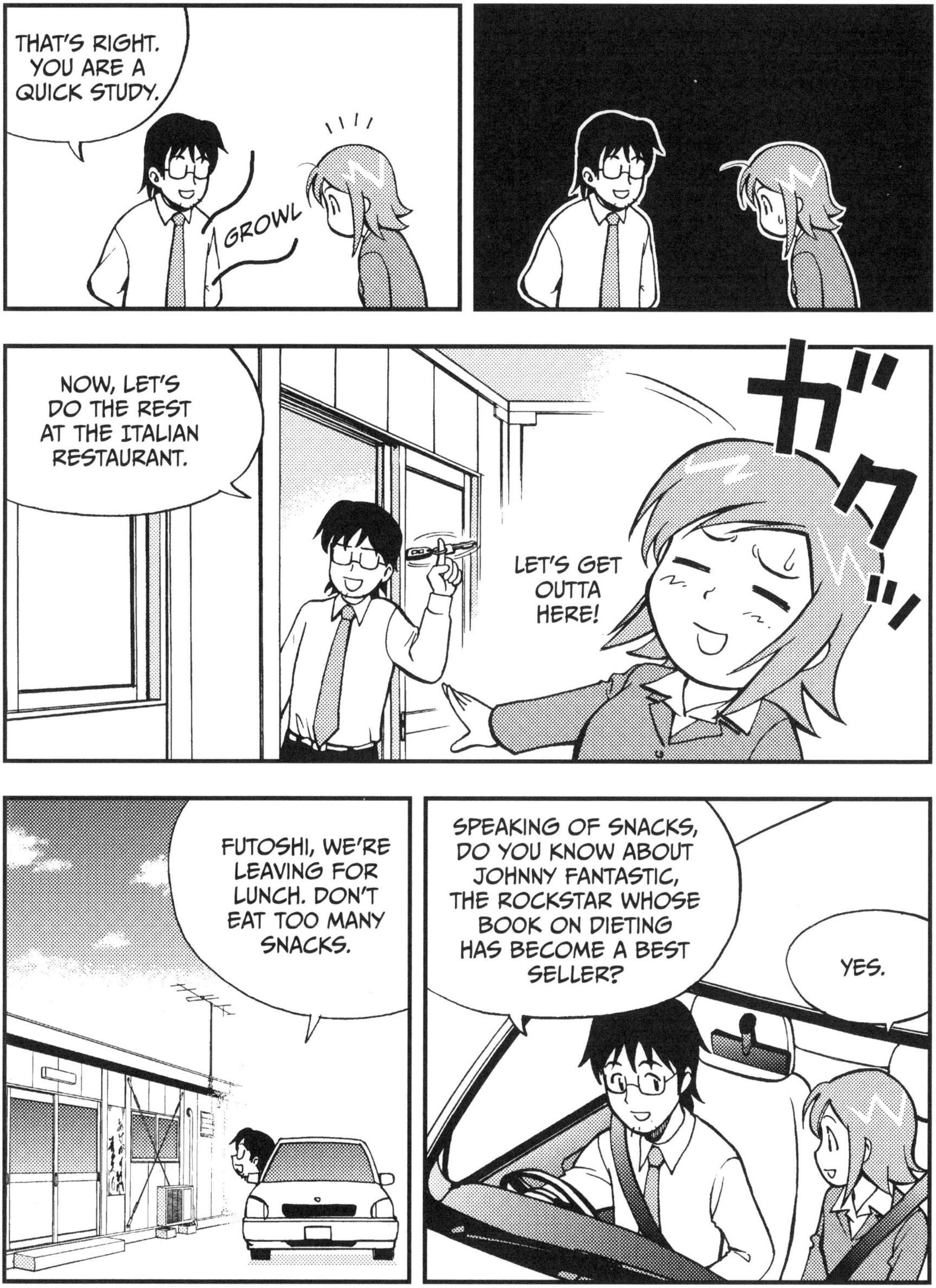
THAT'S RIGHT. YOU ARE A QUICK STUDY.
GROWL
NOW, LET'S DO THE REST AT THE ITALIAN RESTAURANT.
ガクッ
LET'S GET OUTTA HERE!
FUTOSHI, WE'RE LEAVING FOR LUNCH. DON'T EAT TOO MANY SNACKS.
SPEAKING OF SNACKS, DO YOU KNOW ABOUT JOHNNY FANTASTIC, THE ROCKSTAR WHOSE BOOK ON DIETING HAS BECOME A BEST SELLER?
YES.

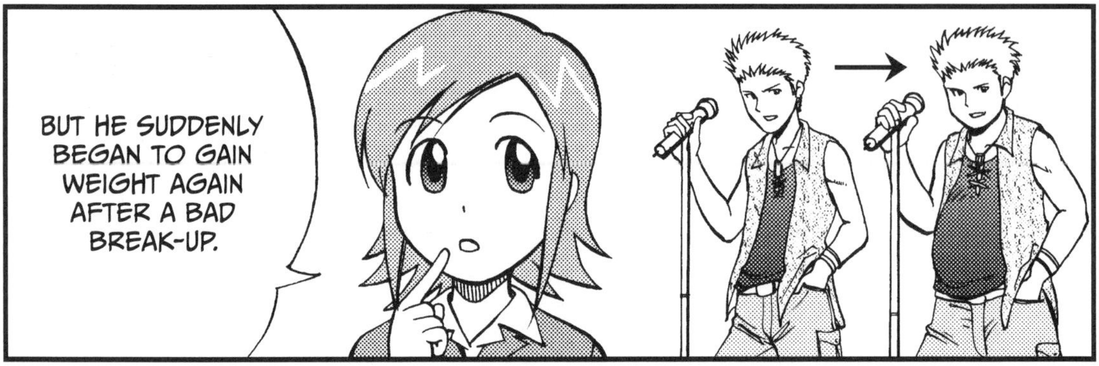
BUT HE SUDDENLY BEGAN TO GAIN WEIGHT AGAIN AFTER A BAD BREAK-UP.

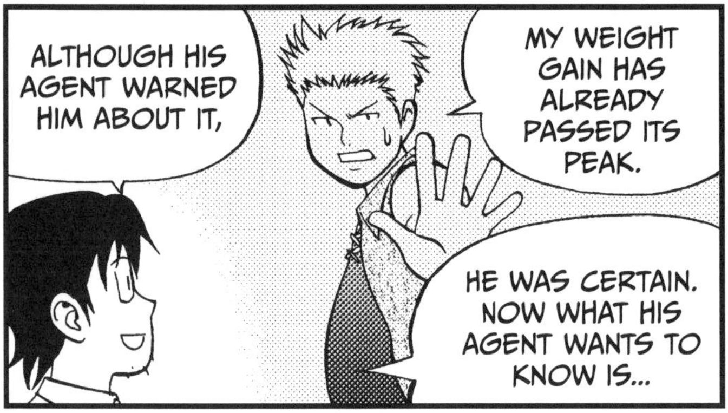
ALTHOUGH HIS AGENT WARNED HIM ABOUT IT,
MY WEIGHT GAIN HAS ALREADY PASSED ITS PEAK.
HE WAS CERTAIN. NOW WHAT HIS AGENT WANTS TO KNOW IS...

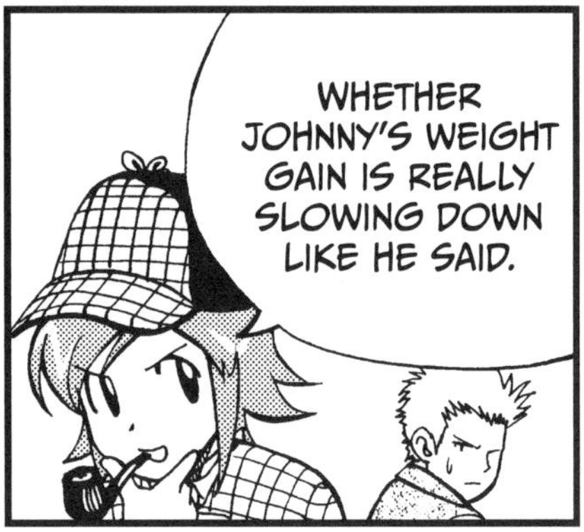
WHETHER JOHNNY'S WEIGHT GAIN IS REALLY SLOWING DOWN LIKE HE SAID.

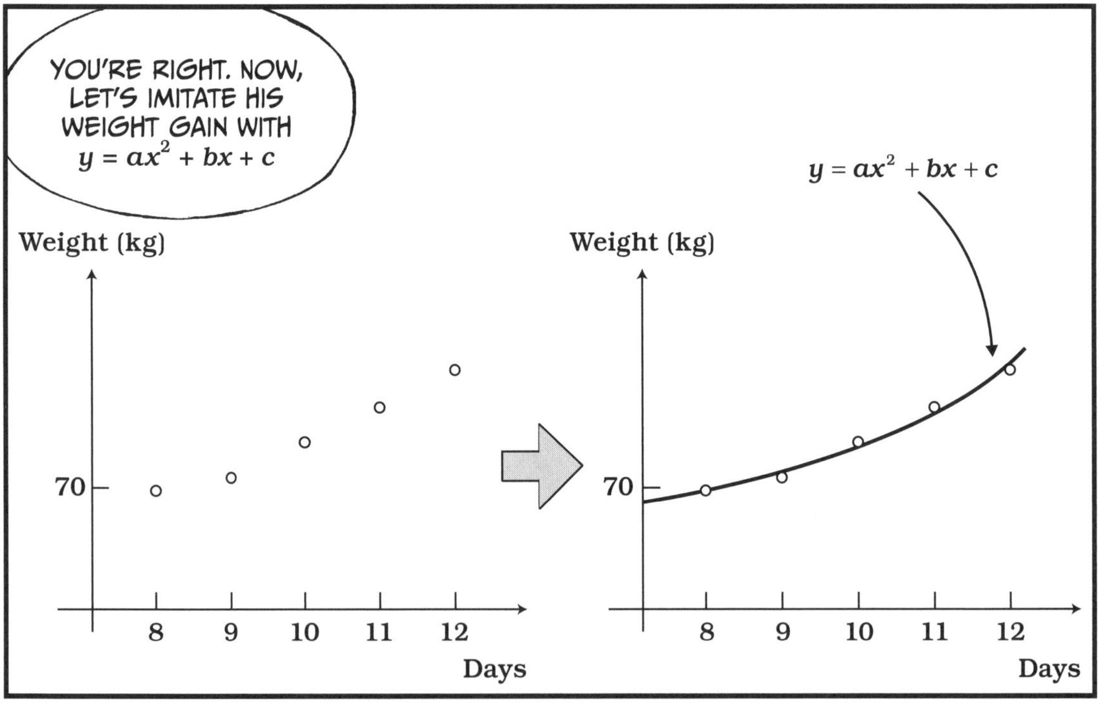
YOU'RE RIGHT. NOW, LET'S IMITATE HIS WEIGHT GAIN WITH $y = ax^2 + bx + c$
Weight (kg)
70
8
9
10
11
12
Days
$y = ax^2 + bx + c$
Weight (kg)
70
8
9
10
11
12
Days

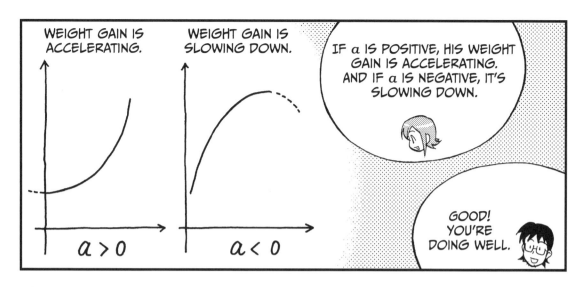

WEIGHT GAIN IS ACCELERATING.

WEIGHT GAIN IS SLOWING DOWN.

IF a IS POSITIVE, HIS WEIGHT GAIN IS ACCELERATING. AND IF a IS NEGATIVE, IT'S SLOWING DOWN.

$a > 0$

$a < 0$

GOOD! YOU'RE DOING WELL.

VROOM...

THERE ARE LOTS OF TIGHT CURVES AROUND HERE.

LET'S ASSUME YOU WANT TO KNOW HOW TIGHT EACH CURVE IS.

EH, I DON'T REALLY CARE ABOUT THAT.

WE CAN APPROXIMATE EACH CURVE WITH A CIRCLE.

...

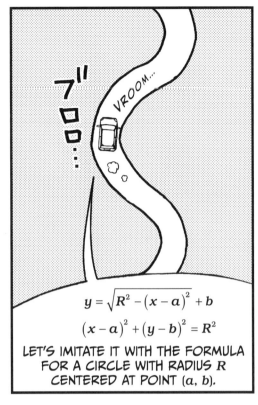

$$y = \sqrt{R^2 - (x-a)^2} + b$$

$$(x-a)^2 + (y-b)^2 = R^2$$

LET'S IMITATE IT WITH THE FORMULA FOR A CIRCLE WITH RADIUS R CENTERED AT POINT (a, b).

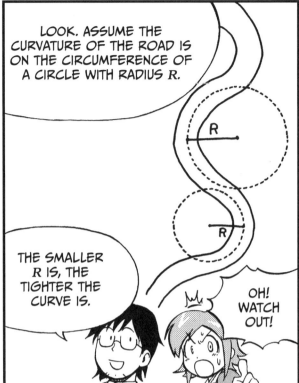

LOOK. ASSUME THE CURVATURE OF THE ROAD IS ON THE CIRCUMFERENCE OF A CIRCLE WITH RADIUS R.

THE SMALLER R IS, THE TIGHTER THE CURVE IS.

OH! WATCH OUT!

ARE YOU ALL RIGHT?

I THINK SO...

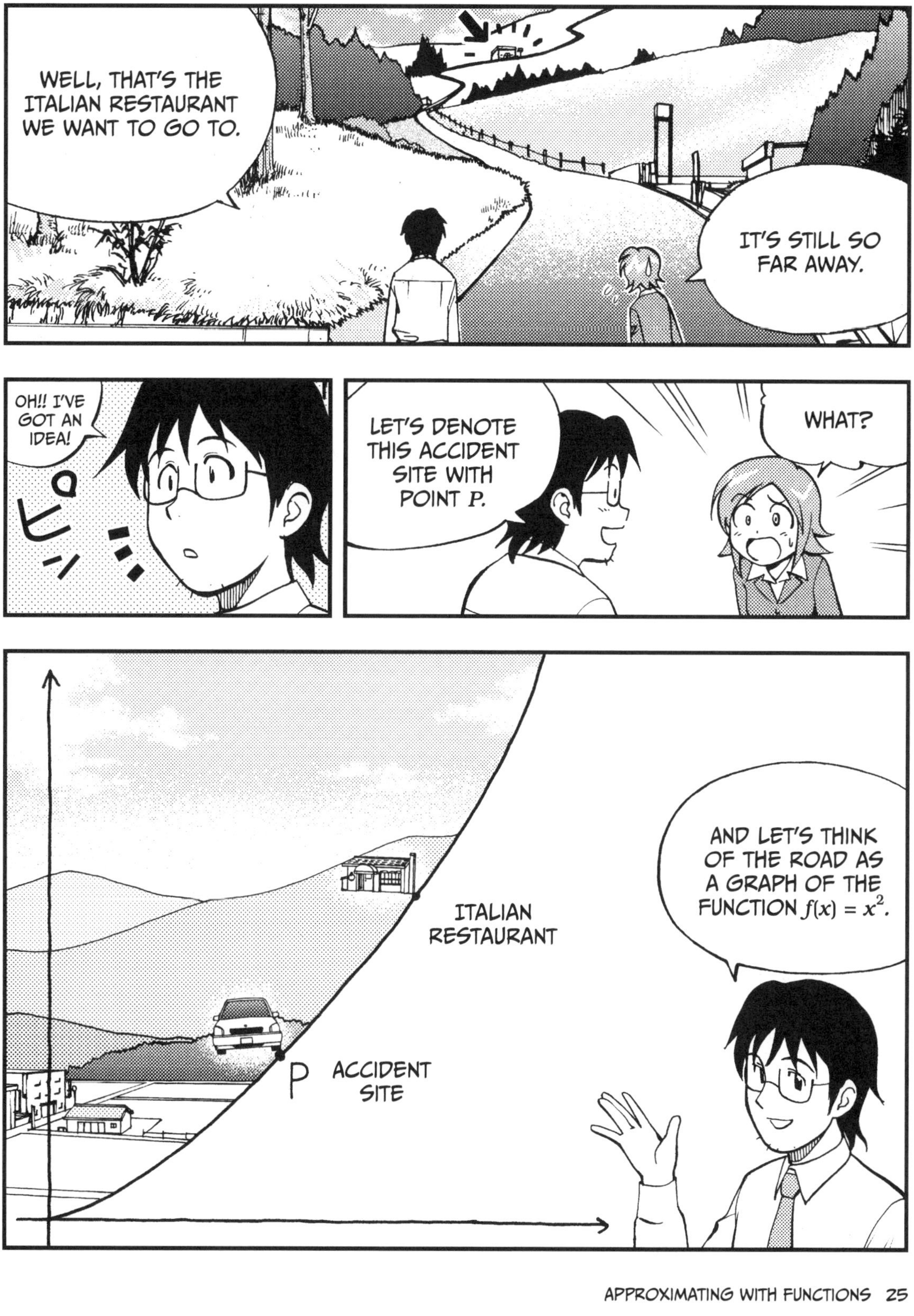
WELL, THAT'S THE ITALIAN RESTAURANT WE WANT TO GO TO.
IT'S STILL SO FAR AWAY.
OH!! I'VE GOT AN IDEA!
LET'S DENOTE THIS ACCIDENT SITE WITH POINT P.
WHAT?
ITALIAN RESTAURANT
P
ACCIDENT SITE
AND LET'S THINK OF THE ROAD AS A GRAPH OF THE FUNCTION $f(x) = x^2$.

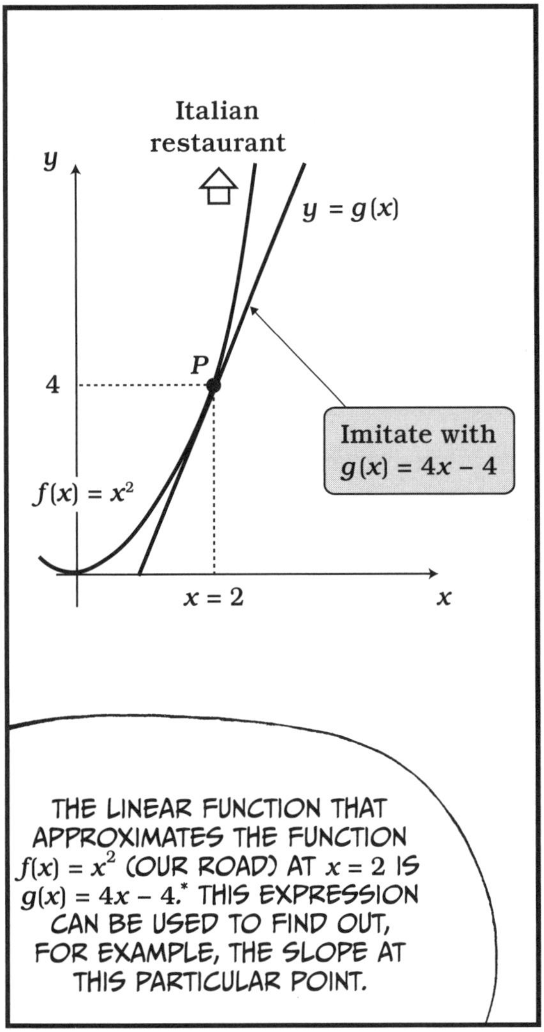
Italian restaurant
y
$y = g(x)$
4
P
Imitate with $g(x) = 4x - 4$
$f(x) = x^2$
$x = 2$
x
THE LINEAR FUNCTION THAT APPROXIMATES THE FUNCTION $f(x) = x^2$ (OUR ROAD) AT $x = 2$ IS $g(x) = 4x - 4$.* THIS EXPRESSION CAN BE USED TO FIND OUT, FOR EXAMPLE, THE SLOPE AT THIS PARTICULAR POINT.

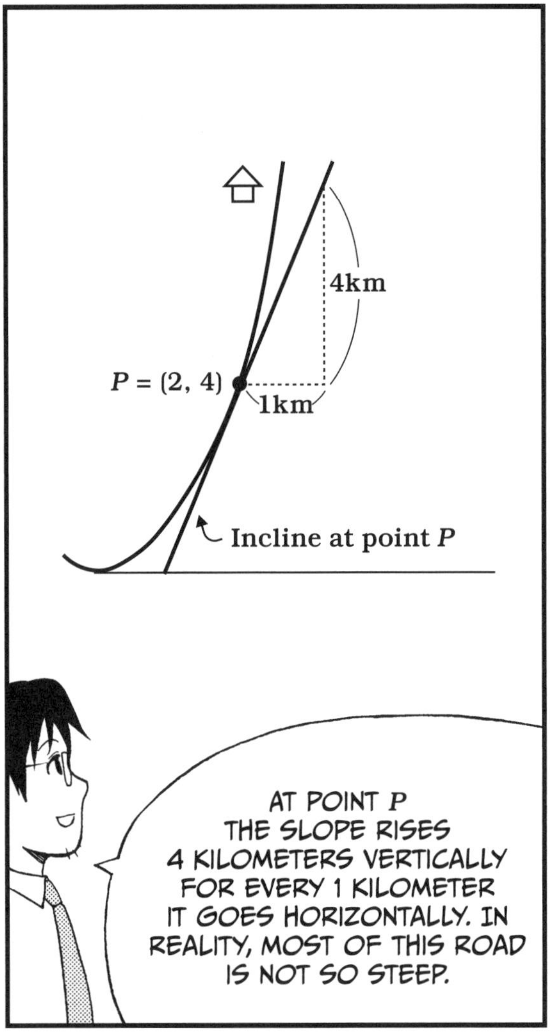
4km
$P = (2, 4)$
1km
Incline at point P
AT POINT P THE SLOPE RISES 4 KILOMETERS VERTICALLY FOR EVERY 1 KILOMETER IT GOES HORIZONTALLY. IN REALITY, MOST OF THIS ROAD IS NOT SO STEEP.

* THE REASON IS GIVEN ON PAGE 39.

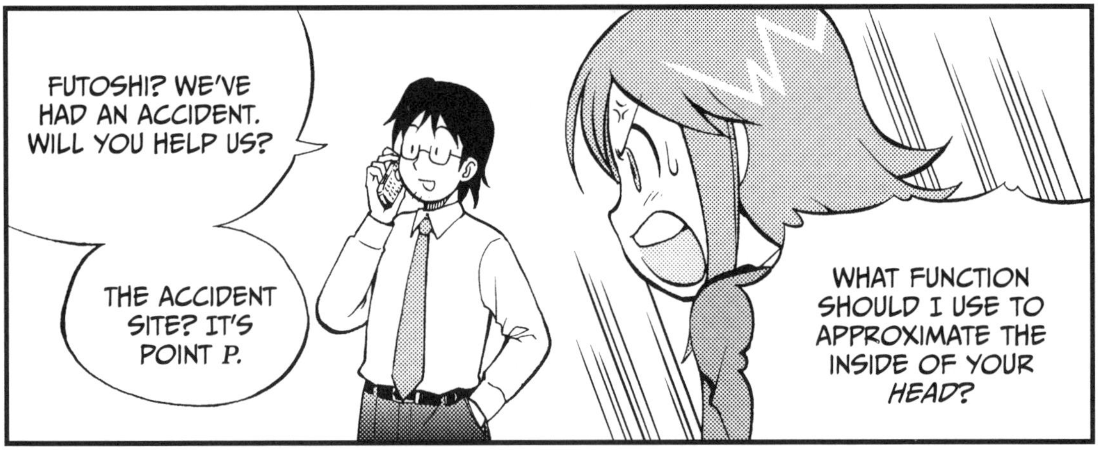
FUTOSHI? WE'VE HAD AN ACCIDENT. WILL YOU HELP US?
THE ACCIDENT SITE? IT'S POINT P.
WHAT FUNCTION SHOULD I USE TO APPROXIMATE THE INSIDE OF YOUR HEAD?

CALCULATING THE RELATIVE ERROR

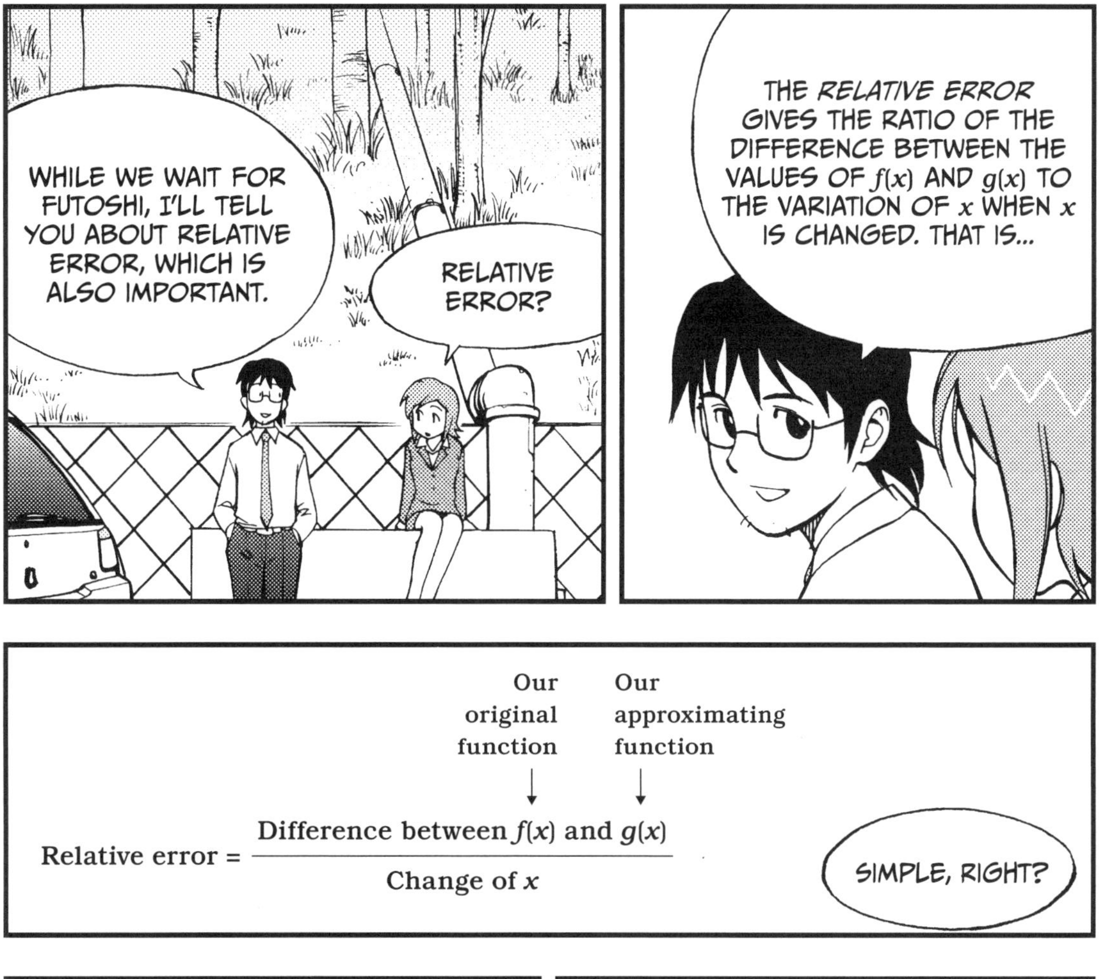

WHILE WE WAIT FOR FUTOSHI, I'LL TELL YOU ABOUT RELATIVE ERROR, WHICH IS ALSO IMPORTANT.
RELATIVE ERROR?
THE *RELATIVE ERROR* GIVES THE RATIO OF THE DIFFERENCE BETWEEN THE VALUES OF $f(x)$ AND $g(x)$ TO THE VARIATION OF x WHEN x IS CHANGED. THAT IS...

Our original function ↓ (f(x)); Our approximating function ↓ (g(x))

$$\text{Relative error} = \frac{\text{Difference between } f(x) \text{ and } g(x)}{\text{Change of } x}$$

SIMPLE, RIGHT?

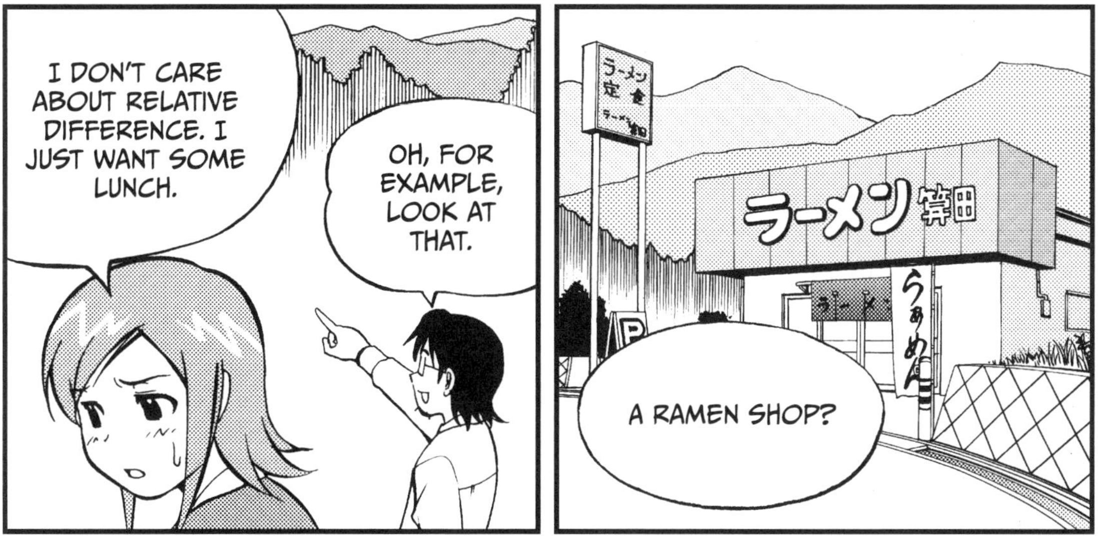

I DON'T CARE ABOUT RELATIVE DIFFERENCE. I JUST WANT SOME LUNCH.
OH, FOR EXAMPLE, LOOK AT THAT.
ラーメン
A RAMEN SHOP?

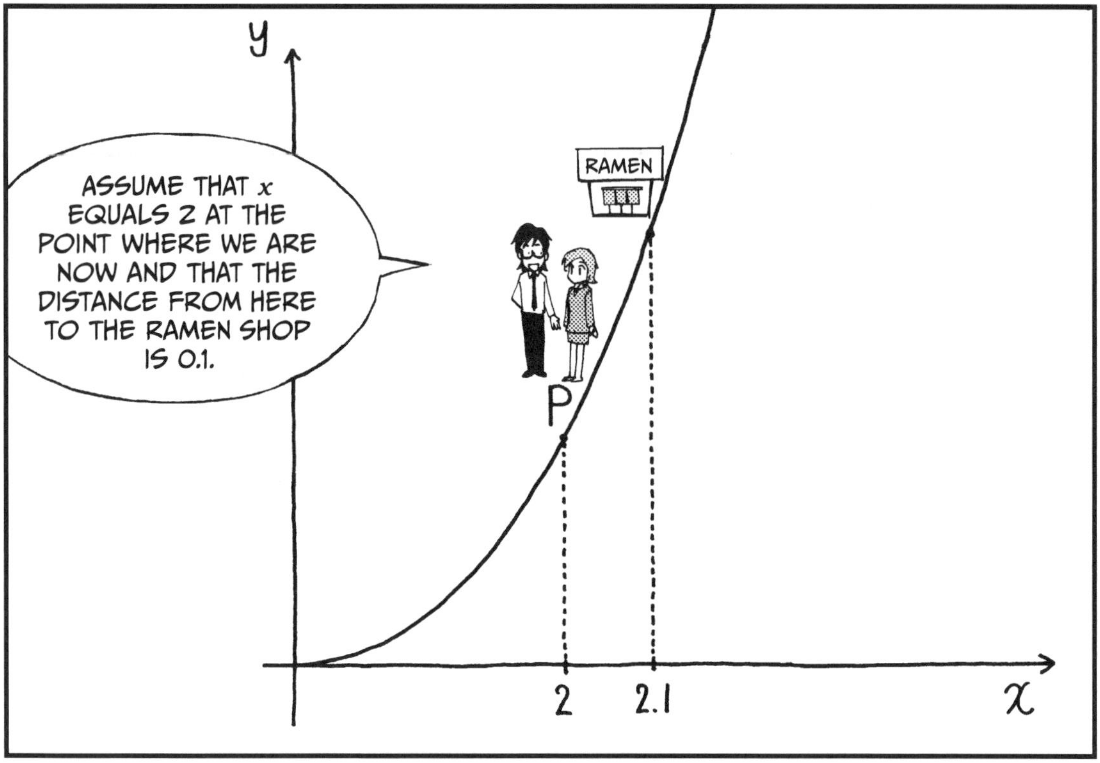
ASSUME THAT x EQUALS 2 AT THE POINT WHERE WE ARE NOW AND THAT THE DISTANCE FROM HERE TO THE RAMEN SHOP IS 0.1.
RAMEN
y
P
2
2.1
x

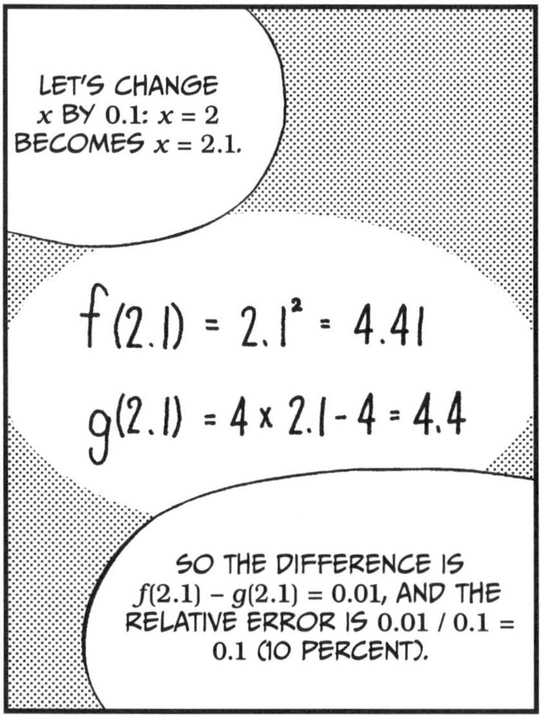
LET'S CHANGE x BY 0.1: $x = 2$ BECOMES $x = 2.1$.
$f(2.1) = 2.1^2 = 4.41$
$g(2.1) = 4 \times 2.1 - 4 = 4.4$
SO THE DIFFERENCE IS $f(2.1) - g(2.1) = 0.01$, AND THE RELATIVE ERROR IS 0.01 / 0.1 = 0.1 (10 PERCENT).

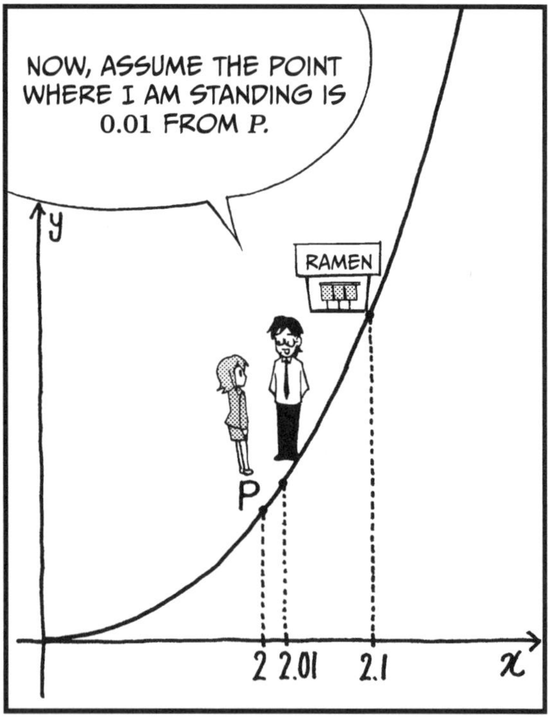
NOW, ASSUME THE POINT WHERE I AM STANDING IS 0.01 FROM P.
RAMEN
y
P
2
2.01
2.1
x

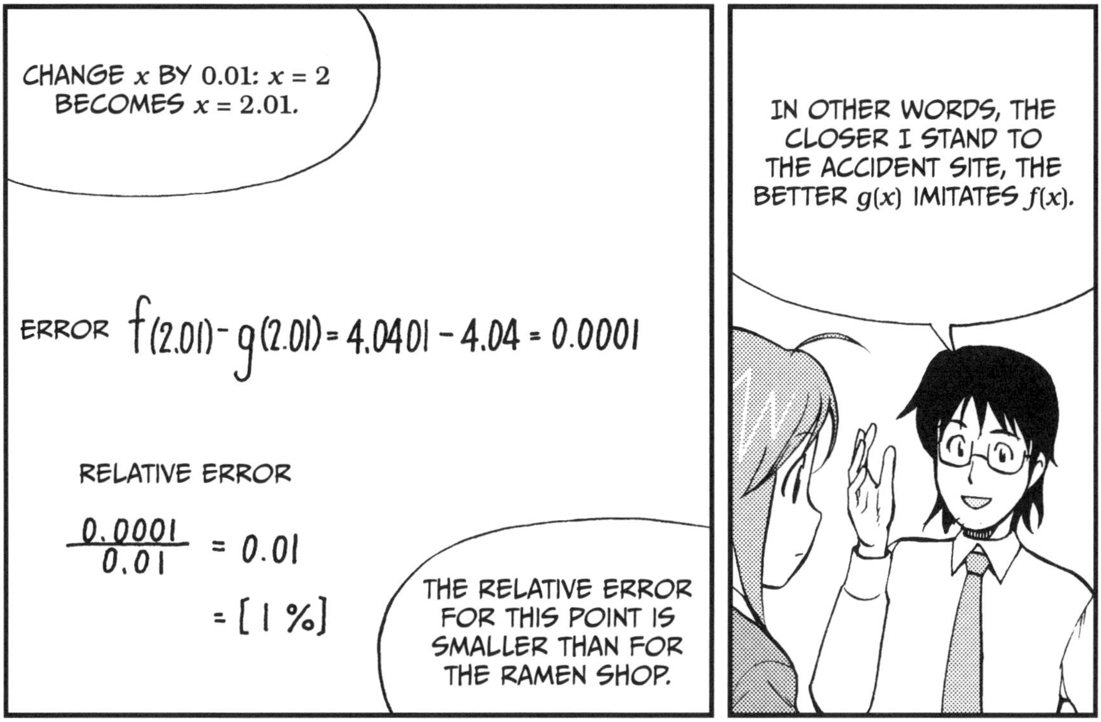

CHANGE x BY 0.01: $x = 2$ BECOMES $x = 2.01$.

ERROR $f(2.01) - g(2.01) = 4.0401 - 4.04 = 0.0001$

RELATIVE ERROR

$$\frac{0.0001}{0.01} = 0.01$$

$$= [1\%]$$

THE RELATIVE ERROR FOR THIS POINT IS SMALLER THAN FOR THE RAMEN SHOP.

IN OTHER WORDS, THE CLOSER I STAND TO THE ACCIDENT SITE, THE BETTER $g(x)$ IMITATES $f(x)$.

As the variation approaches 0, the relative error also approaches 0.

Variation of x from 2	$f(x)$	$g(x)$	Error	Relative error
1	9	8	1	100.0%
0.1	4.41	4.4	0.01	10.0%
0.01	4.0401	4.04	0.0001	1.0%
0.001	4.004001	4.004	0.000001	0.1%
↓				↓
0				0

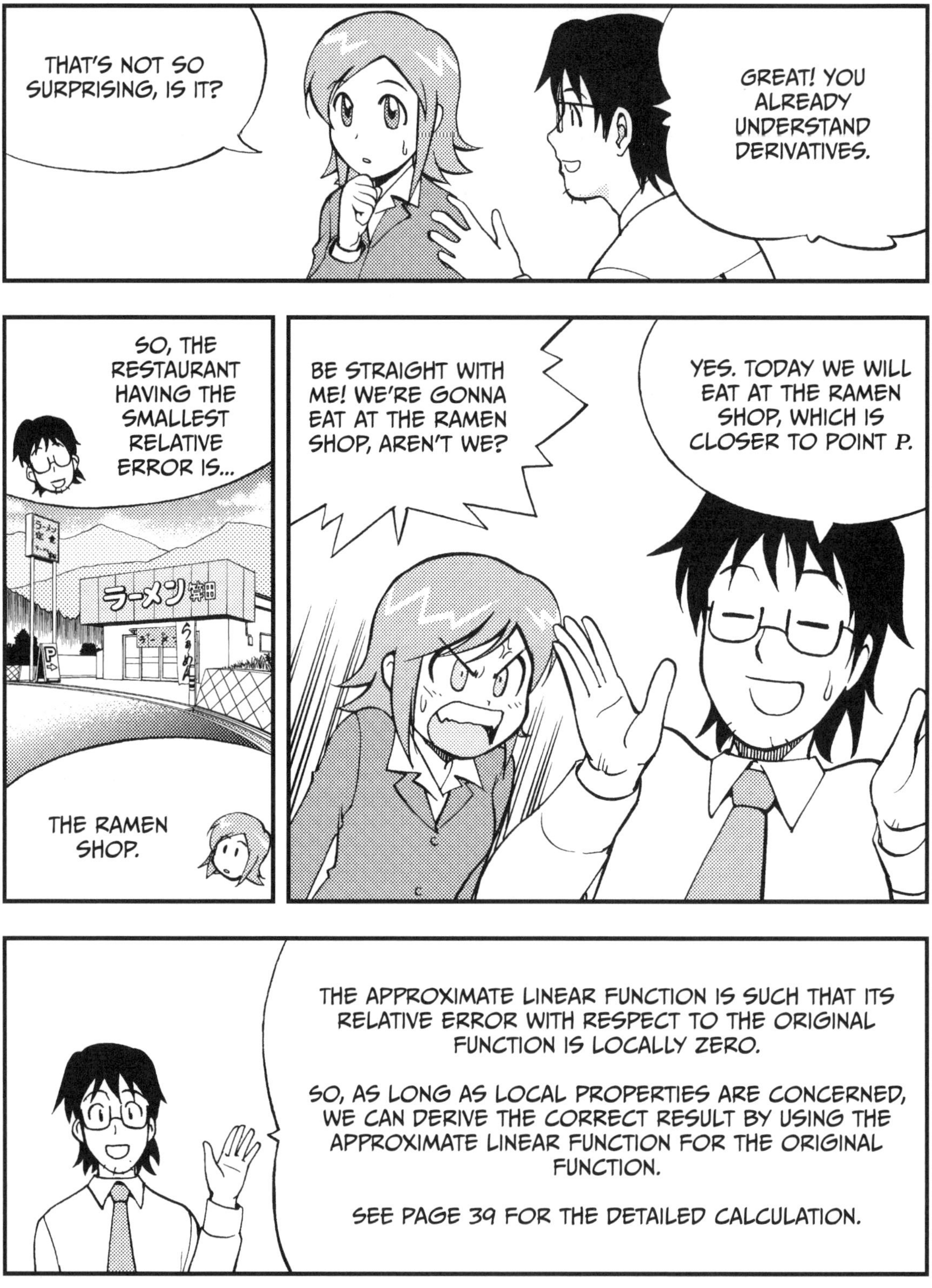
THAT'S NOT SO SURPRISING, IS IT?
GREAT! YOU ALREADY UNDERSTAND DERIVATIVES.
SO, THE RESTAURANT HAVING THE SMALLEST RELATIVE ERROR IS...
ラーメン
THE RAMEN SHOP.
BE STRAIGHT WITH ME! WE'RE GONNA EAT AT THE RAMEN SHOP, AREN'T WE?
YES. TODAY WE WILL EAT AT THE RAMEN SHOP, WHICH IS CLOSER TO POINT P.
THE APPROXIMATE LINEAR FUNCTION IS SUCH THAT ITS RELATIVE ERROR WITH RESPECT TO THE ORIGINAL FUNCTION IS LOCALLY ZERO.
SO, AS LONG AS LOCAL PROPERTIES ARE CONCERNED, WE CAN DERIVE THE CORRECT RESULT BY USING THE APPROXIMATE LINEAR FUNCTION FOR THE ORIGINAL FUNCTION.
SEE PAGE 39 FOR THE DETAILED CALCULATION.

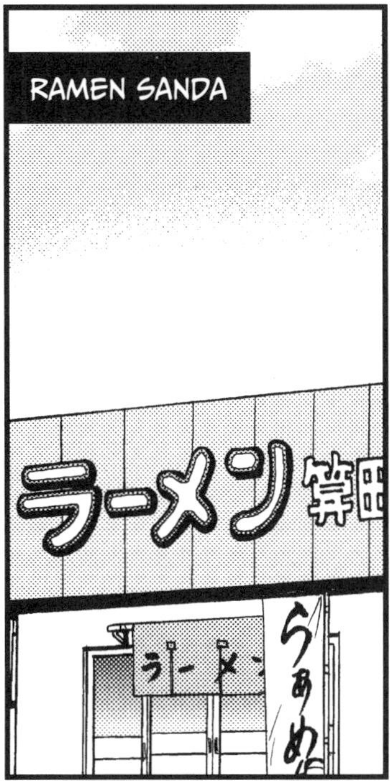
RAMEN SANDA
ラーメン算田
ラーメン

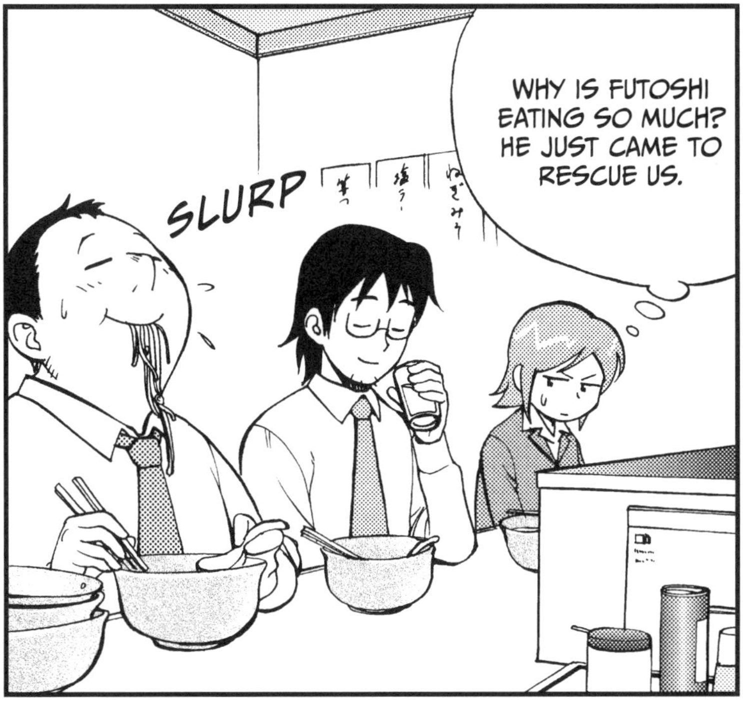
SLURP
WHY IS FUTOSHI EATING SO MUCH? HE JUST CAME TO RESCUE US.

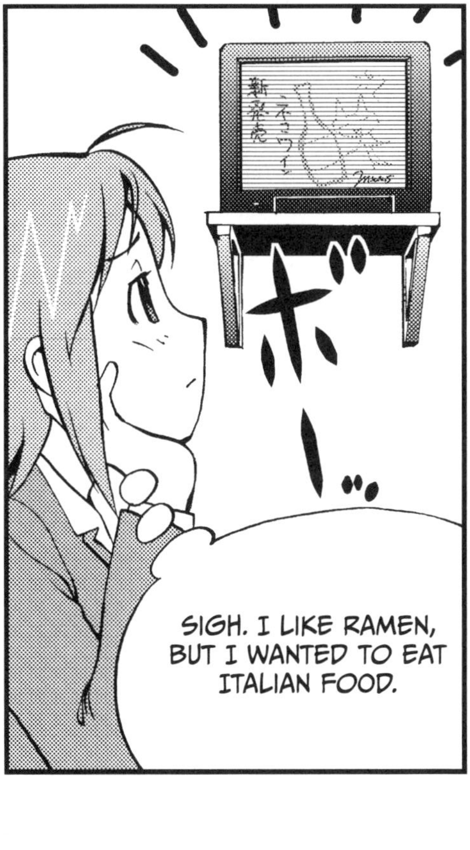
ボー…
SIGH. I LIKE RAMEN, BUT I WANTED TO EAT ITALIAN FOOD.

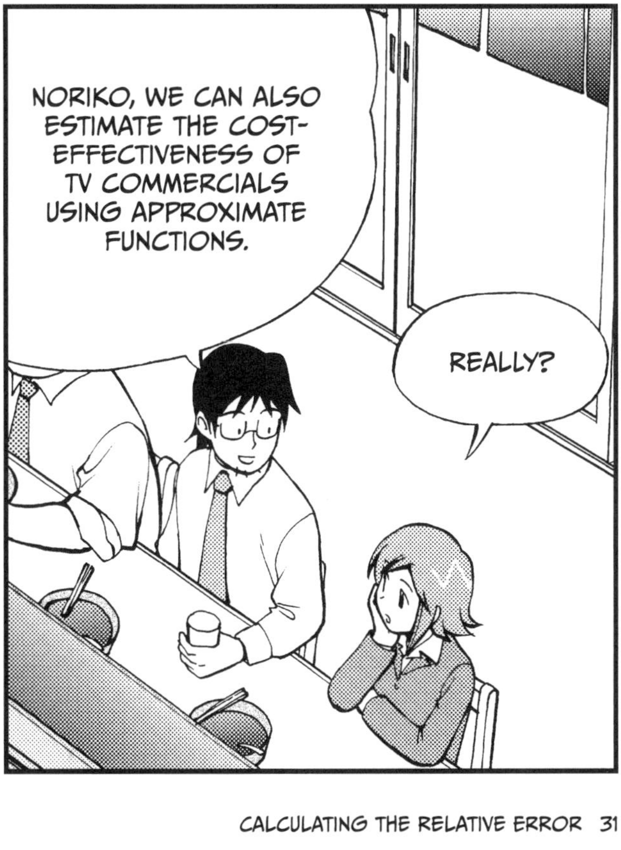
NORIKO, WE CAN ALSO ESTIMATE THE COST-EFFECTIVENESS OF TV COMMERCIALS USING APPROXIMATE FUNCTIONS.
REALLY?

THE DERIVATIVE IN ACTION!

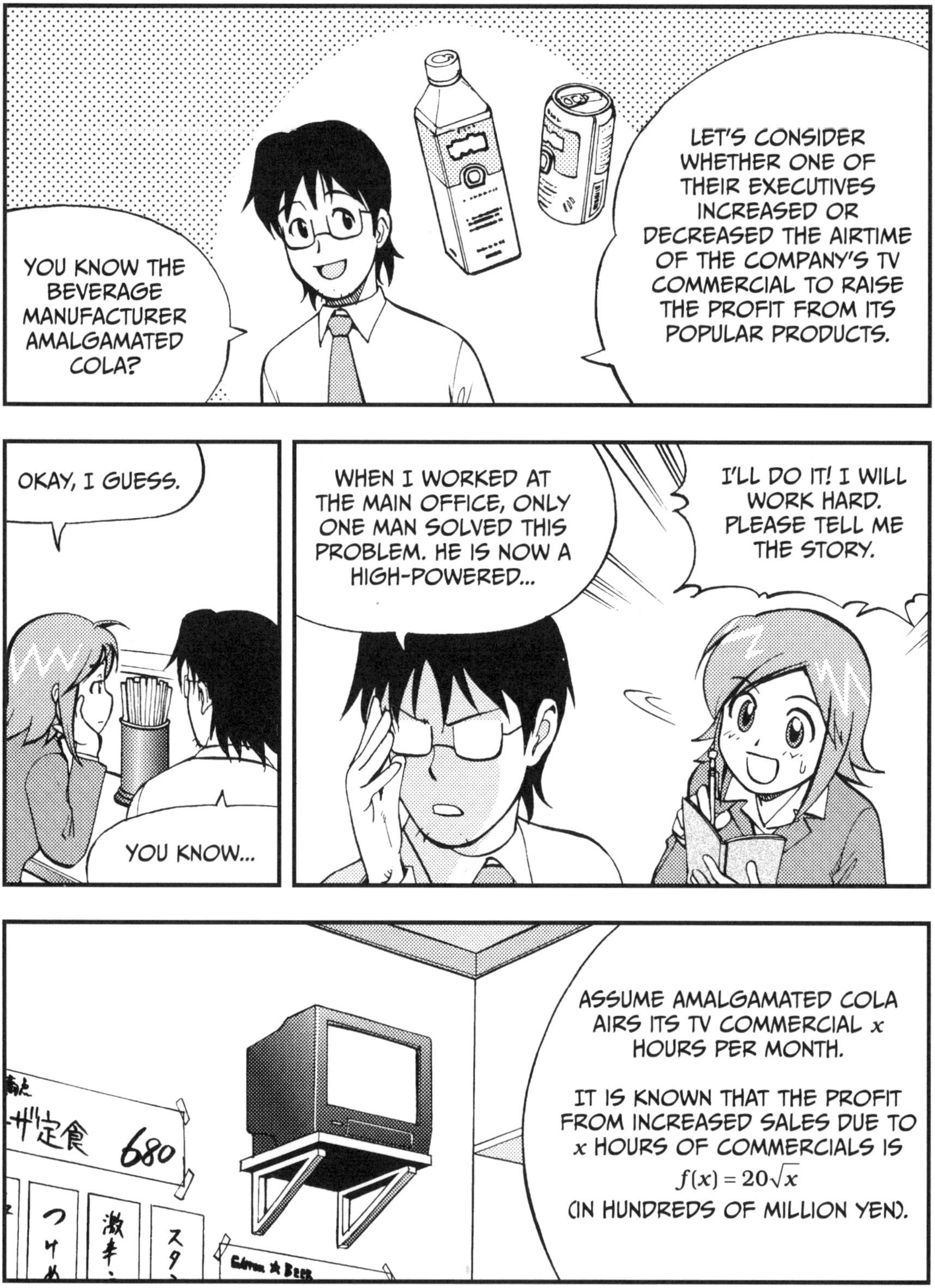
YOU KNOW THE BEVERAGE MANUFACTURER AMALGAMATED COLA?
LET'S CONSIDER WHETHER ONE OF THEIR EXECUTIVES INCREASED OR DECREASED THE AIRTIME OF THE COMPANY'S TV COMMERCIAL TO RAISE THE PROFIT FROM ITS POPULAR PRODUCTS.
OKAY, I GUESS.
YOU KNOW...
WHEN I WORKED AT THE MAIN OFFICE, ONLY ONE MAN SOLVED THIS PROBLEM. HE IS NOW A HIGH-POWERED...
I'LL DO IT! I WILL WORK HARD. PLEASE TELL ME THE STORY.
ASSUME AMALGAMATED COLA AIRS ITS TV COMMERCIAL x HOURS PER MONTH.
IT IS KNOWN THAT THE PROFIT FROM INCREASED SALES DUE TO x HOURS OF COMMERCIALS IS $f(x) = 20\sqrt{x}$ (IN HUNDREDS OF MILLION YEN).

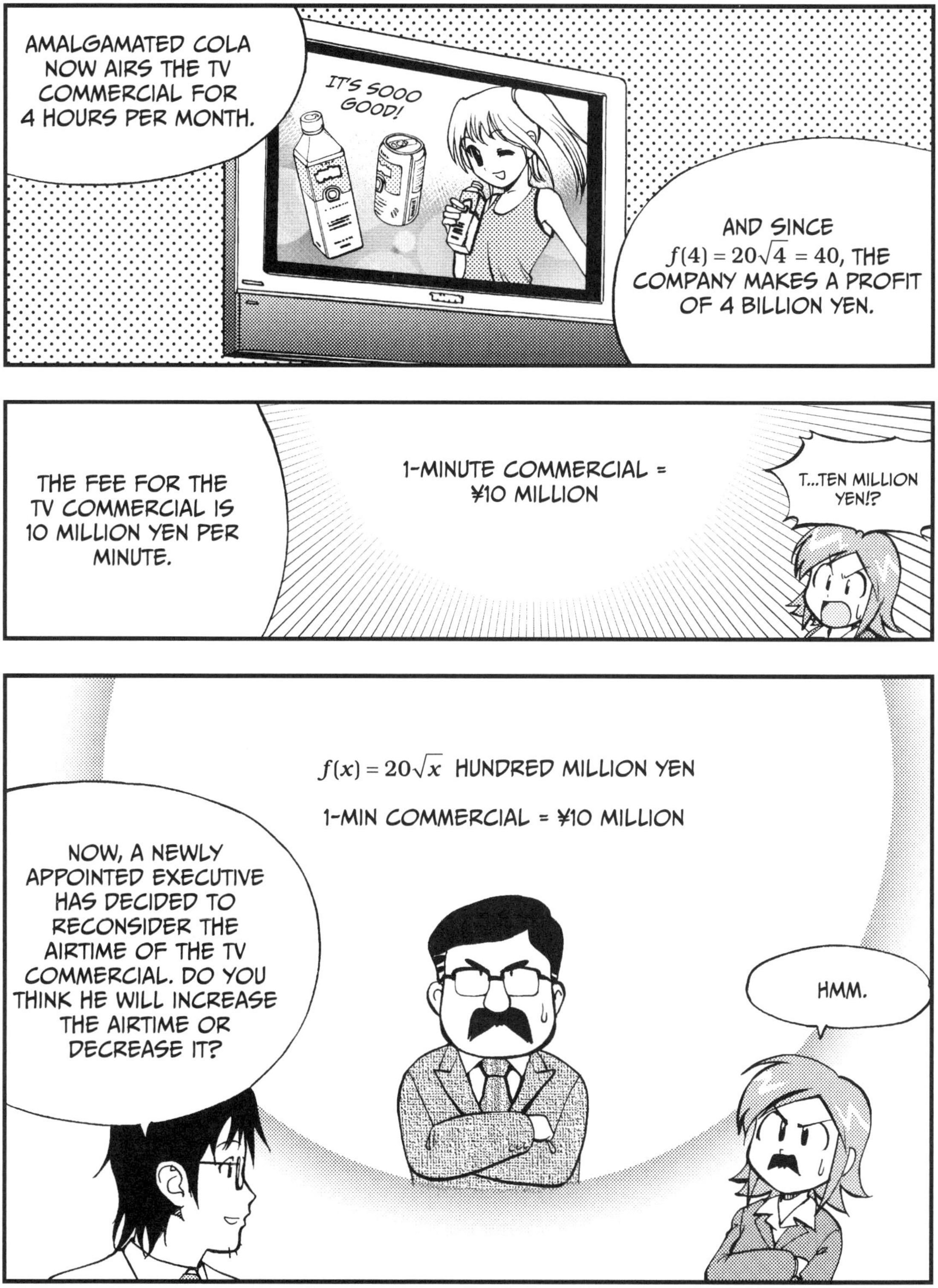
AMALGAMATED COLA NOW AIRS THE TV COMMERCIAL FOR 4 HOURS PER MONTH.
IT'S SOOO GOOD!
AND SINCE $f(4) = 20\sqrt{4} = 40$, THE COMPANY MAKES A PROFIT OF 4 BILLION YEN.
THE FEE FOR THE TV COMMERCIAL IS 10 MILLION YEN PER MINUTE.
1-MINUTE COMMERCIAL = ¥10 MILLION
T...TEN MILLION YEN!?
$f(x) = 20\sqrt{x}$ HUNDRED MILLION YEN
1-MIN COMMERCIAL = ¥10 MILLION
NOW, A NEWLY APPOINTED EXECUTIVE HAS DECIDED TO RECONSIDER THE AIRTIME OF THE TV COMMERCIAL. DO YOU THINK HE WILL INCREASE THE AIRTIME OR DECREASE IT?
HMM.

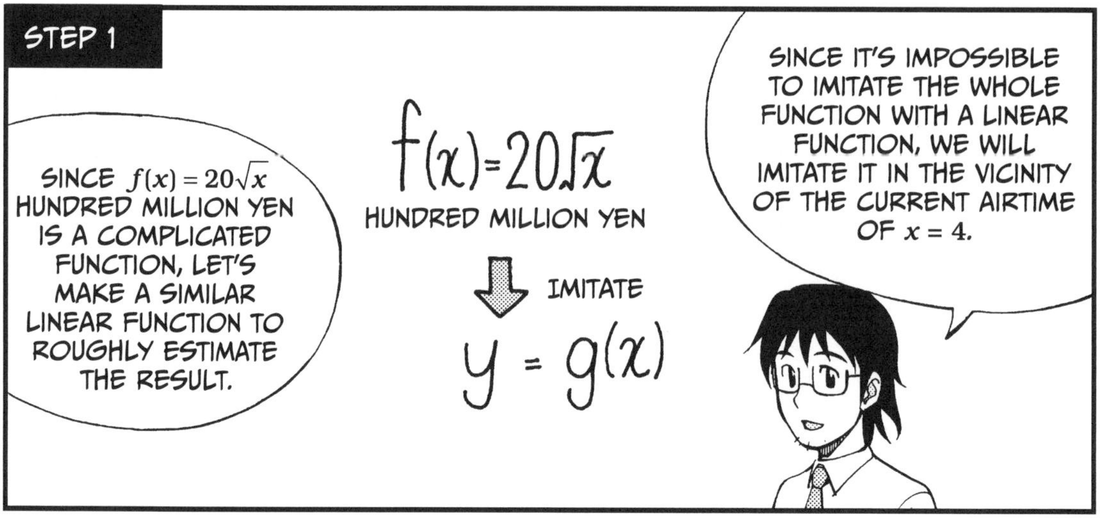
STEP 1
SINCE $f(x) = 20\sqrt{x}$ HUNDRED MILLION YEN IS A COMPLICATED FUNCTION, LET'S MAKE A SIMILAR LINEAR FUNCTION TO ROUGHLY ESTIMATE THE RESULT.
f(x)=20√x
HUNDRED MILLION YEN
IMITATE
y = g(x)
SINCE IT'S IMPOSSIBLE TO IMITATE THE WHOLE FUNCTION WITH A LINEAR FUNCTION, WE WILL IMITATE IT IN THE VICINITY OF THE CURRENT AIRTIME OF $x = 4$.

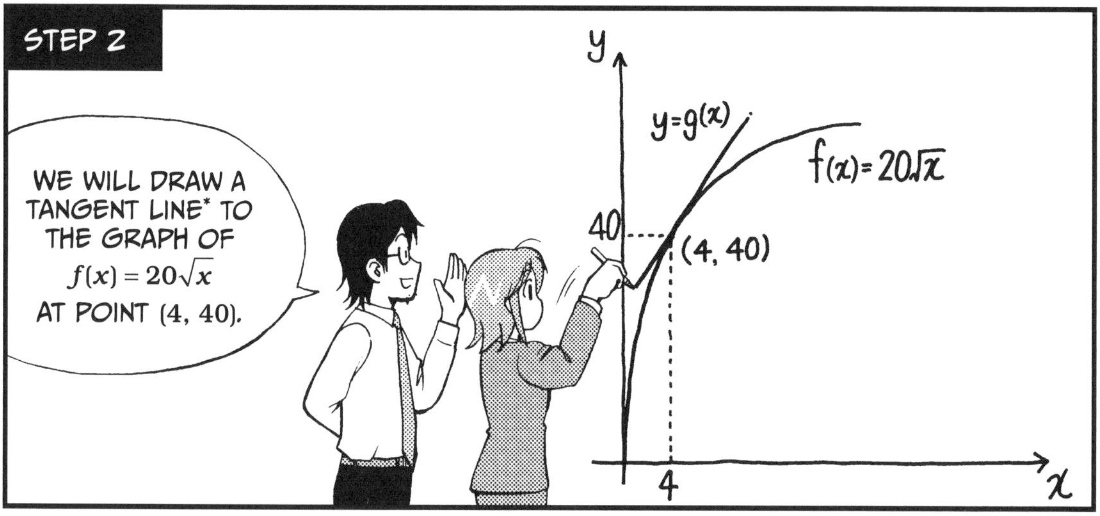
STEP 2
WE WILL DRAW A TANGENT LINE* TO THE GRAPH OF $f(x) = 20\sqrt{x}$ AT POINT (4, 40).
y
y=g(x)
f(x)=20√x
40
(4, 40)
4
x

* Here is the calculation of the tangent line. (See also the explanation of the derivative on page 39.)

For $f(x) = 20\sqrt{x}$, $f'(4)$ is given as follows.

$$\frac{f(4+\varepsilon)-f(4)}{\varepsilon} = \frac{20\sqrt{4+\varepsilon}-20\times 2}{\varepsilon} = 20\frac{\left(\sqrt{4+\varepsilon}-2\right)\times\left(\sqrt{4+\varepsilon}+2\right)}{\varepsilon\times\left(\sqrt{4+\varepsilon}+2\right)}$$

$$= 20\frac{4+\varepsilon-4}{\varepsilon\left(\sqrt{4+\varepsilon}+2\right)} = \frac{20}{\sqrt{4+\varepsilon}+2} \quad ❶$$

When ε approaches 0, the denominator of ❶ $\sqrt{4+\varepsilon}+2 \rightarrow 4$.

Therefore, ❶ $\rightarrow 20 \div 4 = 5$.

Thus, the approximate linear function $g(x) = 5(x-4)+40 = 5x+20$

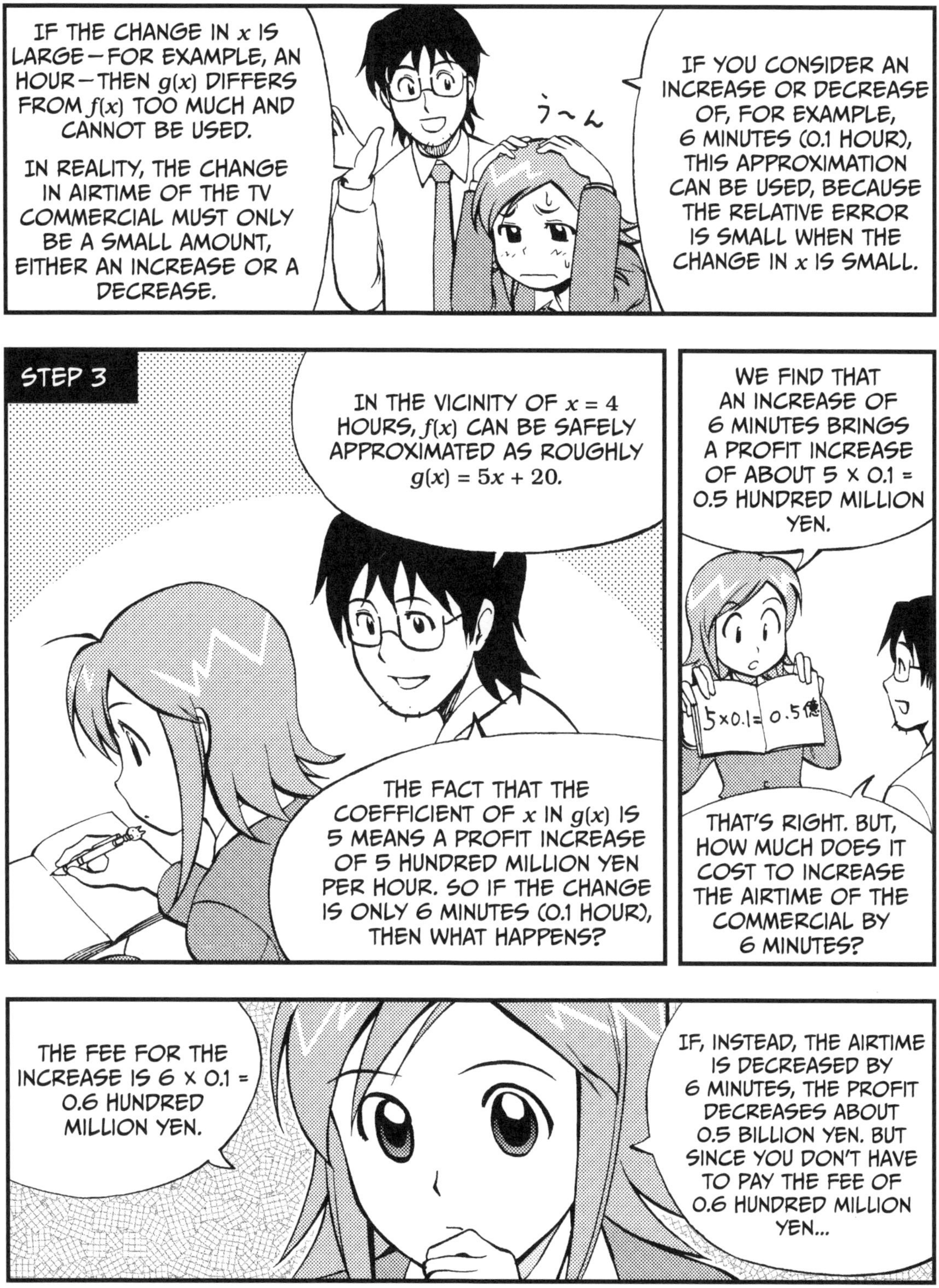
IF THE CHANGE IN x IS LARGE—FOR EXAMPLE, AN HOUR—THEN $g(x)$ DIFFERS FROM $f(x)$ TOO MUCH AND CANNOT BE USED.
IN REALITY, THE CHANGE IN AIRTIME OF THE TV COMMERCIAL MUST ONLY BE A SMALL AMOUNT, EITHER AN INCREASE OR A DECREASE.
う〜ん
IF YOU CONSIDER AN INCREASE OR DECREASE OF, FOR EXAMPLE, 6 MINUTES (0.1 HOUR), THIS APPROXIMATION CAN BE USED, BECAUSE THE RELATIVE ERROR IS SMALL WHEN THE CHANGE IN x IS SMALL.
STEP 3
IN THE VICINITY OF $x = 4$ HOURS, $f(x)$ CAN BE SAFELY APPROXIMATED AS ROUGHLY $g(x) = 5x + 20$.
THE FACT THAT THE COEFFICIENT OF x IN $g(x)$ IS 5 MEANS A PROFIT INCREASE OF 5 HUNDRED MILLION YEN PER HOUR. SO IF THE CHANGE IS ONLY 6 MINUTES (0.1 HOUR), THEN WHAT HAPPENS?
WE FIND THAT AN INCREASE OF 6 MINUTES BRINGS A PROFIT INCREASE OF ABOUT 5 × 0.1 = 0.5 HUNDRED MILLION YEN.
5×0.1=0.5億
THAT'S RIGHT. BUT, HOW MUCH DOES IT COST TO INCREASE THE AIRTIME OF THE COMMERCIAL BY 6 MINUTES?
THE FEE FOR THE INCREASE IS 6 × 0.1 = 0.6 HUNDRED MILLION YEN.
IF, INSTEAD, THE AIRTIME IS DECREASED BY 6 MINUTES, THE PROFIT DECREASES ABOUT 0.5 BILLION YEN. BUT SINCE YOU DON'T HAVE TO PAY THE FEE OF 0.6 HUNDRED MILLION YEN...

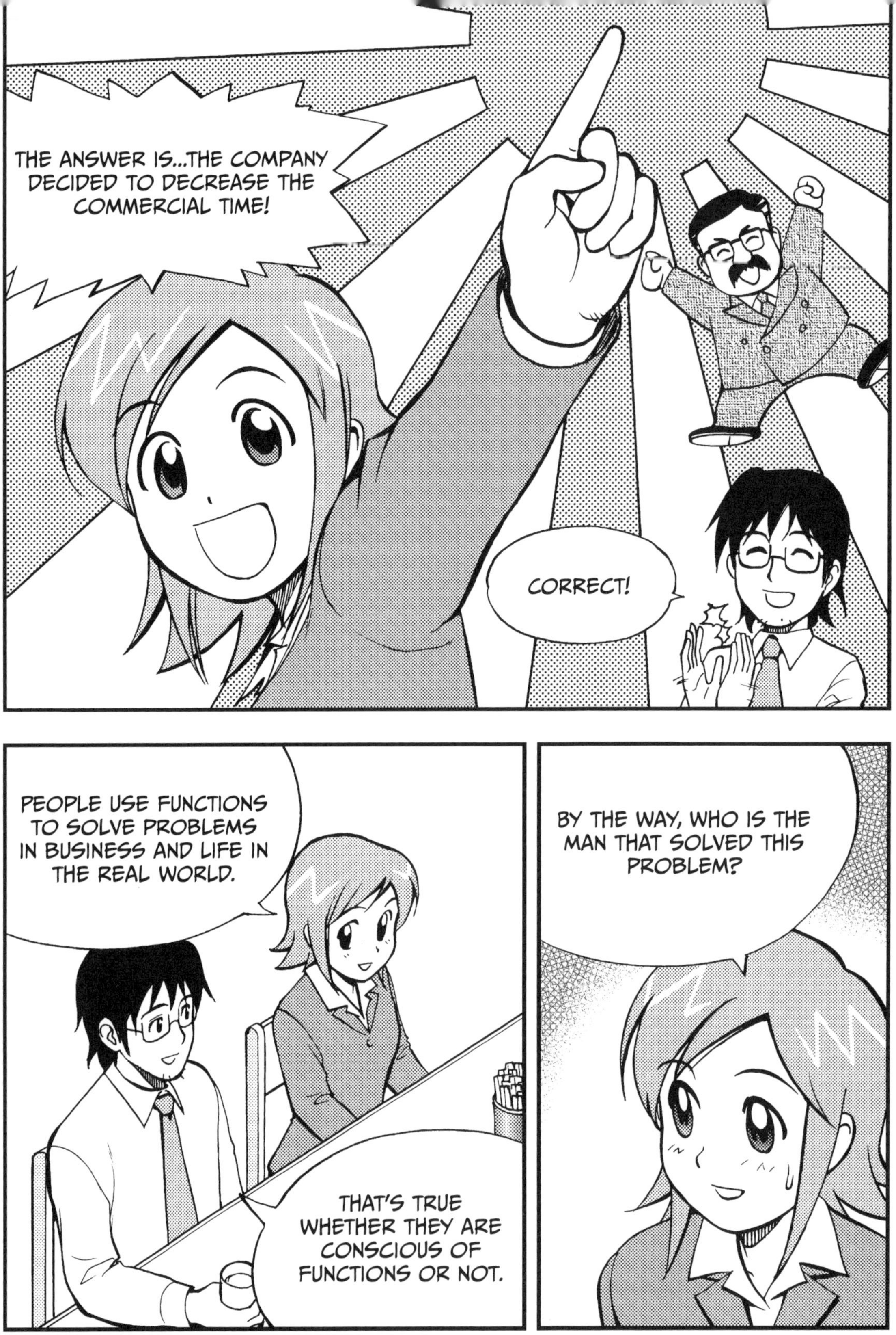
THE ANSWER IS...THE COMPANY DECIDED TO DECREASE THE COMMERCIAL TIME!
CORRECT!
PEOPLE USE FUNCTIONS TO SOLVE PROBLEMS IN BUSINESS AND LIFE IN THE REAL WORLD.
THAT'S TRUE WHETHER THEY ARE CONSCIOUS OF FUNCTIONS OR NOT.
BY THE WAY, WHO IS THE MAN THAT SOLVED THIS PROBLEM?

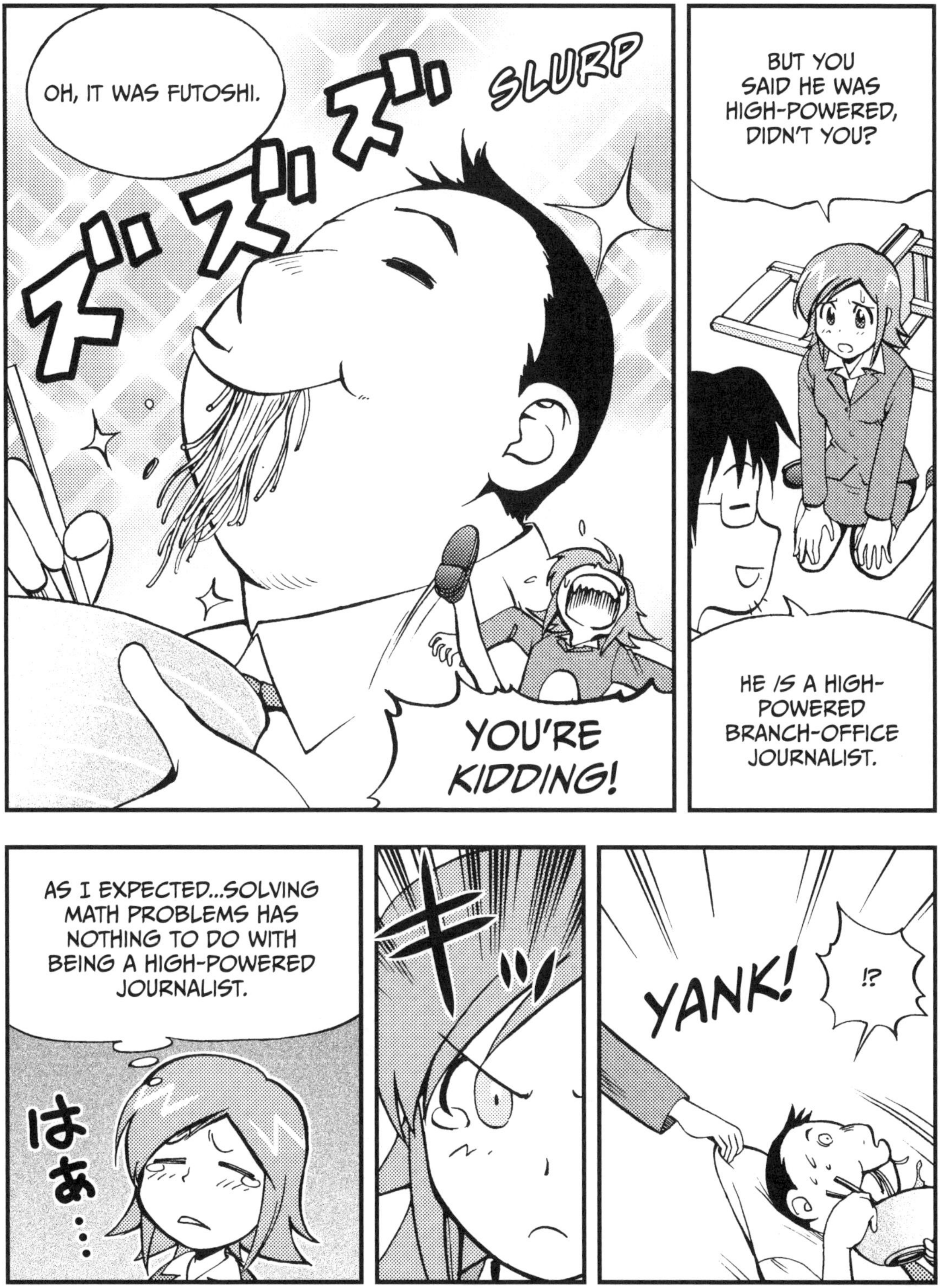
OH, IT WAS FUTOSHI.
ズズズズ
SLURP
YOU'RE KIDDING!
BUT YOU SAID HE WAS HIGH-POWERED, DIDN'T YOU?
HE IS A HIGH-POWERED BRANCH-OFFICE JOURNALIST.
AS I EXPECTED...SOLVING MATH PROBLEMS HAS NOTHING TO DO WITH BEING A HIGH-POWERED JOURNALIST.
はぁ…
YANK!
!?

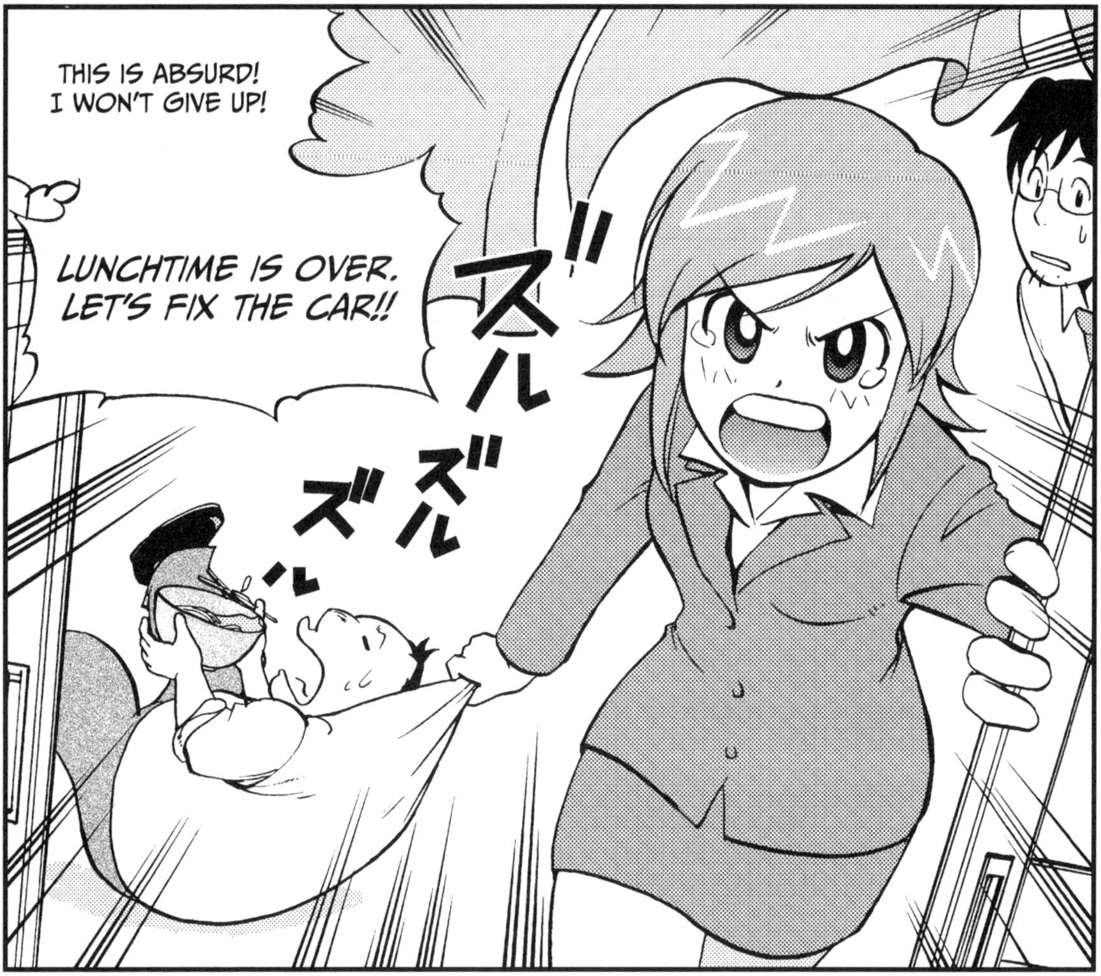
THIS IS ABSURD! I WON'T GIVE UP!
LUNCHTIME IS OVER. LET'S FIX THE CAR!!
ズ ズルズルズル

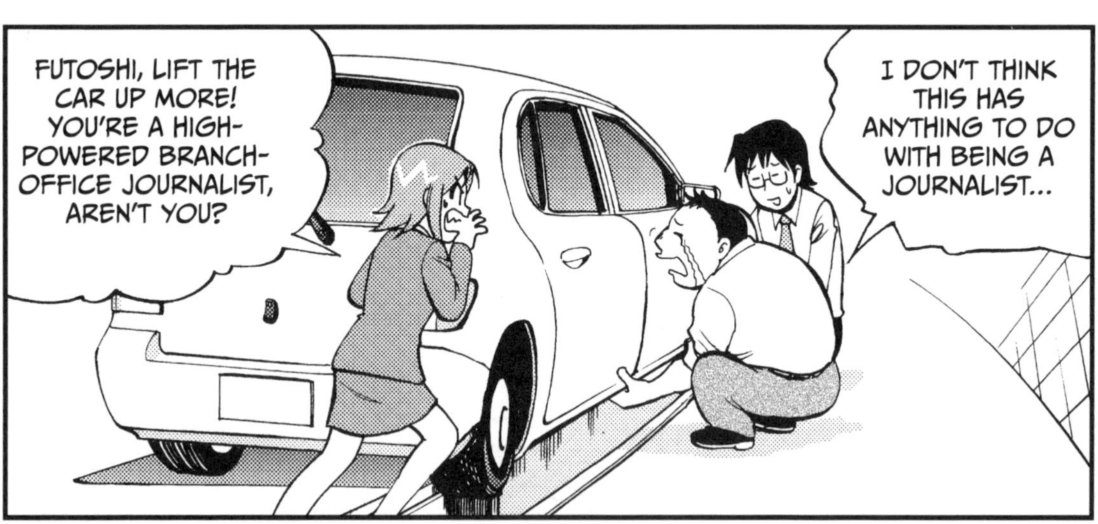
FUTOSHI, LIFT THE CAR UP MORE! YOU'RE A HIGH-POWERED BRANCH-OFFICE JOURNALIST, AREN'T YOU?
I DON'T THINK THIS HAS ANYTHING TO DO WITH BEING A JOURNALIST...

CALCULATING THE DERIVATIVE

Let's find the imitating linear function $g(x) = kx + l$ of function $f(x)$ at $x = a$.
We need to find slope k.

> ❶ $g(x) = k(x - a) + f(a)$ ($g(x)$ coincides with $f(a)$ when $x = a$.)

Now, let's calculate the relative error when x changes from $x = a$ to $x = a + \varepsilon$.

$$\text{Relative error} = \frac{\text{Difference between } f \text{ and } g \text{ after } x \text{ has changed}}{\text{Change of } x \text{ from } x = a}$$

$$= \frac{f(a + \varepsilon) - g(a + \varepsilon)}{\varepsilon}$$

$$= \frac{f(a + \varepsilon) - (k\varepsilon + f(a))}{\varepsilon}$$

> $g(a + \varepsilon) = k(a + \varepsilon - a) + f(a)$
> $= k\varepsilon + f(a)$

$$= \frac{f(a + \varepsilon) - f(a)}{\varepsilon} - k \xrightarrow[\varepsilon \to 0]{} 0$$

> When ε approaches 0, the relative error also approaches 0.

$$k = \lim_{\varepsilon \to 0} \frac{f(a + \varepsilon) - f(a)}{\varepsilon}$$

> $\dfrac{f(a + \varepsilon) - f(a)}{\varepsilon}$ approaches k when $\varepsilon \to 0$.

(The *lim* notation expresses the operation that obtains the value when ε approaches 0.)

Linear function ❶, or $g(x)$, with this k, is an approximate function of $f(x)$. k is called the *differential coefficient* of $f(x)$ at $x = a$.

$$\lim_{\varepsilon \to 0} \frac{f(a + \varepsilon) - f(a)}{\varepsilon}$$

Slope of the line tangent to $y = f(x)$ at any point $(a, f(a))$.

We make symbol f' by attaching a prime to f.

$$f'(a) = \lim_{\varepsilon \to 0} \frac{f(a + \varepsilon) - f(a)}{\varepsilon}$$

$f'(a)$ is the slope of the line tangent to $y = f(x)$ at $x = a$.

Letter a can be replaced with x.

Since f' can been seen as a function of x, it is called "the function derived from function f," or the *derivative* of function f.

CALCULATING THE DERIVATIVE OF A CONSTANT, LINEAR, OR QUADRATIC FUNCTION

1. Let's find the derivative of constant function $f(x) = \alpha$. The differential coefficient of $f(x)$ at $x = a$ is

$$\lim_{\varepsilon \to 0} \frac{f(a + \varepsilon) - f(a)}{\varepsilon} = \lim_{\varepsilon \to 0} \frac{\alpha - \alpha}{\varepsilon} = \lim_{\varepsilon \to 0} 0 = 0$$

Thus, the derivative of $f(x)$ is $f'(x) = 0$. This makes sense, since our function is constant—the rate of change is 0.

NOTE The *differential coefficient* of $f(x)$ at $x = a$ is often simply called the derivative of $f(x)$ at $x = a$, or just $f'(a)$.

2. Let's calculate the derivative of linear function $f(x) = \alpha x + \beta$. The derivative of $f(x)$ at $x = \alpha$ is

$$\lim_{\varepsilon \to 0} \frac{f(\alpha + \varepsilon) - f(a)}{\varepsilon} = \lim_{\varepsilon \to 0} \frac{\alpha(a + \varepsilon) + \beta - (\alpha a + \beta)}{\varepsilon} = \lim_{\varepsilon \to 0} \alpha = \alpha$$

Thus, the derivative of $f(x)$ is $f'(x) = \alpha$, a constant value. This result should also be intuitive—linear functions have a constant rate of change by definition.

3. Let's find the derivative of $f(x) = x^2$, which appeared in the story. The differential coefficient of $f(x)$ at $x = a$ is

$$\lim_{\varepsilon \to 0} \frac{f(a + \varepsilon) - f(a)}{\varepsilon} = \lim_{\varepsilon \to 0} \frac{(a + \varepsilon)^2 - a^2}{\varepsilon} = \lim_{\varepsilon \to 0} \frac{2a\varepsilon + \varepsilon^2}{\varepsilon} = \lim_{\varepsilon \to 0} (2a + \varepsilon) = 2a$$

Thus, the differential coefficient of $f(x)$ at $x = a$ is $2a$, or $f'(a) = 2a$. Therefore, the derivative of $f(x)$ is $f'(x) = 2x$.

SUMMARY

*   The calculation of a limit that appears in calculus is simply a formula calculating an error.
*   A limit is used to obtain a derivative.
*   The derivative is the slope of the tangent line at a given point.
*   The derivative is nothing but the rate of change.

The derivative of $f(x)$ at $x = a$ is calculated by

$$\lim_{\varepsilon \to 0} \frac{f(a + \varepsilon) - f(a)}{\varepsilon}$$

$g(x) = f'(a)(x - a) + f(a)$ is then the *approximate linear function* of $f(x)$.
$f'(x)$, which expresses the slope of the line tangent to $f(x)$ at the point $(x, f(x))$, is called the *derivative* of $f(x)$, because it is derived from $f(x)$.

Other than $f'(x)$, the following symbols are also used to denote the derivative of $y = f(x)$.

$$y', \quad \frac{dy}{dx}, \quad \frac{df}{dx}, \quad \frac{d}{dx} f(x)$$

EXERCISES

1. We have function $f(x)$ and linear function $g(x) = 8x + 10$. It is known that the relative error of the two functions approaches 0 when x approaches 5.

    A. Obtain $f(5)$.

    B. Obtain $f'(5)$.

2. For $f(x) = x^3$, obtain its derivative $f'(x)$.

2
LET'S LEARN DIFFERENTIATION TECHNIQUES!

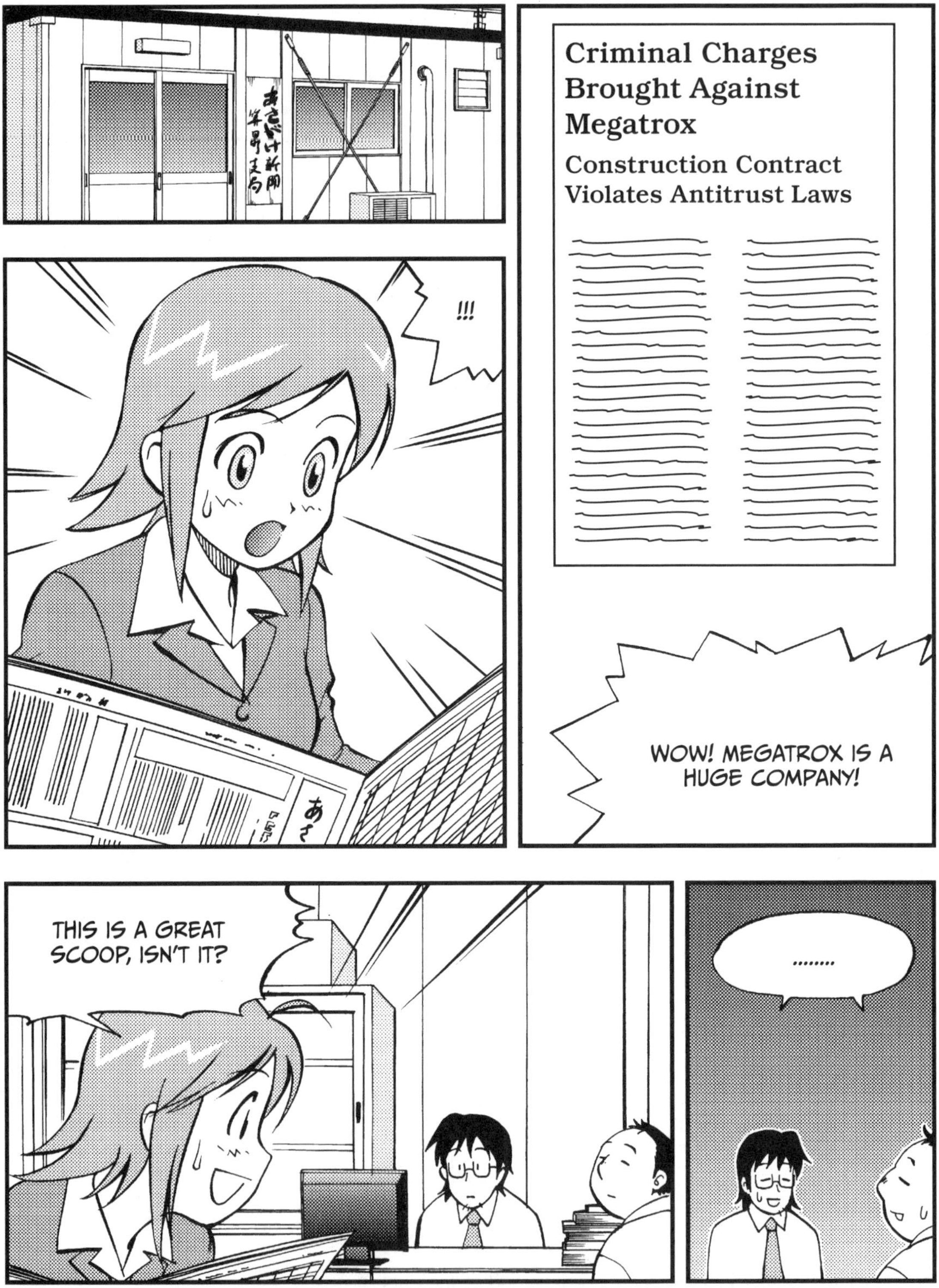
!!!
Criminal Charges Brought Against Megatrox
Construction Contract Violates Antitrust Laws
WOW! MEGATROX IS A HUGE COMPANY!
THIS IS A GREAT SCOOP, ISN'T IT?
....

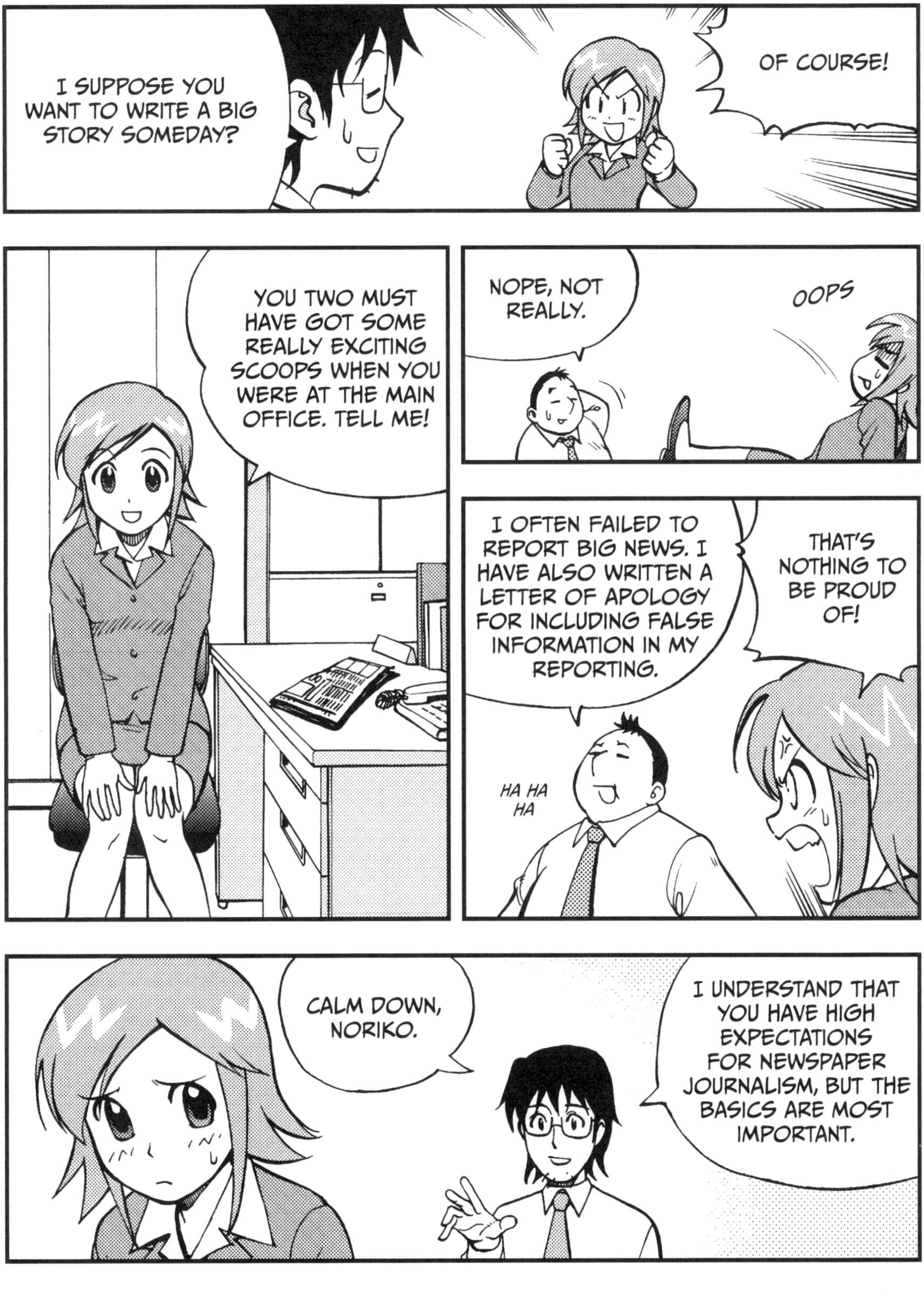
I SUPPOSE YOU WANT TO WRITE A BIG STORY SOMEDAY?
OF COURSE!
YOU TWO MUST HAVE GOT SOME REALLY EXCITING SCOOPS WHEN YOU WERE AT THE MAIN OFFICE. TELL ME!
NOPE, NOT REALLY.
OOPS
I OFTEN FAILED TO REPORT BIG NEWS. I HAVE ALSO WRITTEN A LETTER OF APOLOGY FOR INCLUDING FALSE INFORMATION IN MY REPORTING.
HA HA HA
THAT'S NOTHING TO BE PROUD OF!
CALM DOWN, NORIKO.
I UNDERSTAND THAT YOU HAVE HIGH EXPECTATIONS FOR NEWSPAPER JOURNALISM, BUT THE BASICS ARE MOST IMPORTANT.

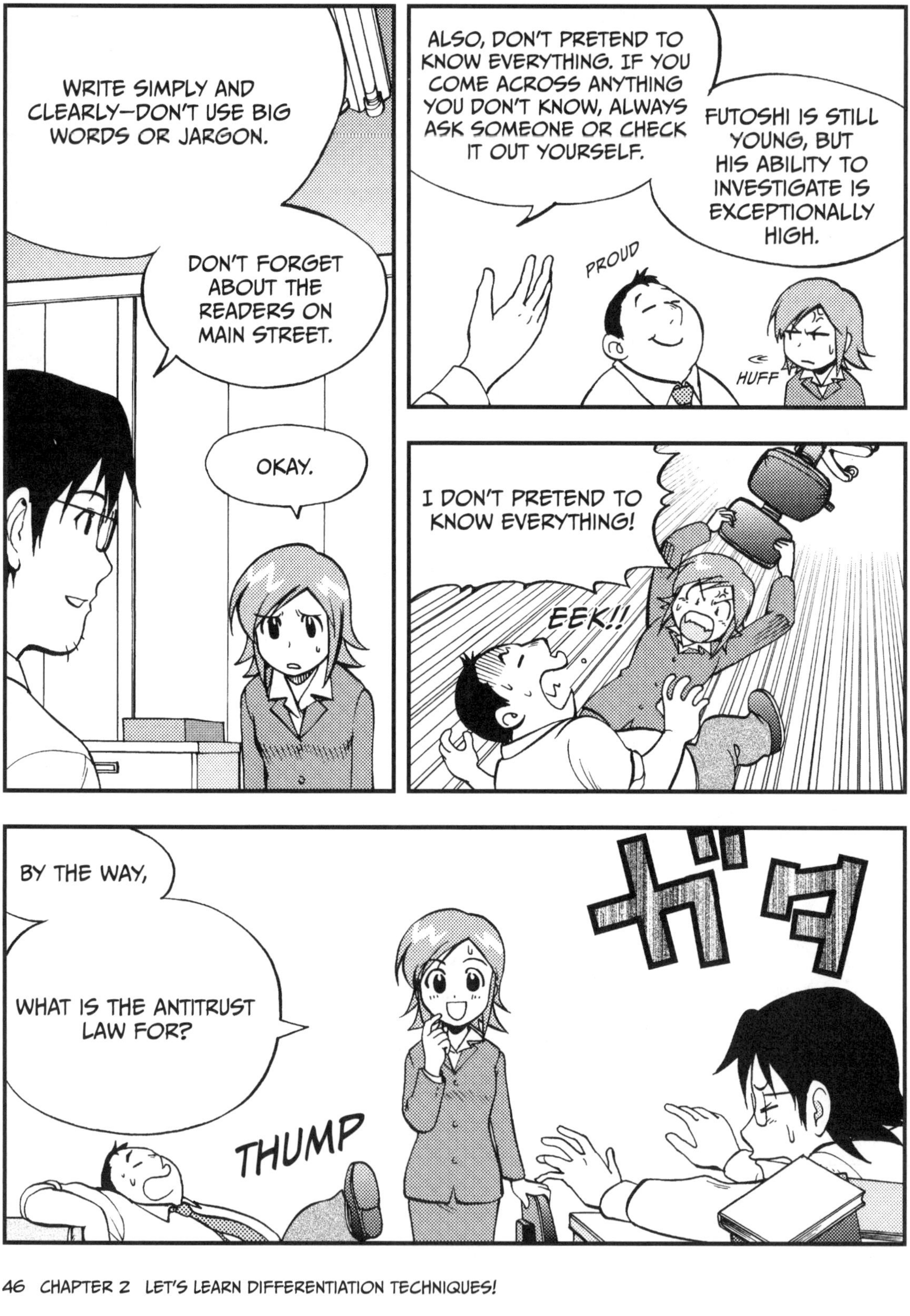

WRITE SIMPLY AND CLEARLY—DON'T USE BIG WORDS OR JARGON.
DON'T FORGET ABOUT THE READERS ON MAIN STREET.
OKAY.
ALSO, DON'T PRETEND TO KNOW EVERYTHING. IF YOU COME ACROSS ANYTHING YOU DON'T KNOW, ALWAYS ASK SOMEONE OR CHECK IT OUT YOURSELF.
FUTOSHI IS STILL YOUNG, BUT HIS ABILITY TO INVESTIGATE IS EXCEPTIONALLY HIGH.
PROUD
HUFF
I DON'T PRETEND TO KNOW EVERYTHING!
EEK!!
BY THE WAY,
WHAT IS THE ANTITRUST LAW FOR?
ガタ
THUMP

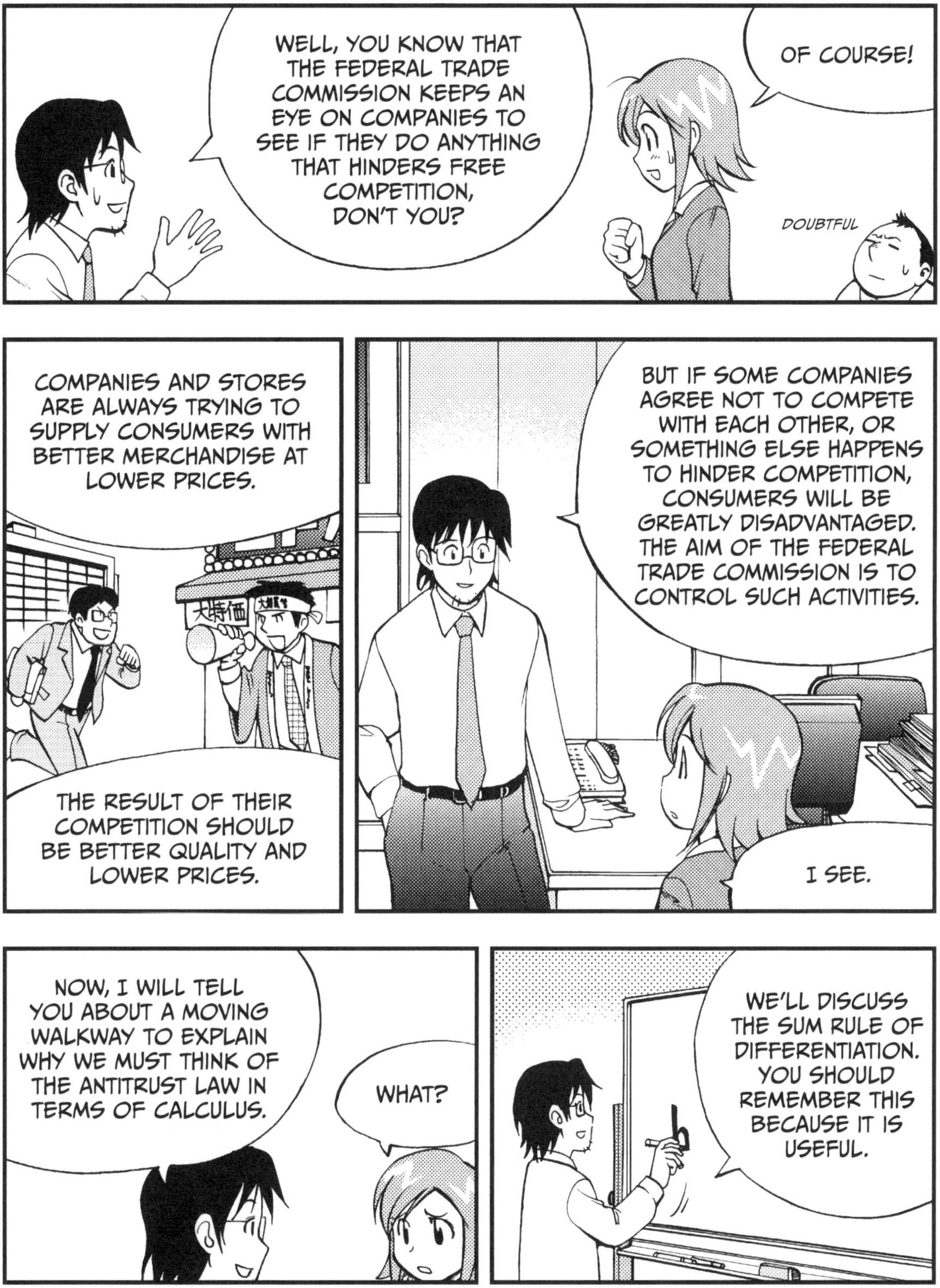
WELL, YOU KNOW THAT THE FEDERAL TRADE COMMISSION KEEPS AN EYE ON COMPANIES TO SEE IF THEY DO ANYTHING THAT HINDERS FREE COMPETITION, DON'T YOU?
OF COURSE!
DOUBTFUL
COMPANIES AND STORES ARE ALWAYS TRYING TO SUPPLY CONSUMERS WITH BETTER MERCHANDISE AT LOWER PRICES.
THE RESULT OF THEIR COMPETITION SHOULD BE BETTER QUALITY AND LOWER PRICES.
BUT IF SOME COMPANIES AGREE NOT TO COMPETE WITH EACH OTHER, OR SOMETHING ELSE HAPPENS TO HINDER COMPETITION, CONSUMERS WILL BE GREATLY DISADVANTAGED. THE AIM OF THE FEDERAL TRADE COMMISSION IS TO CONTROL SUCH ACTIVITIES.
I SEE.
NOW, I WILL TELL YOU ABOUT A MOVING WALKWAY TO EXPLAIN WHY WE MUST THINK OF THE ANTITRUST LAW IN TERMS OF CALCULUS.
WHAT?
WE'LL DISCUSS THE SUM RULE OF DIFFERENTIATION. YOU SHOULD REMEMBER THIS BECAUSE IT IS USEFUL.

THE SUM RULE OF DIFFERENTIATION

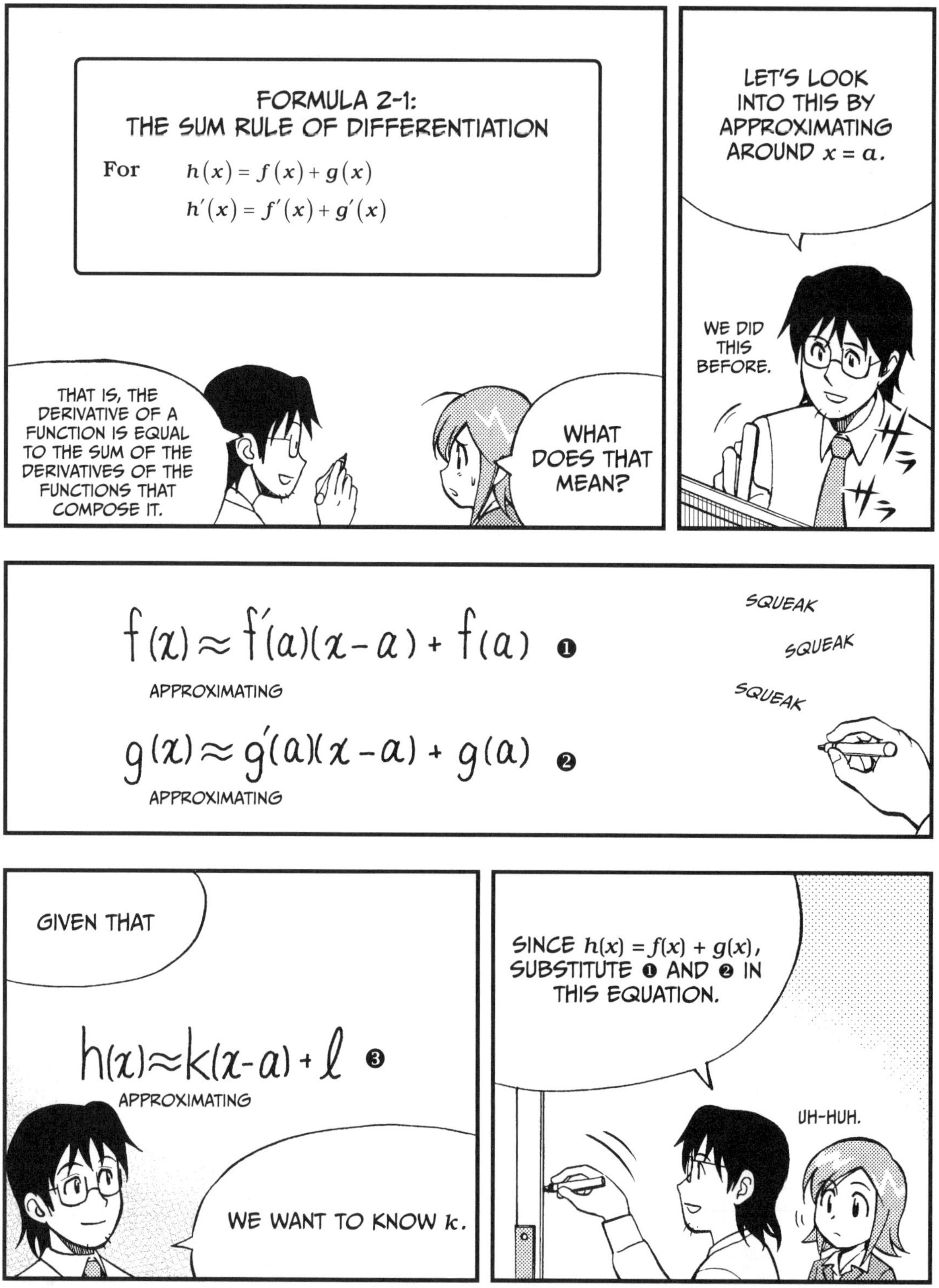
FORMULA 2-1: THE SUM RULE OF DIFFERENTIATION
For $h(x) = f(x) + g(x)$
$h'(x) = f'(x) + g'(x)$
THAT IS, THE DERIVATIVE OF A FUNCTION IS EQUAL TO THE SUM OF THE DERIVATIVES OF THE FUNCTIONS THAT COMPOSE IT.
WHAT DOES THAT MEAN?
LET'S LOOK INTO THIS BY APPROXIMATING AROUND $x = a$.
WE DID THIS BEFORE.
$f(x) \approx f'(a)(x-a) + f(a)$ ❶
APPROXIMATING
$g(x) \approx g'(a)(x-a) + g(a)$ ❷
APPROXIMATING
SQUEAK
SQUEAK
SQUEAK
GIVEN THAT
$h(x) \approx k(x-a) + l$ ❸
APPROXIMATING
WE WANT TO KNOW k.
SINCE $h(x) = f(x) + g(x)$, SUBSTITUTE ❶ AND ❷ IN THIS EQUATION.
UH-HUH.

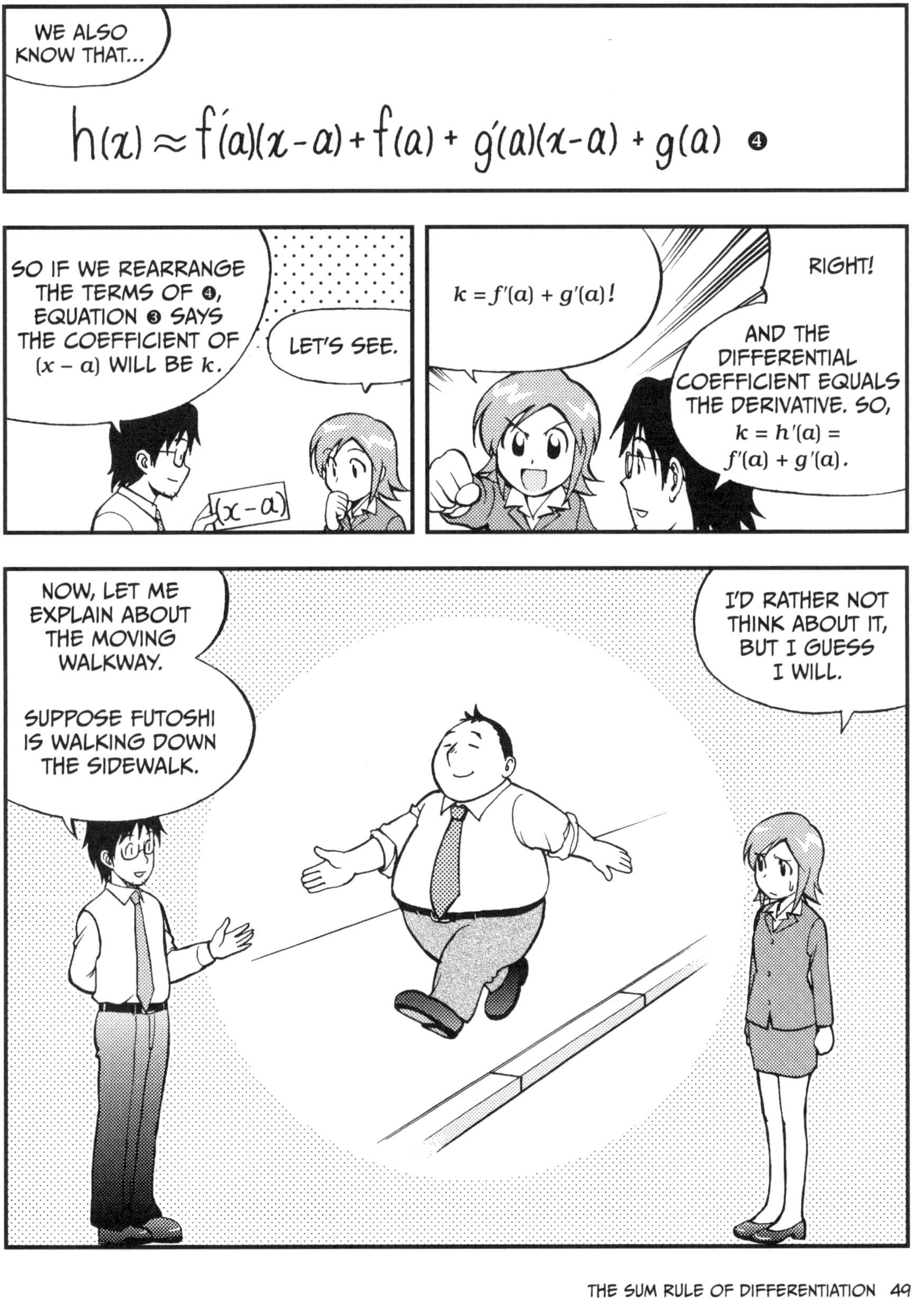
WE ALSO KNOW THAT...
$h(x) \approx f'(a)(x-a) + f(a) + g'(a)(x-a) + g(a)$ ❹
SO IF WE REARRANGE THE TERMS OF ❹, EQUATION ❸ SAYS THE COEFFICIENT OF $(x - a)$ WILL BE k.
LET'S SEE.
$(x-a)$
$k = f'(a) + g'(a)$!
RIGHT!
AND THE DIFFERENTIAL COEFFICIENT EQUALS THE DERIVATIVE. SO, $k = h'(a) = f'(a) + g'(a)$.
NOW, LET ME EXPLAIN ABOUT THE MOVING WALKWAY.
SUPPOSE FUTOSHI IS WALKING DOWN THE SIDEWALK.
I'D RATHER NOT THINK ABOUT IT, BUT I GUESS I WILL.

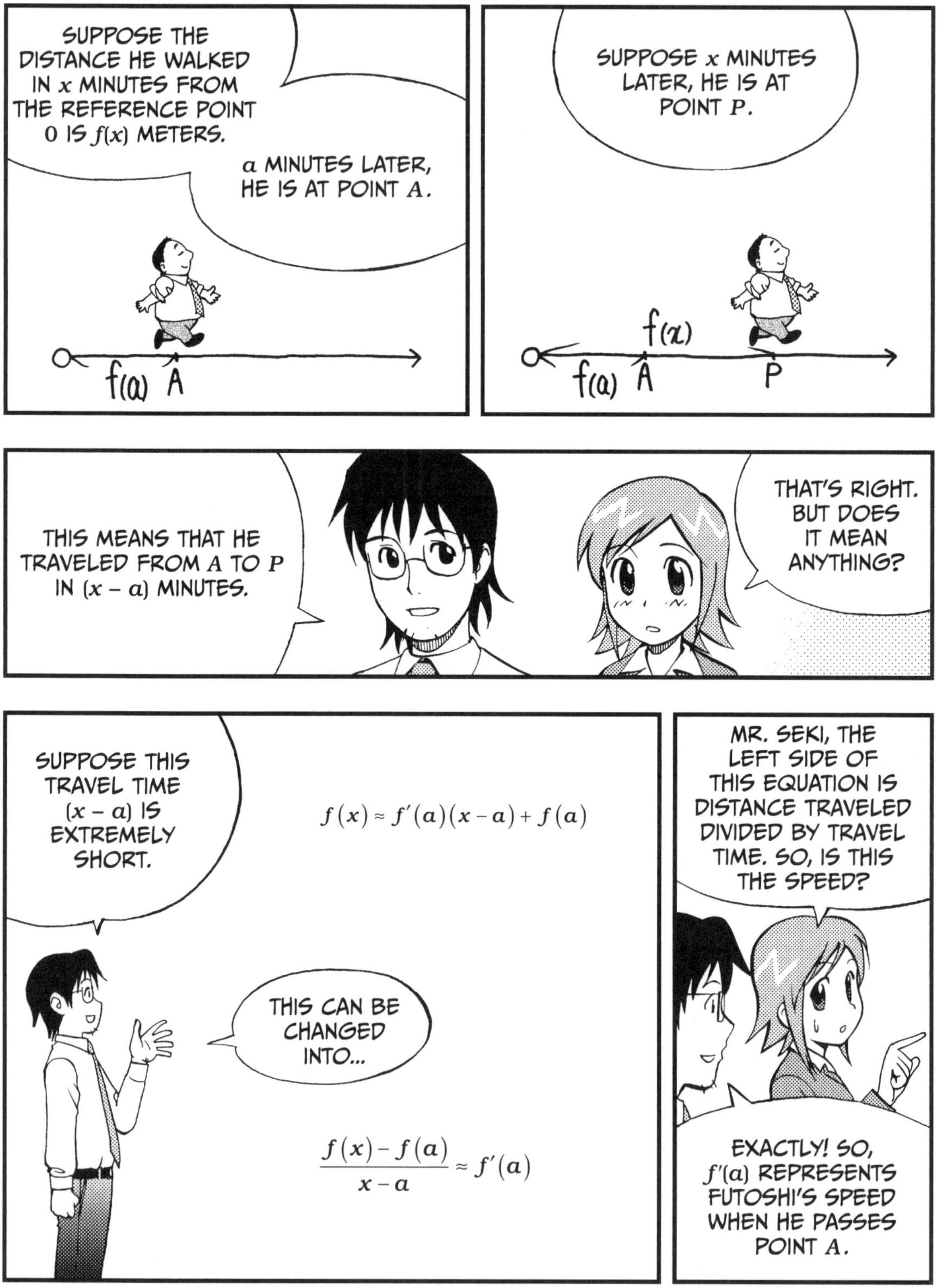
SUPPOSE THE DISTANCE HE WALKED IN x MINUTES FROM THE REFERENCE POINT 0 IS $f(x)$ METERS.
a MINUTES LATER, HE IS AT POINT A.
O
f(a)
A
SUPPOSE x MINUTES LATER, HE IS AT POINT P.
O
f(x)
f(a)
A
P
THIS MEANS THAT HE TRAVELED FROM A TO P IN $(x - a)$ MINUTES.
THAT'S RIGHT. BUT DOES IT MEAN ANYTHING?
SUPPOSE THIS TRAVEL TIME $(x - a)$ IS EXTREMELY SHORT.
$f(x) \approx f'(a)(x-a)+f(a)$
THIS CAN BE CHANGED INTO...
$\frac{f(x)-f(a)}{x-a} \approx f'(a)$
MR. SEKI, THE LEFT SIDE OF THIS EQUATION IS DISTANCE TRAVELED DIVIDED BY TRAVEL TIME. SO, IS THIS THE SPEED?
EXACTLY! SO, $f'(a)$ REPRESENTS FUTOSHI'S SPEED WHEN HE PASSES POINT A.

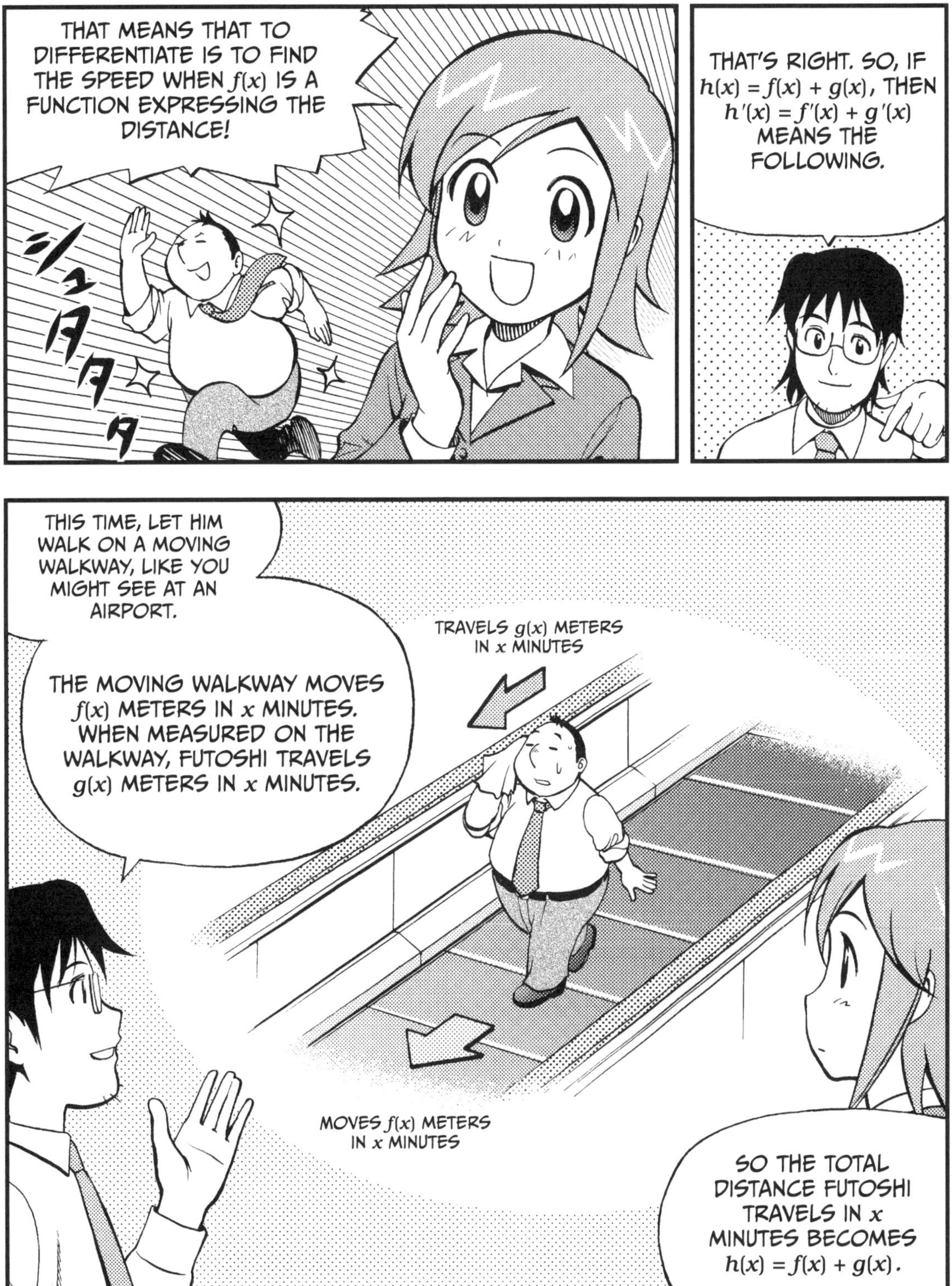
THAT MEANS THAT TO DIFFERENTIATE IS TO FIND THE SPEED WHEN $f(x)$ IS A FUNCTION EXPRESSING THE DISTANCE!
THAT'S RIGHT. SO, IF $h(x) = f(x) + g(x)$, THEN $h'(x) = f'(x) + g'(x)$ MEANS THE FOLLOWING.
THIS TIME, LET HIM WALK ON A MOVING WALKWAY, LIKE YOU MIGHT SEE AT AN AIRPORT.
THE MOVING WALKWAY MOVES $f(x)$ METERS IN x MINUTES. WHEN MEASURED ON THE WALKWAY, FUTOSHI TRAVELS $g(x)$ METERS IN x MINUTES.
TRAVELS $g(x)$ METERS IN x MINUTES
MOVES $f(x)$ METERS IN x MINUTES
SO THE TOTAL DISTANCE FUTOSHI TRAVELS IN x MINUTES BECOMES $h(x) = f(x) + g(x)$.

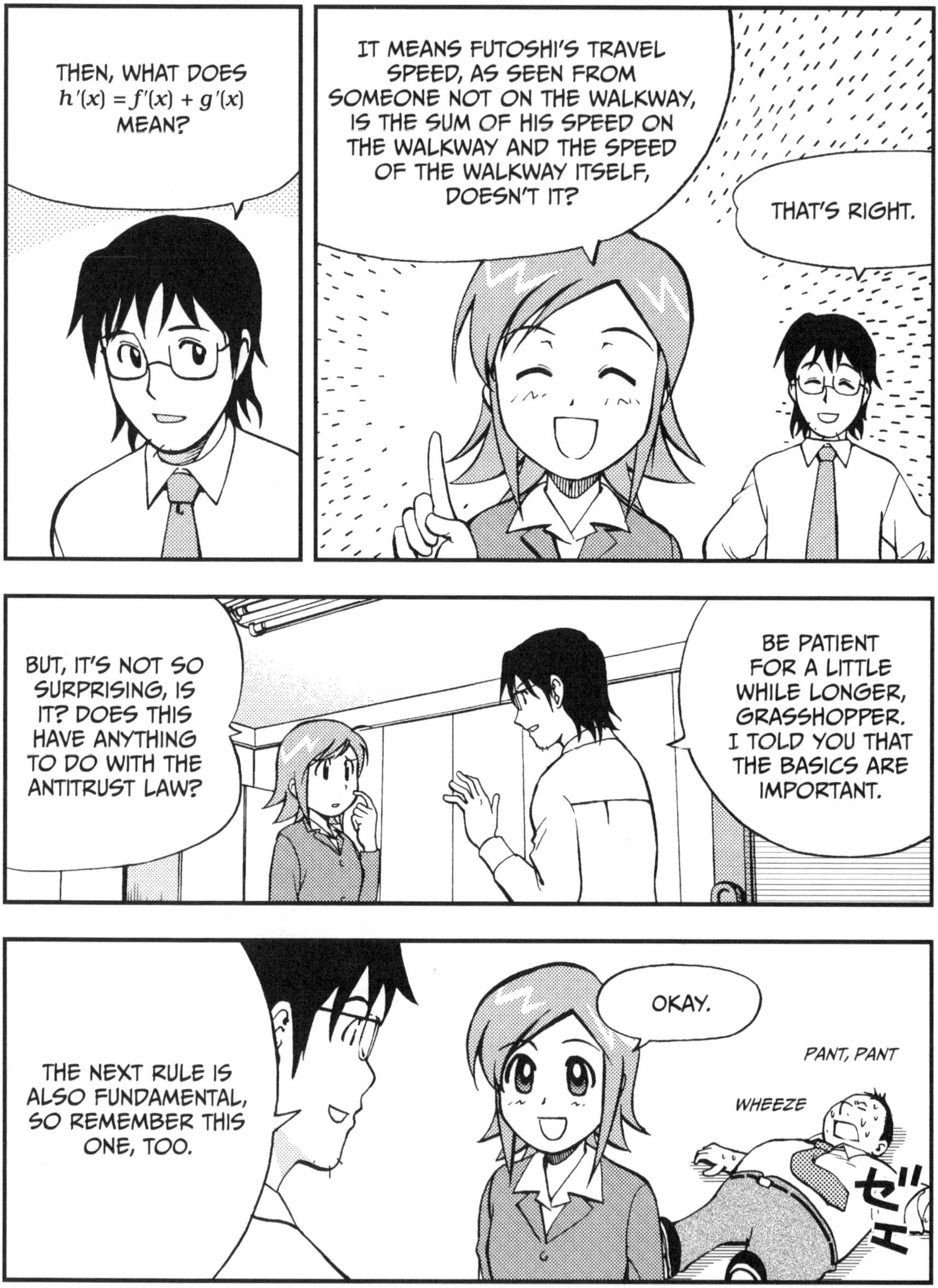
THEN, WHAT DOES $h'(x) = f'(x) + g'(x)$ MEAN?
IT MEANS FUTOSHI'S TRAVEL SPEED, AS SEEN FROM SOMEONE NOT ON THE WALKWAY, IS THE SUM OF HIS SPEED ON THE WALKWAY AND THE SPEED OF THE WALKWAY ITSELF, DOESN'T IT?
THAT'S RIGHT.
BUT, IT'S NOT SO SURPRISING, IS IT? DOES THIS HAVE ANYTHING TO DO WITH THE ANTITRUST LAW?
BE PATIENT FOR A LITTLE WHILE LONGER, GRASSHOPPER. I TOLD YOU THAT THE BASICS ARE IMPORTANT.
THE NEXT RULE IS ALSO FUNDAMENTAL, SO REMEMBER THIS ONE, TOO.
OKAY.
PANT, PANT
WHEEZE
ゼェ

THE PRODUCT RULE OF DIFFERENTIATION

FORMULA 2-2:
THE PRODUCT RULE OF DIFFERENTIATION

For $\qquad h(x) = f(x)g(x)$

$\qquad h'(x) = f'(x)g(x) + f(x)g'(x)$

The derivative of a product is the sum of the products with only one function differentiated.

ONLY ONE FUNCTION?

YES. LET'S CONSIDER $x = a$.

$$f(x) \approx f'(a)(x-a) + f(a)$$

$$g(x) \approx g'(a)(x-a) + g(a)$$

$h(x) = f(x)g(x) \approx k(x-a) + l$

$h(x) \approx \{f'(a)(x-a) + f(a)\} \times \{g'(a)(x-a) + g(a)\}$

$h(x) \approx f'(a)g'(a)(x-a)^2 + f(a)g'(a)(x-a) + f'(a)(x-a)g(a) + f(a)g(a)$

$(x - a)$ IS A SMALL CHANGE. THAT MEANS $(x - a)^2$ IS VERY, VERY SMALL. SINCE WE ARE APPROXIMATING, WE CAN THROW THAT TERM OUT.

$h(x) \approx \{f'(a)g(a) + f(a)g'(a)\}(x-a) + f(a)g(a)$

$k = f'(a)g(a) + f(a)g'(a)$

WE GET THIS.

NOW, I WILL USE DIFFERENTIATION TO EXPLAIN WHY A MONOPOLY SHOULD NOT BE ALLOWED.

HOW DO YOU SOLVE A SOCIAL PROBLEM USING DIFFERENTIATION?

ISN'T IT RATHER AN ISSUE OF MORALITY, JUSTICE, AND TRUTH?

LET'S LOOK AT THE WORLD IN A MORE BUSINESSLIKE MANNER.

A MARKET WHERE MANY COMPANIES SUPPLY PRODUCTS THAT CANNOT BE DISCRIMINATED BETWEEN IS CALLED "A PERFECTLY COMPETITIVE MARKET."

FOR EXAMPLE?

PERFECTLY COMPETITIVE MARKET

LET'S SEE... VIDEO RENTAL SHOPS?

THAT'S RIGHT.* COMPANIES IN A PERFECTLY COMPETITIVE MARKET ACCEPT THE COMMODITY PRICE DETERMINED BY THE MARKET AND CONTINUE TO PRODUCE AND SUPPLY THEIR PRODUCT AS LONG AS THEY MAKE PROFITS.

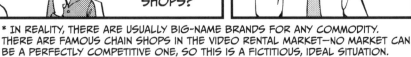

* IN REALITY, THERE ARE USUALLY BIG-NAME BRANDS FOR ANY COMMODITY. THERE ARE FAMOUS CHAIN SHOPS IN THE VIDEO RENTAL MARKET—NO MARKET CAN BE A PERFECTLY COMPETITIVE ONE, SO THIS IS A FICTITIOUS, IDEAL SITUATION.

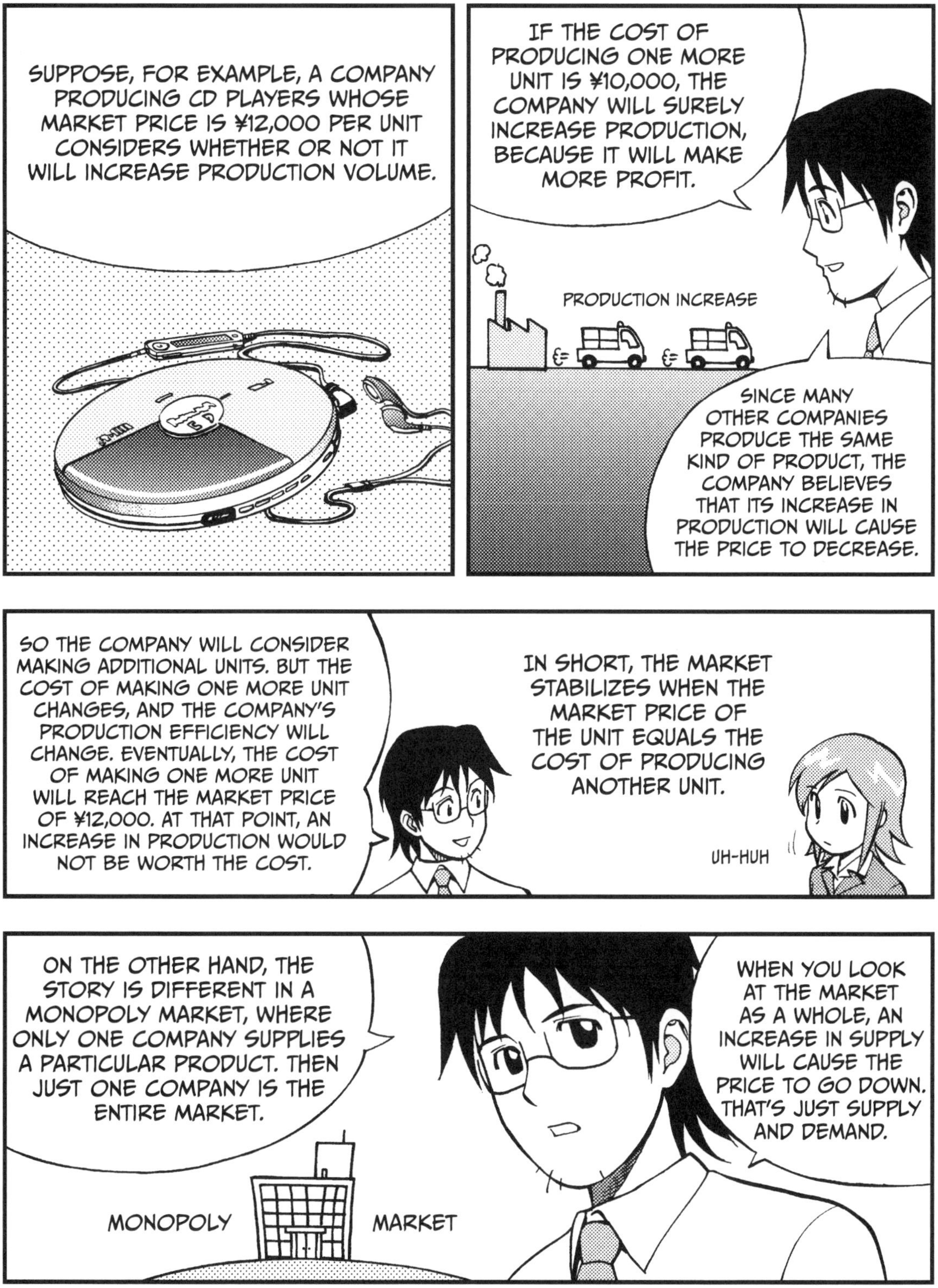
SUPPOSE, FOR EXAMPLE, A COMPANY PRODUCING CD PLAYERS WHOSE MARKET PRICE IS ¥12,000 PER UNIT CONSIDERS WHETHER OR NOT IT WILL INCREASE PRODUCTION VOLUME.
IF THE COST OF PRODUCING ONE MORE UNIT IS ¥10,000, THE COMPANY WILL SURELY INCREASE PRODUCTION, BECAUSE IT WILL MAKE MORE PROFIT.
PRODUCTION INCREASE
SINCE MANY OTHER COMPANIES PRODUCE THE SAME KIND OF PRODUCT, THE COMPANY BELIEVES THAT ITS INCREASE IN PRODUCTION WILL CAUSE THE PRICE TO DECREASE.
SO THE COMPANY WILL CONSIDER MAKING ADDITIONAL UNITS. BUT THE COST OF MAKING ONE MORE UNIT CHANGES, AND THE COMPANY'S PRODUCTION EFFICIENCY WILL CHANGE. EVENTUALLY, THE COST OF MAKING ONE MORE UNIT WILL REACH THE MARKET PRICE OF ¥12,000. AT THAT POINT, AN INCREASE IN PRODUCTION WOULD NOT BE WORTH THE COST.
IN SHORT, THE MARKET STABILIZES WHEN THE MARKET PRICE OF THE UNIT EQUALS THE COST OF PRODUCING ANOTHER UNIT.
UH-HUH
ON THE OTHER HAND, THE STORY IS DIFFERENT IN A MONOPOLY MARKET, WHERE ONLY ONE COMPANY SUPPLIES A PARTICULAR PRODUCT. THEN JUST ONE COMPANY IS THE ENTIRE MARKET.
MONOPOLY
MARKET
WHEN YOU LOOK AT THE MARKET AS A WHOLE, AN INCREASE IN SUPPLY WILL CAUSE THE PRICE TO GO DOWN. THAT'S JUST SUPPLY AND DEMAND.

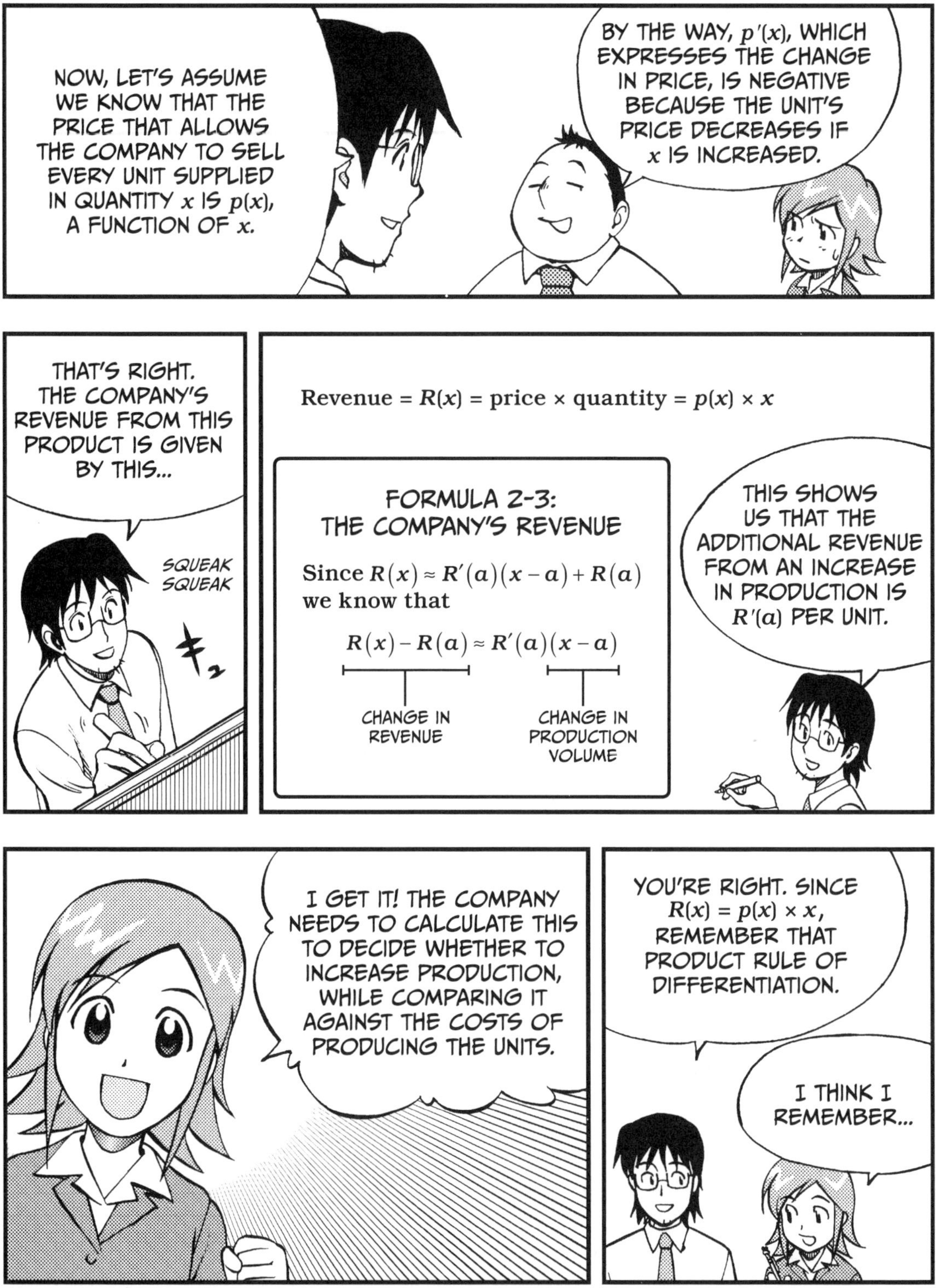
NOW, LET'S ASSUME WE KNOW THAT THE PRICE THAT ALLOWS THE COMPANY TO SELL EVERY UNIT SUPPLIED IN QUANTITY x IS $p(x)$, A FUNCTION OF x.
BY THE WAY, $p'(x)$, WHICH EXPRESSES THE CHANGE IN PRICE, IS NEGATIVE BECAUSE THE UNIT'S PRICE DECREASES IF x IS INCREASED.
THAT'S RIGHT. THE COMPANY'S REVENUE FROM THIS PRODUCT IS GIVEN BY THIS...
SQUEAK SQUEAK
Revenue = $R(x)$ = price × quantity = $p(x) \times x$
FORMULA 2-3: THE COMPANY'S REVENUE
Since $R(x) \approx R'(a)(x-a) + R(a)$ we know that
$R(x) - R(a) \approx R'(a)(x-a)$
CHANGE IN REVENUE
CHANGE IN PRODUCTION VOLUME
THIS SHOWS US THAT THE ADDITIONAL REVENUE FROM AN INCREASE IN PRODUCTION IS $R'(a)$ PER UNIT.
I GET IT! THE COMPANY NEEDS TO CALCULATE THIS TO DECIDE WHETHER TO INCREASE PRODUCTION, WHILE COMPARING IT AGAINST THE COSTS OF PRODUCING THE UNITS.
YOU'RE RIGHT. SINCE $R(x) = p(x) \times x$, REMEMBER THAT PRODUCT RULE OF DIFFERENTIATION.
I THINK I REMEMBER...

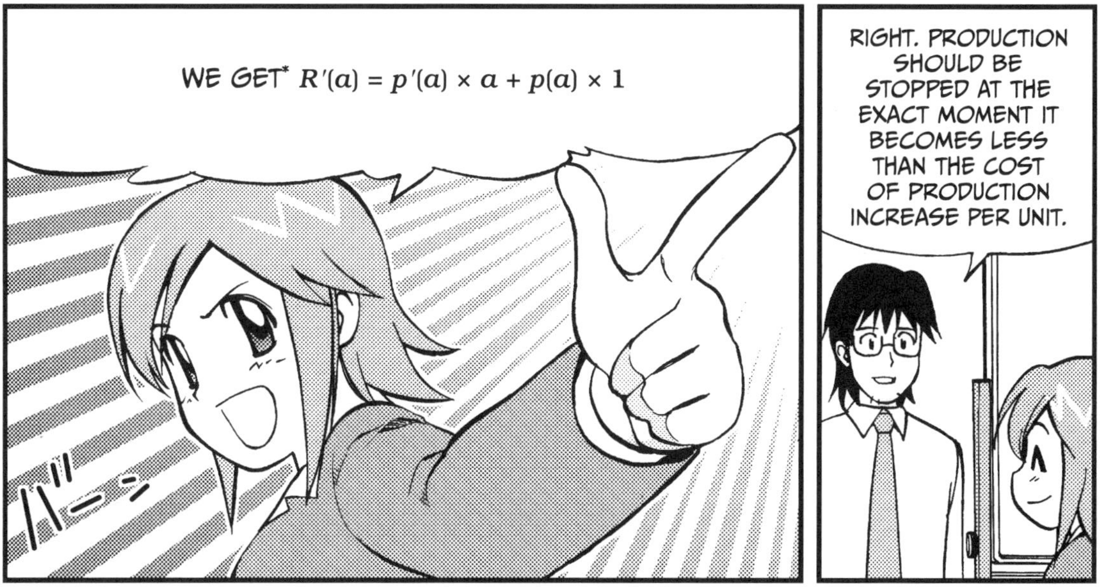

WE GET* $R'(a) = p'(a) \times a + p(a) \times 1$
バーン
RIGHT. PRODUCTION SHOULD BE STOPPED AT THE EXACT MOMENT IT BECOMES LESS THAN THE COST OF PRODUCTION INCREASE PER UNIT.

* THE DERIVATIVE OF x IS 1 (SEE PAGE 40 FOR MORE ON DIFFERENTIATING LINEAR FUNCTIONS).

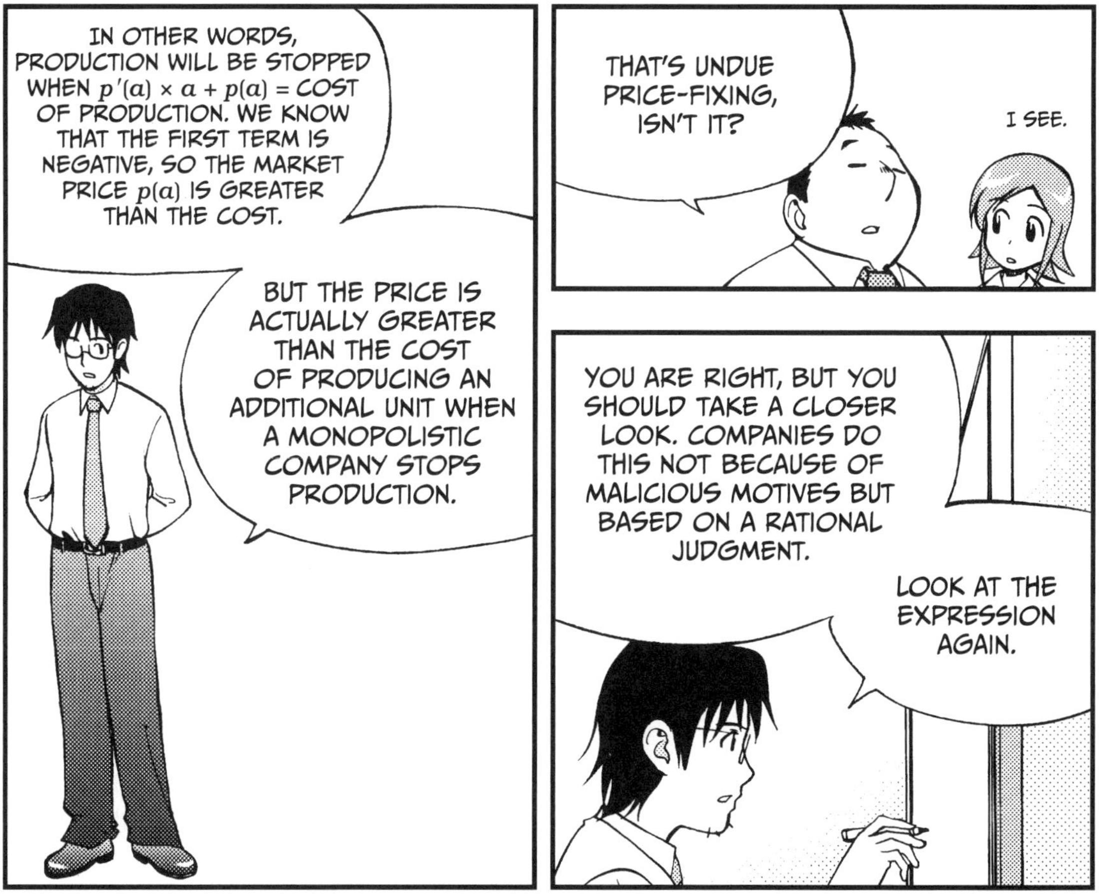

IN OTHER WORDS, PRODUCTION WILL BE STOPPED WHEN $p'(a) \times a + p(a)$ = COST OF PRODUCTION. WE KNOW THAT THE FIRST TERM IS NEGATIVE, SO THE MARKET PRICE $p(a)$ IS GREATER THAN THE COST.
BUT THE PRICE IS ACTUALLY GREATER THAN THE COST OF PRODUCING AN ADDITIONAL UNIT WHEN A MONOPOLISTIC COMPANY STOPS PRODUCTION.
THAT'S UNDUE PRICE-FIXING, ISN'T IT?
I SEE.
YOU ARE RIGHT, BUT YOU SHOULD TAKE A CLOSER LOOK. COMPANIES DO THIS NOT BECAUSE OF MALICIOUS MOTIVES BUT BASED ON A RATIONAL JUDGMENT.
LOOK AT THE EXPRESSION AGAIN.

Sales increase (per unit) when production is increased a little more:

$$R'(a) = p'(a)a + p(a)$$

The two terms in the last expression mean the following:

$p(a)$ represents the revenue from selling a units

$p'(a)a$ = Rate of price decrease × Amount of production
= A heavy loss due to price decrease influencing all units

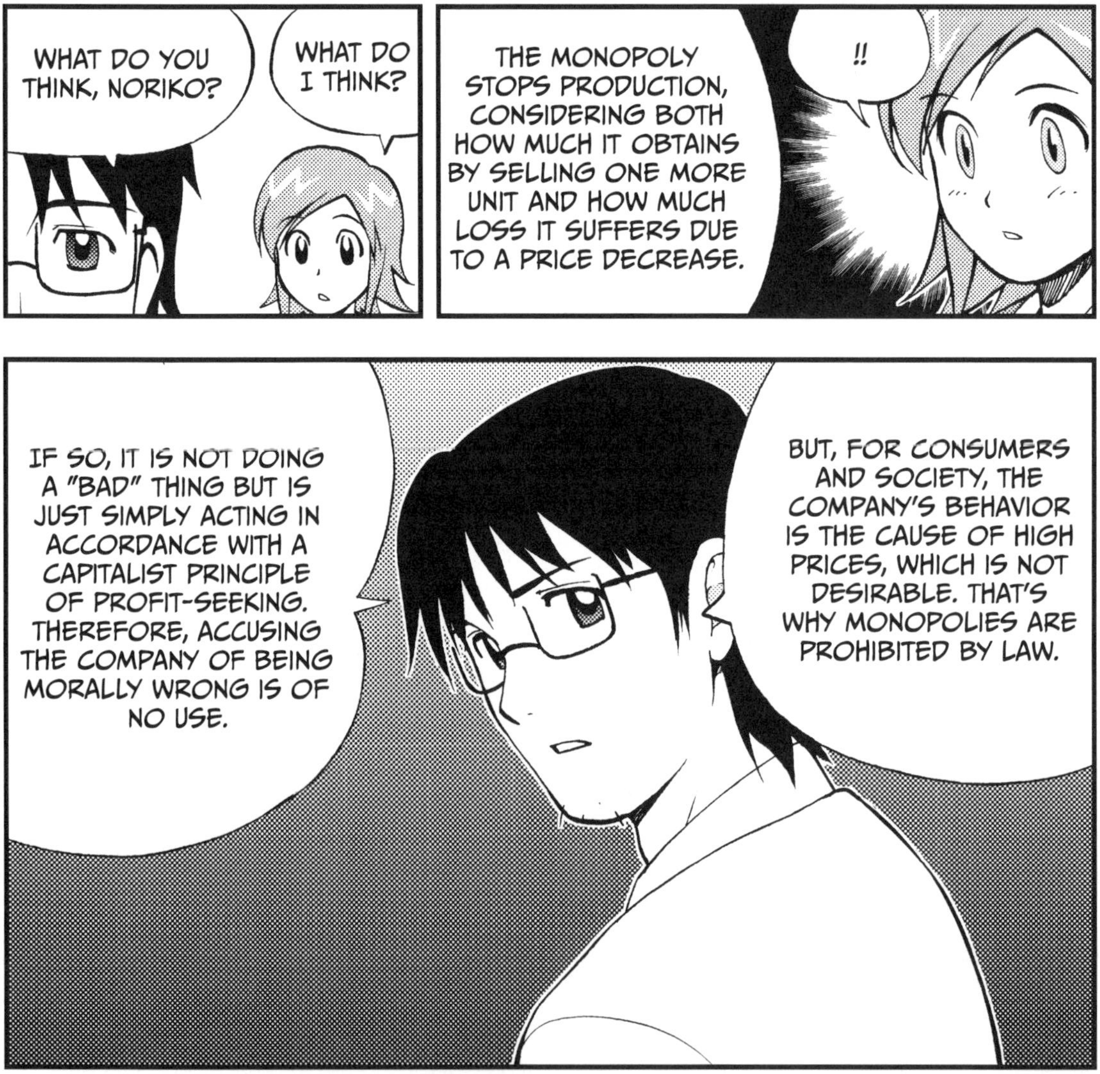
WHAT DO YOU THINK, NORIKO?
WHAT DO I THINK?
THE MONOPOLY STOPS PRODUCTION, CONSIDERING BOTH HOW MUCH IT OBTAINS BY SELLING ONE MORE UNIT AND HOW MUCH LOSS IT SUFFERS DUE TO A PRICE DECREASE.
!!
IF SO, IT IS NOT DOING A "BAD" THING BUT IS JUST SIMPLY ACTING IN ACCORDANCE WITH A CAPITALIST PRINCIPLE OF PROFIT-SEEKING. THEREFORE, ACCUSING THE COMPANY OF BEING MORALLY WRONG IS OF NO USE.
BUT, FOR CONSUMERS AND SOCIETY, THE COMPANY'S BEHAVIOR IS THE CAUSE OF HIGH PRICES, WHICH IS NOT DESIRABLE. THAT'S WHY MONOPOLIES ARE PROHIBITED BY LAW.

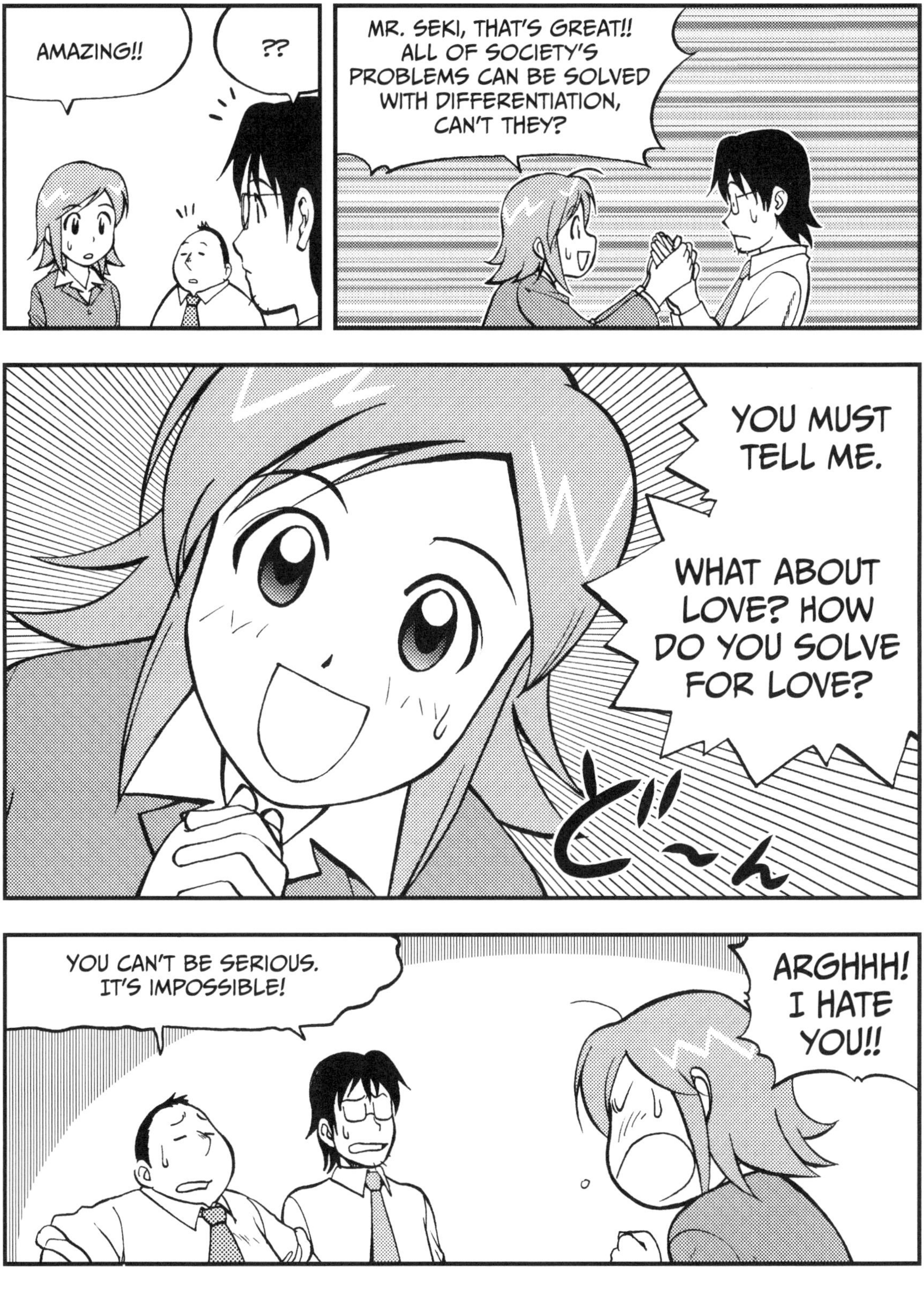
AMAZING!!
??
MR. SEKI, THAT'S GREAT!! ALL OF SOCIETY'S PROBLEMS CAN BE SOLVED WITH DIFFERENTIATION, CAN'T THEY?
YOU MUST TELL ME.
WHAT ABOUT LOVE? HOW DO YOU SOLVE FOR LOVE?
YOU CAN'T BE SERIOUS. IT'S IMPOSSIBLE!
ARGHHH! I HATE YOU!!

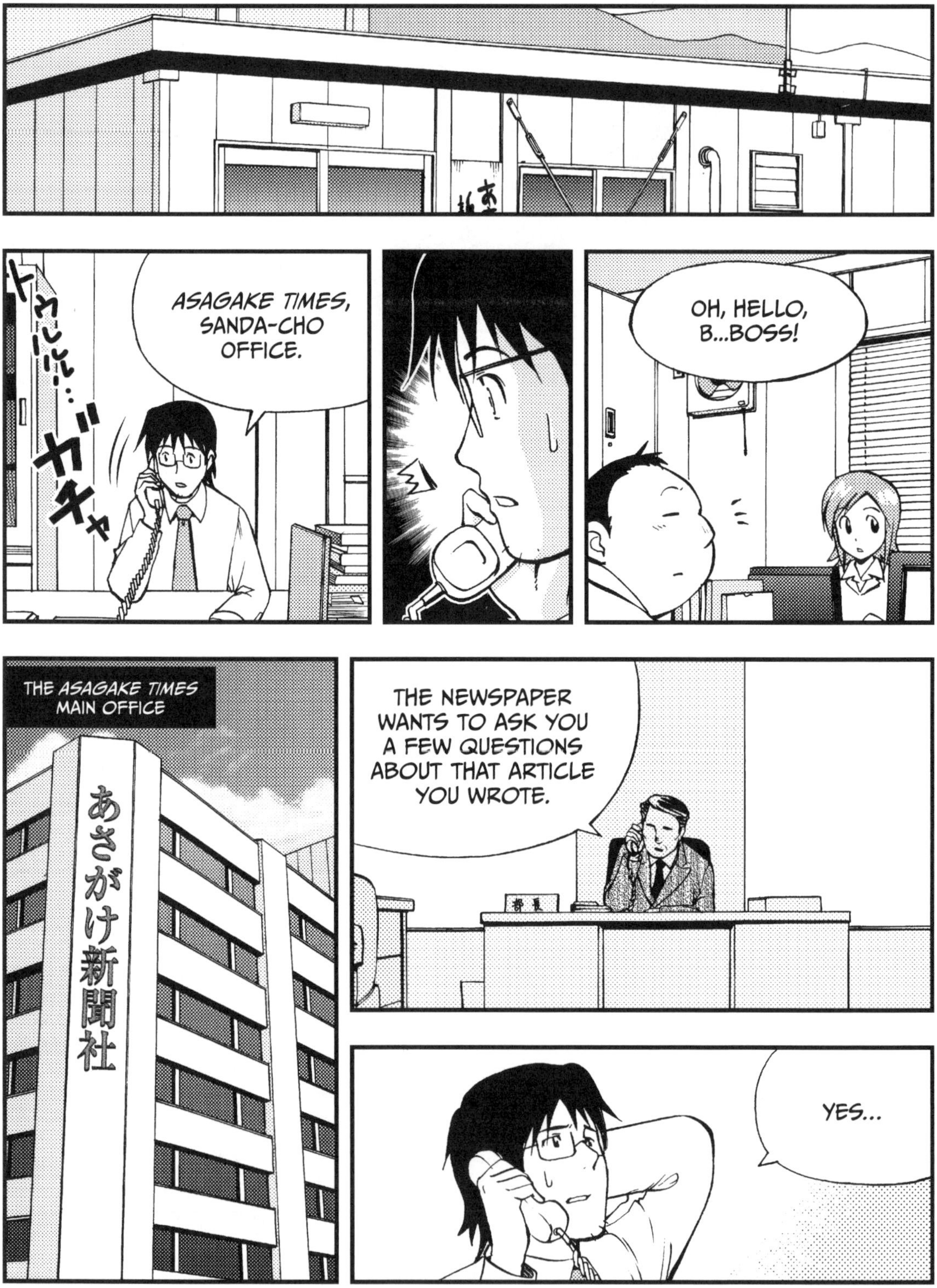
トゥルルル…
ASAGAKE TIMES, SANDA-CHO OFFICE.
ガチャ
OH, HELLO, B...BOSS!
THE ASAGAKE TIMES MAIN OFFICE
あさがけ新聞社
THE NEWSPAPER WANTS TO ASK YOU A FEW QUESTIONS ABOUT THAT ARTICLE YOU WROTE.
YES...

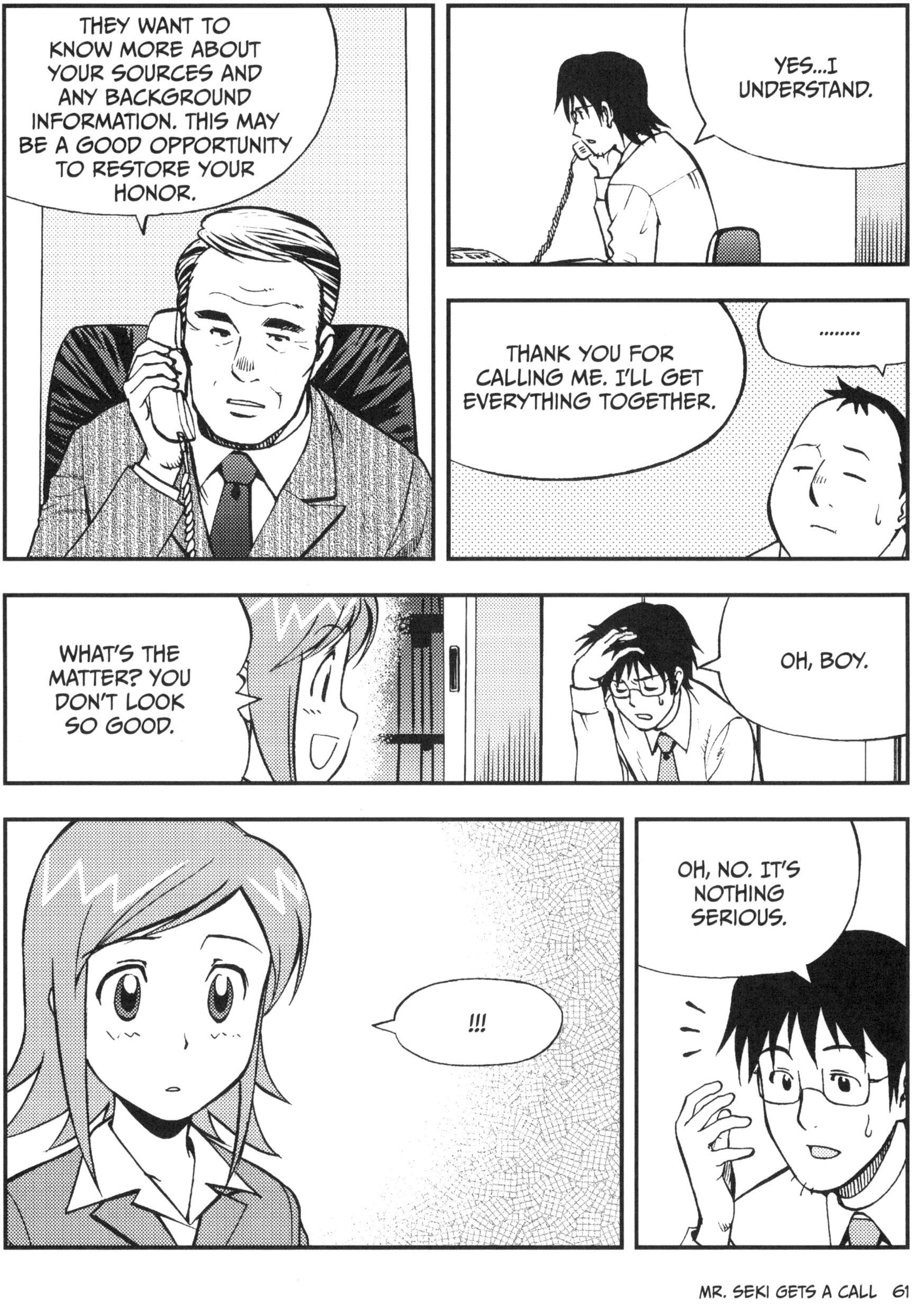
THEY WANT TO KNOW MORE ABOUT YOUR SOURCES AND ANY BACKGROUND INFORMATION. THIS MAY BE A GOOD OPPORTUNITY TO RESTORE YOUR HONOR.
YES...I UNDERSTAND.
THANK YOU FOR CALLING ME. I'LL GET EVERYTHING TOGETHER.
....
WHAT'S THE MATTER? YOU DON'T LOOK SO GOOD.
OH, BOY.
!!!
OH, NO. IT'S NOTHING SERIOUS.

DIFFERENTIATING POLYNOMIALS

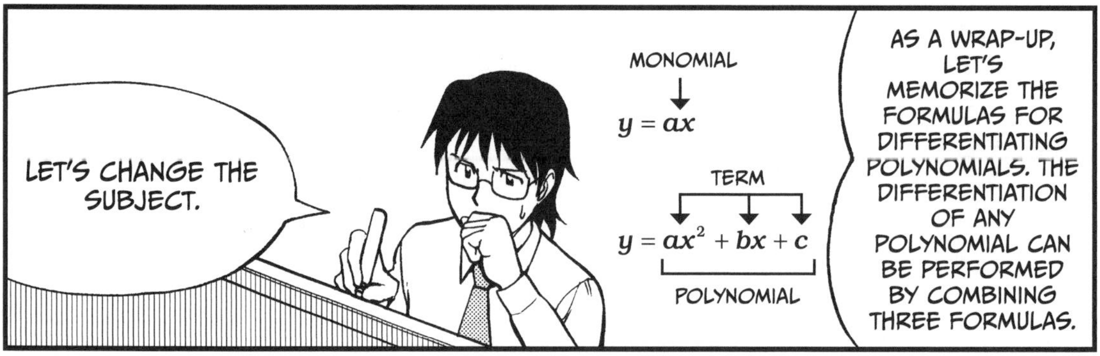
LET'S CHANGE THE SUBJECT.
MONOMIAL
y = ax
TERM
y = ax² + bx + c
POLYNOMIAL
AS A WRAP-UP, LET'S MEMORIZE THE FORMULAS FOR DIFFERENTIATING POLYNOMIALS. THE DIFFERENTIATION OF ANY POLYNOMIAL CAN BE PERFORMED BY COMBINING THREE FORMULAS.

FORMULA 2-4: THE DERIVATIVE OF AN nTH-DEGREE FUNCTION

The derivative of $h(x) = x^n$ is $h'(x) = nx^{n-1}$

How do we get this general rule? We use the product rule of differentiation repeatedly.

For $h(x) = x^2$, since $h(x) = x \times x, h'(x) = x \times 1 + 1 \times x = 2x$ THIS RESULT IS USED

The formula is correct in this case.

For $h(x) = x^3$, since $h(x) = x^2 \times x$, $h'(x) = \left(x^2\right)' \times x + x^2 \times (x)' = (2x)x + x^2 \times 1 = 3x^2$

The formula is correct in this case, too.

For $h(x) = x^4$, since $h(x) = x^3 \times x$, $h'(x) = \left(x^3\right)' \times x + x^3 \times (x)' = 3x^2 \times x + x^3 \times 1 = 4x^3$

Again, the formula is correct. This continues forever. Any polynomial can be differentiated by combining the three formulas!

FORMULA 2-5: THE DIFFERENTIATION FORMULAS OF SUM RULE, CONSTANT MULTIPLICATION, AND x^n

❶ Sum rule: $\{f(x) + g(x)\}' = f'(x) + g'(x)$

❷ Constant multiplication: $\{\alpha f(x)\}' = \alpha f'(x)$

❸ Power rule (x^n): $\{x^n\}' = nx^{n-1}$

Let's see it in action! Differentiate $h(x) = x^3 + 2x^2 + 5x + 3$

$$h'(x) = \{x^3 + 2x^2 + 5x + 3\}' = \underbrace{\left(x^3\right)' + \left(2x^2\right)' + (5x)' + (3)'}_{\text{rule ❶}}$$

$$= \underbrace{\left(x^3\right)' + 2\left(x^2\right)' + 5(x)'}_{\text{rule ❷}} = 3x^2 + 2(2x) + 5 \times 1 = \underbrace{3x^2 + 4x + 5}_{\text{rule ❸}}$$

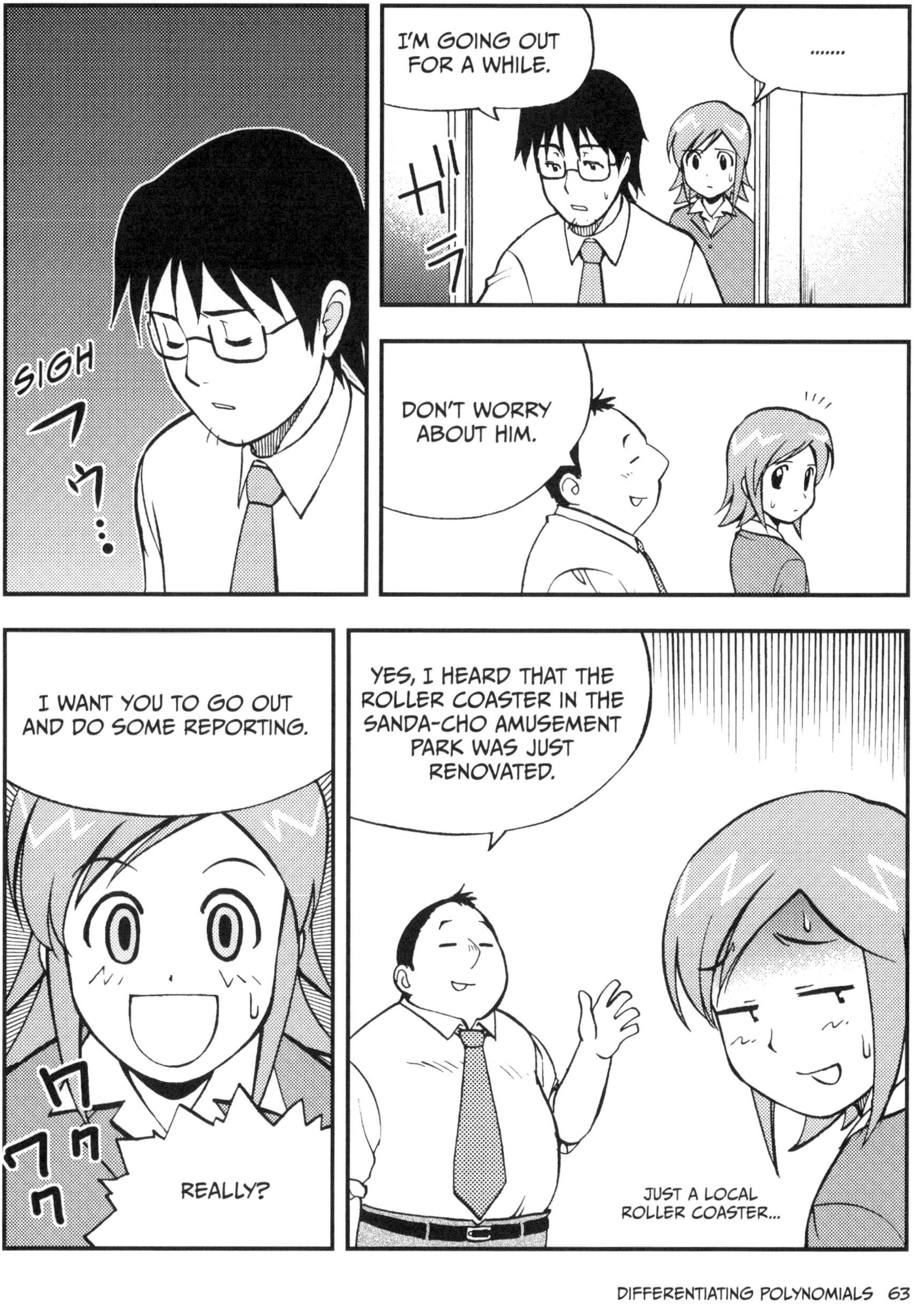
SIGH
I'M GOING OUT FOR A WHILE.
....
DON'T WORRY ABOUT HIM.
I WANT YOU TO GO OUT AND DO SOME REPORTING.
REALLY?
YES, I HEARD THAT THE ROLLER COASTER IN THE SANDA-CHO AMUSEMENT PARK WAS JUST RENOVATED.
JUST A LOCAL ROLLER COASTER...

FINDING MAXIMA AND MINIMA

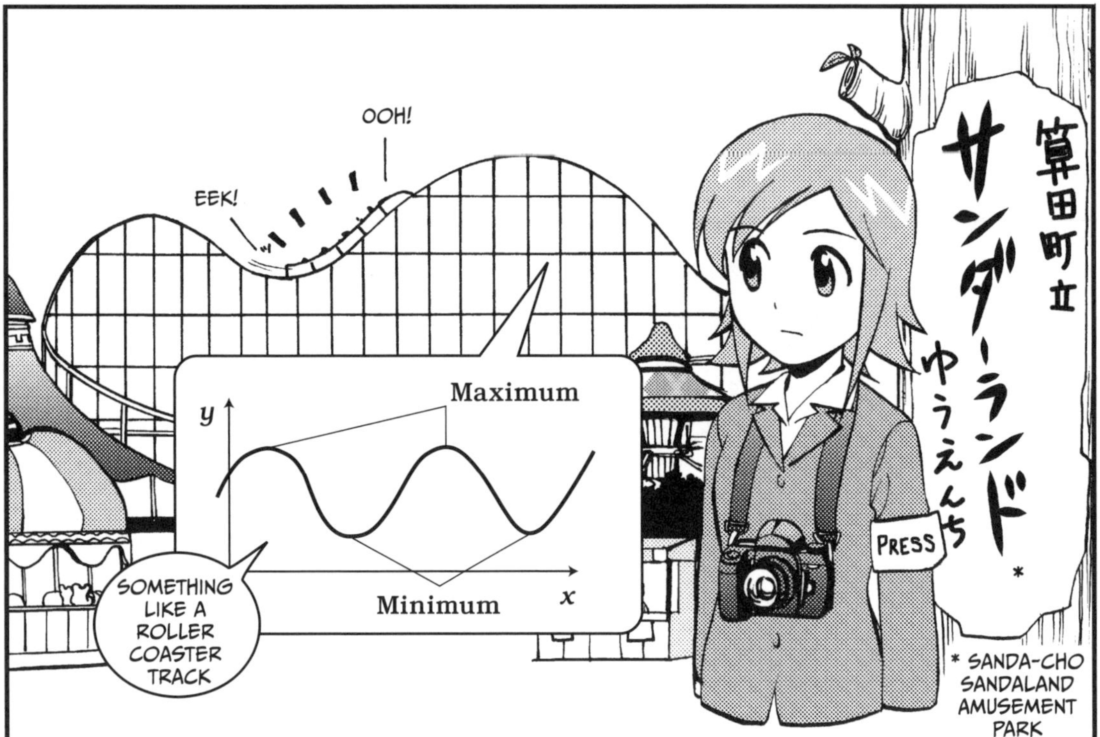
EEK!
OOH!
y
Maximum
Minimum
x
SOMETHING LIKE A ROLLER COASTER TRACK
PRESS
算田町立
サンダーランド
ゆうえんち*
* SANDA-CHO SANDALAND AMUSEMENT PARK

Maxima and *minima* are where a function changes from a decrease to an increase or vice versa. Thus they are important for examining the properties of a function.

Since a maximum or minimum is often the absolute maximum or minimum, respectively, it is an important point for obtaining an optimum solution.

> **THEOREM 2-1: THE CONDITIONS FOR EXTREMA**
>
> If $y = f(x)$ has a maximum or minimum at $x = a$, then $f'(a) = 0$.

This means that we can find maxima or minima by finding values of a that satisfy $f'(a) = 0$. These values are also called *extrema*.

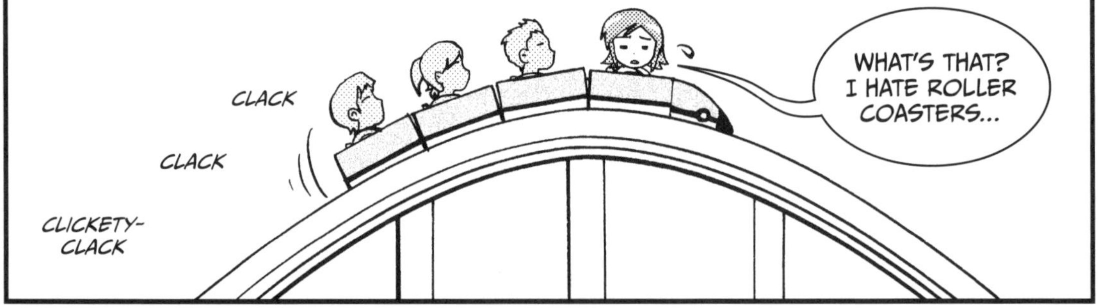
CLICKETY-CLACK
CLACK
CLACK
WHAT'S THAT? I HATE ROLLER COASTERS...

Assume $f'(a) > 0$.

Since $f(x) \approx f'(a)(x - a) + f(a)$ near $x = a$, $f'(a) > 0$ means that the approximate linear function is increasing at $x = a$. Thus, so is $f(x)$.

In other words, the roller coaster is ascending, and it is not at the top or at the bottom.

Similarly, $y = f(x)$ is descending when $f'(a) < 0$, and it is not at the top or the bottom, either.

If $y = f(x)$ is ascending or descending when $f'(a) > 0$ or $f'(a) < 0$, respectively, we can only have $f'(a) = 0$ at the top or bottom.

In fact, the approximate linear function $y = f'(a)(x - a) + f(a) = 0 \times (x - a) + f(a)$ is a horizontal constant function when $f'(a) = 0$, which fits our understanding of maxima and minima.

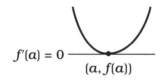

$f'(a) = 0$ ——— $(a, f(a))$

$(a, f(a))$

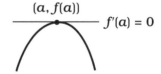

——— $f'(a) = 0$

THIS DISCUSSION CAN BE SUMMARIZED INTO THE FOLLOWING THEOREM.

THEOREM 2-2: THE CRITERIA FOR INCREASING AND DECREASING

$y = f(x)$ is increasing around $x = a$ when $f'(a) > 0$.

$y = f(x)$ is decreasing around $x = a$ when $f'(a) < 0$.

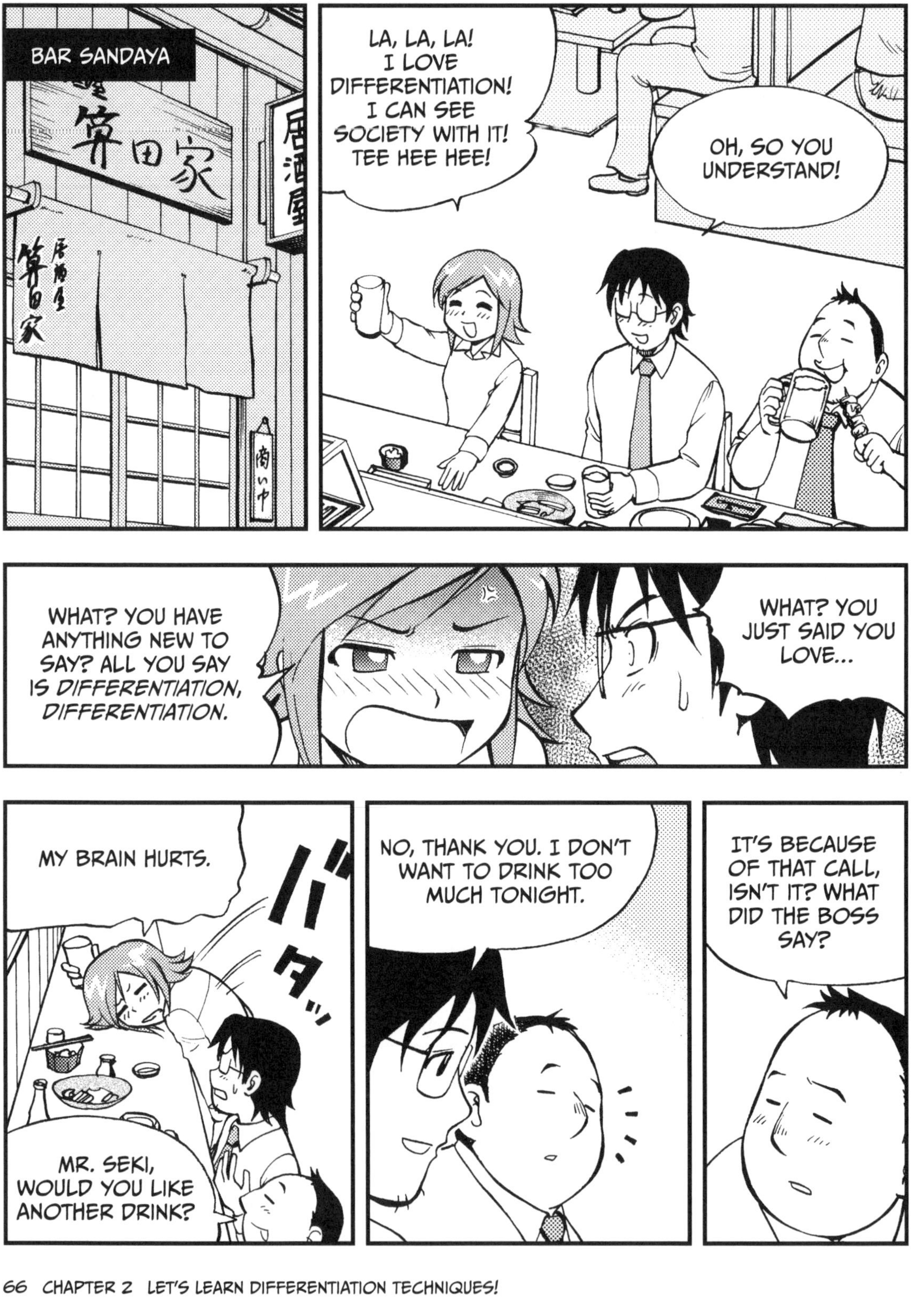
BAR SANDAYA
居酒屋
LA, LA, LA! I LOVE DIFFERENTIATION! I CAN SEE SOCIETY WITH IT! TEE HEE HEE!
OH, SO YOU UNDERSTAND!
WHAT? YOU HAVE ANYTHING NEW TO SAY? ALL YOU SAY IS *DIFFERENTIATION, DIFFERENTIATION.*
WHAT? YOU JUST SAID YOU LOVE...
MY BRAIN HURTS.
バタッ
MR. SEKI, WOULD YOU LIKE ANOTHER DRINK?
NO, THANK YOU. I DON'T WANT TO DRINK TOO MUCH TONIGHT.
IT'S BECAUSE OF THAT CALL, ISN'T IT? WHAT DID THE BOSS SAY?

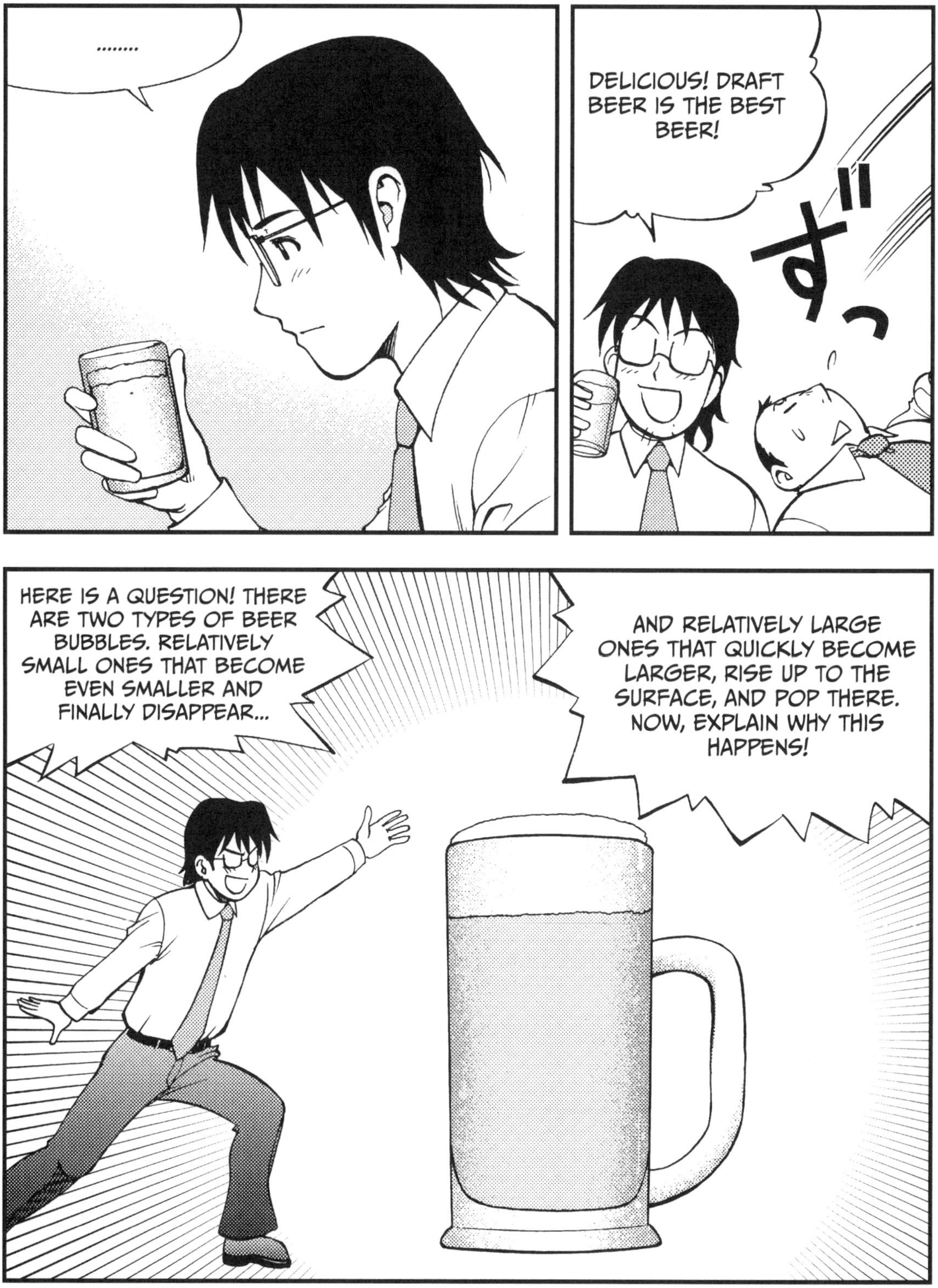
....
DELICIOUS! DRAFT BEER IS THE BEST BEER!
ずっ
HERE IS A QUESTION! THERE ARE TWO TYPES OF BEER BUBBLES. RELATIVELY SMALL ONES THAT BECOME EVEN SMALLER AND FINALLY DISAPPEAR...
AND RELATIVELY LARGE ONES THAT QUICKLY BECOME LARGER, RISE UP TO THE SURFACE, AND POP THERE. NOW, EXPLAIN WHY THIS HAPPENS!

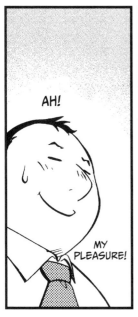

AH!
MY PLEASURE!
SINCE CARBON DIOXIDE IN CARBONATED DRINKS, SUCH AS BEER, IS SUPERSATURATED, IT IS MORE STABLE AS A GAS THAN WHEN IT IS DISSOLVED IN FLUID.
CO_2
CO_2
CO_2
SO, THE ENERGY OF A BUBBLE *DECREASES* IN PROPORTION TO ITS VOLUME ($\frac{4}{3}\pi r^3$, WITH r BEING THE RADIUS).
GAS (BUBBLE)
FLUID
SURFACE TENSION ACTS
ON THE OTHER HAND, SURFACE TENSION ACTS ON THE BOUNDARY SURFACE BETWEEN THE BUBBLE AND THE FLUID, TRYING TO REDUCE THE SURFACE AREA.
THEREFORE, THE ENERGY OF THE BUBBLE DUE TO THIS FORCE *INCREASES* IN PROPORTION TO THE SURFACE AREA, $4\pi r^2$.
CONSIDERING THESE TWO EFFECTS, THE ENERGY $E(r)$ OF A BUBBLE OF RADIUS r CAN BE EXPRESSED
VOLUME OF A SPHERE
SURFACE AREA OF A SPHERE
$E(r) = -a\left(\frac{4}{3}\pi r^3\right) + b\left(4\pi r^2\right)$
TERM FOR THE VOLUME
TERM FOR THE AREA
AS SHOWN HERE.

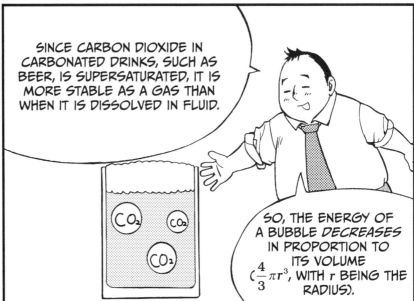

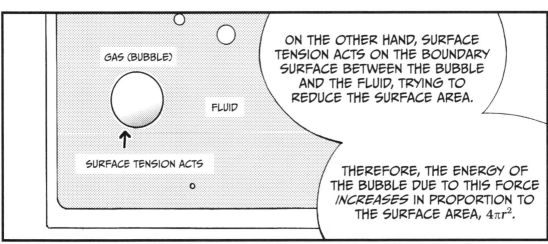

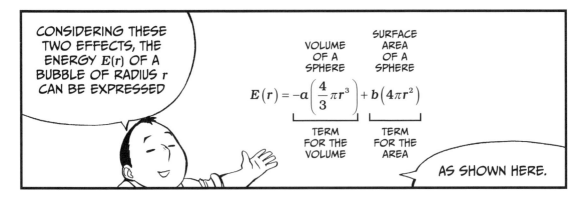

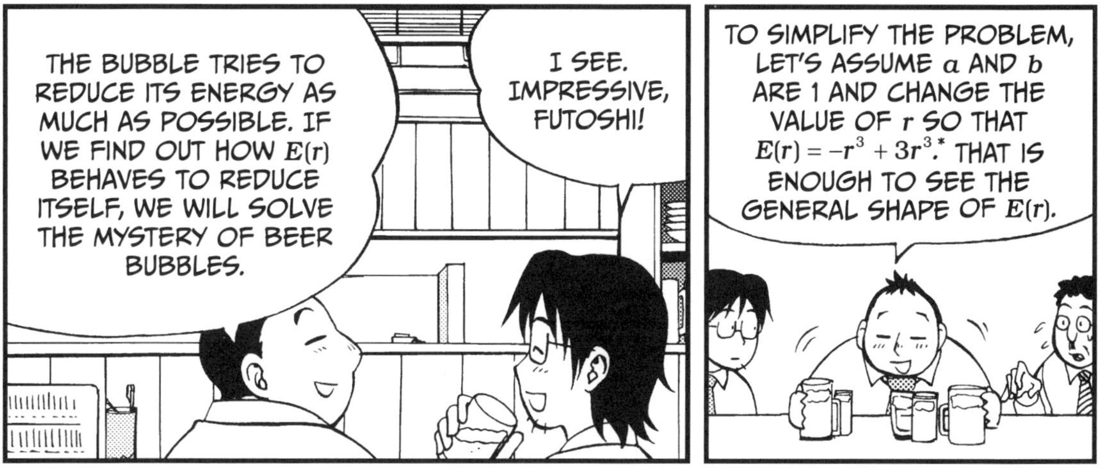
THE BUBBLE TRIES TO REDUCE ITS ENERGY AS MUCH AS POSSIBLE. IF WE FIND OUT HOW $E(r)$ BEHAVES TO REDUCE ITSELF, WE WILL SOLVE THE MYSTERY OF BEER BUBBLES.
I SEE. IMPRESSIVE, FUTOSHI!
TO SIMPLIFY THE PROBLEM, LET'S ASSUME a AND b ARE 1 AND CHANGE THE VALUE OF r SO THAT $E(r) = -r^3 + 3r^3$.* THAT IS ENOUGH TO SEE THE GENERAL SHAPE OF $E(r)$.

* THIS IS CALLED *NORMALIZING A VARIABLE*. WE'VE SIMPLY MULTIPLIED EACH TERM BY $3/(4\pi)$.

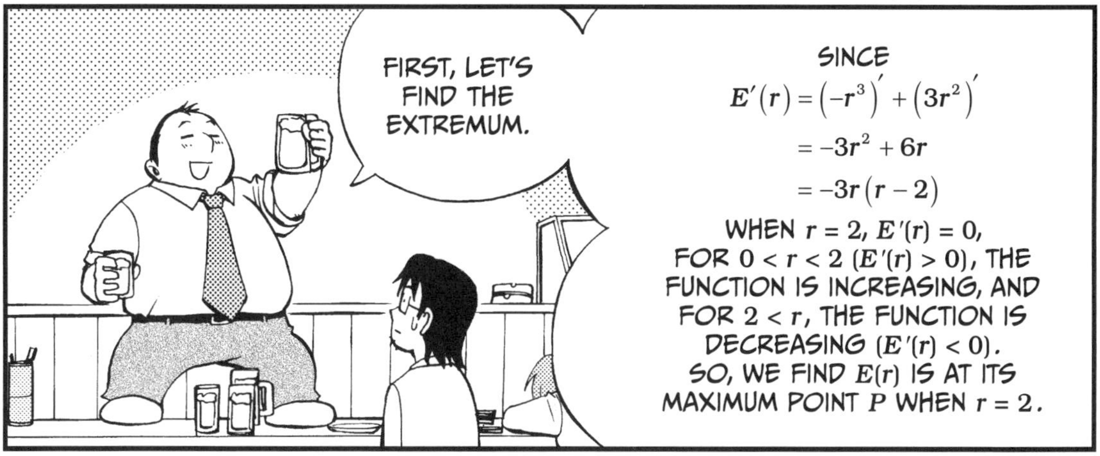
FIRST, LET'S FIND THE EXTREMUM.
SINCE
$E'(r) = \left(-r^3\right)' + \left(3r^2\right)'$
$= -3r^2 + 6r$
$= -3r\left(r-2\right)$
WHEN $r = 2$, $E'(r) = 0$, FOR $0 < r < 2$ ($E'(r) > 0$), THE FUNCTION IS INCREASING, AND FOR $2 < r$, THE FUNCTION IS DECREASING ($E'(r) < 0$). SO, WE FIND $E(r)$ IS AT ITS MAXIMUM POINT P WHEN $r = 2$.

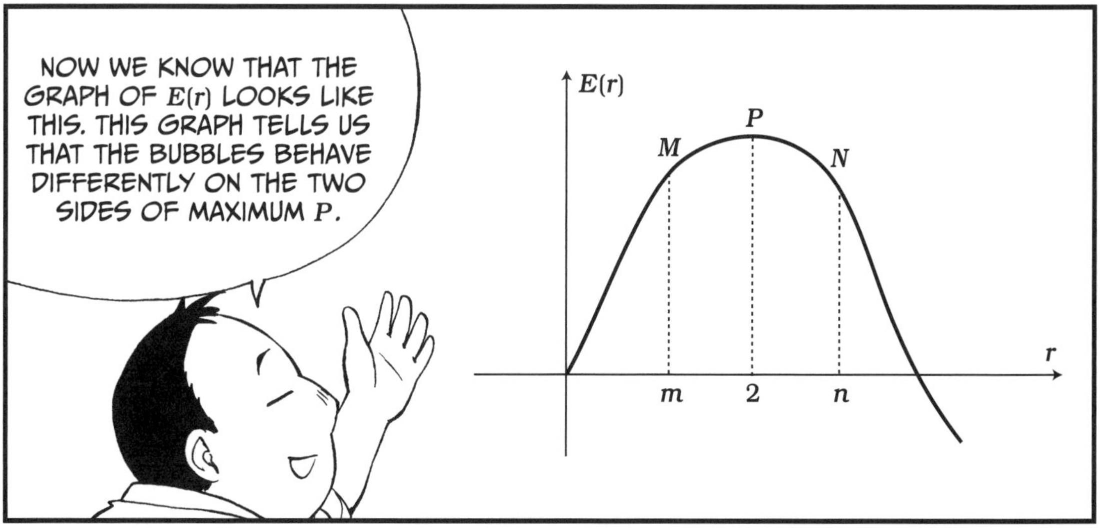
NOW WE KNOW THAT THE GRAPH OF $E(r)$ LOOKS LIKE THIS. THIS GRAPH TELLS US THAT THE BUBBLES BEHAVE DIFFERENTLY ON THE TWO SIDES OF MAXIMUM P.
$E(r)$
P
M
N
r
m
2
n

A BUBBLE THAT HAS THE RADIUS AND ENERGY OF POINT *M* SHOULD REDUCE ITS RADIUS UNTIL IT IS SMALLER THAN *m* TO MAKE ITS ENERGY *E(r)* SMALLER. THE BUBBLE WILL CONTINUE TO BECOME SMALLER UNTIL IT FINALLY DISAPPEARS.

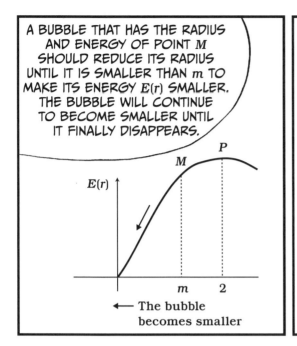

← The bubble becomes smaller

ON THE OTHER HAND, A BUBBLE THAT HAS THE RADIUS AND ENERGY OF POINT *N* SHOULD INCREASE ITS RADIUS TO MAKE ITS ENERGY *E(r)* SMALLER. THE BUBBLE WILL CONTINUE TO GROW LARGER AND TO RISE UP INSIDE THE BEER.

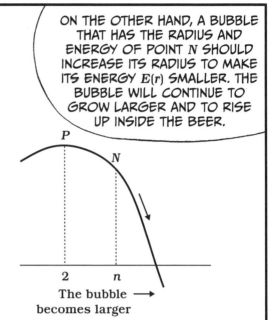

The bubble becomes larger →

BRAVO!

CLAP
CLAP

?!

YANK

HEH-HEH...FUTOSHI.

N...NORIKO!

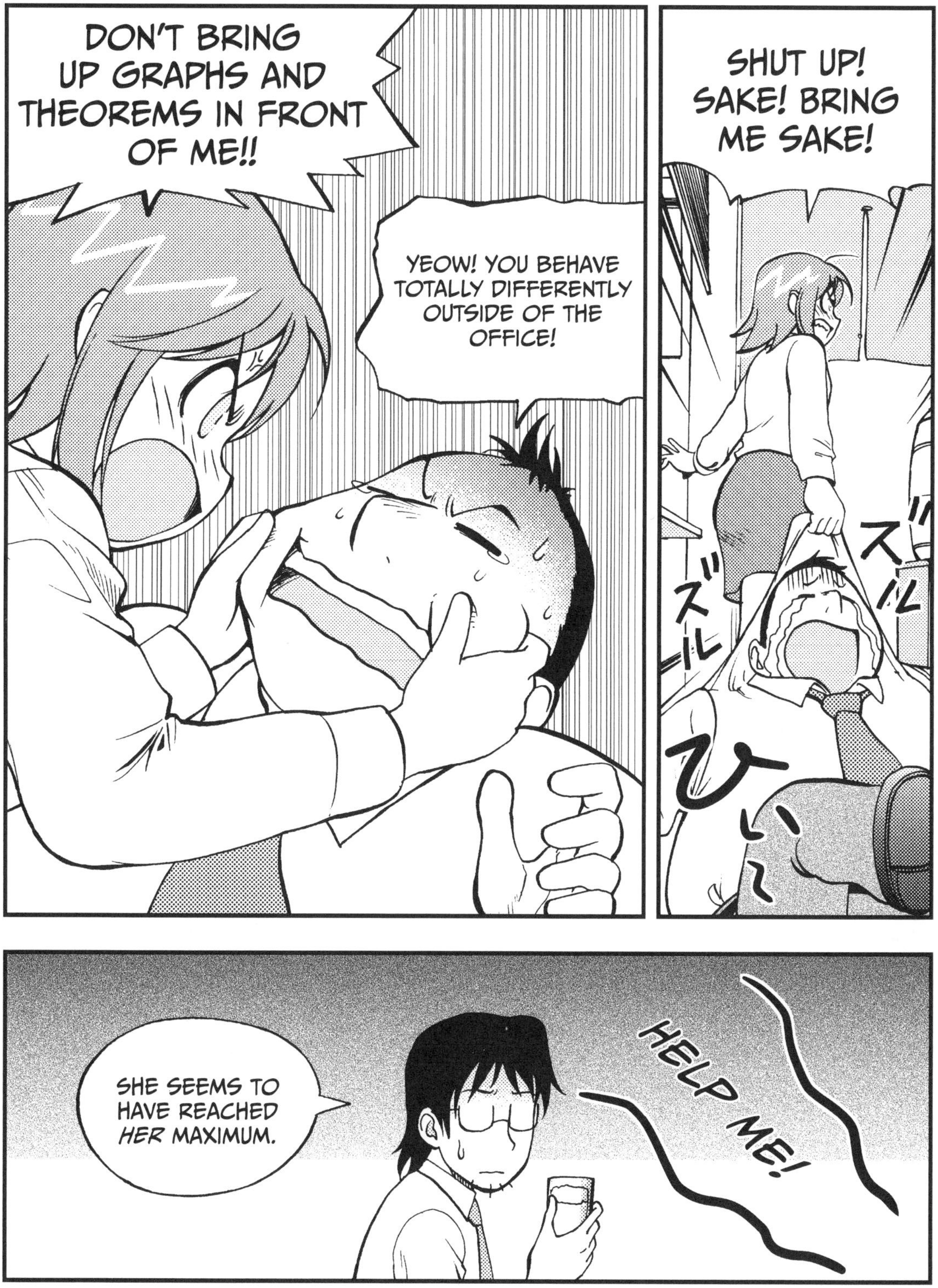
DON'T BRING UP GRAPHS AND THEOREMS IN FRONT OF ME!!
YEOW! YOU BEHAVE TOTALLY DIFFERENTLY OUTSIDE OF THE OFFICE!
SHUT UP! SAKE! BRING ME SAKE!
ズルズル
ひぃ〜
SHE SEEMS TO HAVE REACHED *HER* MAXIMUM.
HELP ME!

USING THE MEAN VALUE THEOREM

We saw before that the derivative is the coefficient of x in the approximate linear function that imitates function $f(x)$ in the vicinity of $x = a$.

That is,

$$f(x) \approx f'(a)(x-a) + f(a) \quad \text{(when } x \text{ is very close to } a\text{)}$$

But the linear function only "pretends to be" or "imitates" $f(x)$, and for b, which is near a, we generally have

❶ $\quad f(b) \neq f'(a)(b-a) + f(a)$

So, this is not exactly an equation.

FOR THOSE WHO CANNOT STAND FOR THIS, WE HAVE THE FOLLOWING THEOREM.

THEOREM 2-3: THE MEAN VALUE THEOREM

For a, b $(a < b)$, and c, which satisfy $a < c < b$, there exists a number c that satisfies

$$f(b) = f'(c)(b-a) + f(a)$$

In other words, we can make expression **❶** hold with an equal sign not with $f'(a)$ but with $f'(c)$, where c is a value existing somewhere between a and b.[*]

WHY IS THIS?

---

[*] That is, there must be a value for x between a and b (which we'll call c) that has a tangent line matching the slope of a line connecting points A and B.

Let's draw a line through point $A = (a, f(a))$ and point $B = (b, f(b))$ to form line segment AB.

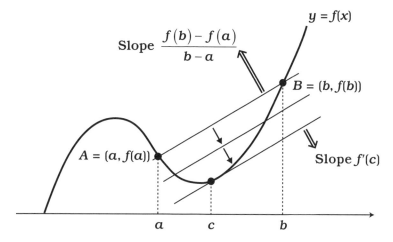

Slope $\dfrac{f(b) - f(a)}{b - a}$

$y = f(x)$

$B = (b, f(b))$

$A = (a, f(a))$

Slope $f'(c)$

$a \qquad c \qquad b$

We know the slope is simply $\Delta y \,/\, \Delta x$:

❷ Slope of $AB = \dfrac{f(b) - f(a)}{b - a}$

Now, move line AB parallel to its initial state as shown in the figure.

The line eventually comes to a point beyond which it separates from the graph. Denote this point by $(c, f(c))$.

At this moment, the line is a tangent line, and its slope is $f'(c)$.

Since the line has been moved parallel to the initial state, this slope has not been changed from slope ❷.

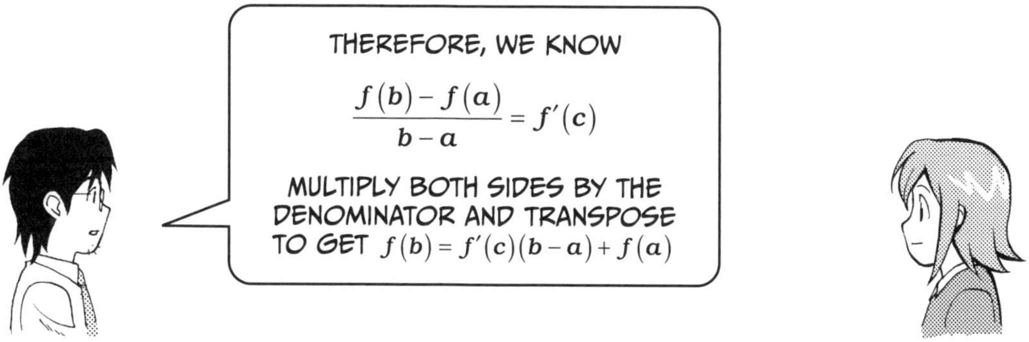

THEREFORE, WE KNOW

$$\dfrac{f(b) - f(a)}{b - a} = f'(c)$$

MULTIPLY BOTH SIDES BY THE DENOMINATOR AND TRANSPOSE TO GET $f(b) = f'(c)(b - a) + f(a)$

USING THE QUOTIENT RULE OF DIFFERENTIATION

Let's find the formula for the derivative of $h(x) = \dfrac{g(x)}{f(x)}$

First, we find the derivative of function $p(x) = \dfrac{1}{f(x)}$, which is the reciprocal of $f(x)$.

If we know this, we'll be able to apply the product rule to $h(x)$.

Using simple algebra, we see that $f(x)\,p(x) = 1$ always holds.

$$1 = f(x)p(x) \approx \{f'(a)(x-a) + f(a)\}\{p'(a)(x-a) + p(a)\}$$

Since these two are equal, their derivatives must be equal as well.

$$0 = p(x)f'(x) + p'(x)f(x)$$

Thus, we have $p'(x) = -\dfrac{p(x)f'(x)}{f(x)}$.

Since $p(a) = \dfrac{1}{f(a)}$, substituting this for $p(a)$ in the numerator gives

$p'(a) = \dfrac{-f'(a)}{f(a)^2}$.

For $h(x) = \dfrac{g(x)}{f(x)}$ in general, we consider $h(x) = g(x) \times \dfrac{1}{f(x)} = g(x)p(x)$

and use the product rule and the above formula.

$$h'(x) = g'(x)p(x) + g(x)p'(x) = g'(x)\frac{1}{f(x)} - g(x)\frac{f'(x)}{f(x)^2}$$

$$= \frac{g'(x)f(x) - g(x)f'(x)}{f(x)^2}$$

Therefore, we obtain the following formula.

FORMULA 2-6: THE QUOTIENT RULE OF DIFFERENTIATION

$$h'(x) = \frac{g'(x)f(x) - g(x)f'(x)}{f(x)^2}$$

CALCULATING DERIVATIVES OF COMPOSITE FUNCTIONS

Let's obtain the formula for the derivative of $h(x) = g(f(x))$.

Near $x = a$,

$$f(x) - f(a) \approx f'(a)(x - a)$$

And near $y = b$,

$$g(y) - g(b) \approx g'(b)(y - b)$$

We now substitute $b = f(a)$ and $y = f(x)$ in the last expression.
Near $x = a$,

$$g(f(x)) - g(f(a)) \approx g'(f(a))(f(x) - f(a))$$

Replace $f(x) - f(a)$ in the right side with the right side of the first expression.

$$g(f(x)) - g(f(a)) \approx g'(f(a)) f'(a)(x - a)$$

Since $g(f(x)) = h(x)$, the coefficient of $(x - a)$ in this expression gives us $h'(a) = g'(f(a)) f'(a)$.

We thus obtain the following formula.

> **FORMULA 2-7: THE DERIVATIVES OF COMPOSITE FUNCTIONS**
>
> $$h'(a) = g'(f(x)) f'(x)$$

CALCULATING DERIVATIVES OF INVERSE FUNCTIONS

Let's use the above formula to find the formula for the derivative of $x = g(y)$, the inverse function of $y = f(x)$.

Since $x = g(f(x))$ for any x, differentiating both sides of this expression gives $1 = g'(f(x)) f'(x)$.

Thus, $1 = g'(y) f'(x)$, and we obtain the following formula.

> **FORMULA 2-8: THE DERIVATIVES OF INVERSE FUNCTIONS**
>
> $$g'(y) = \frac{1}{f'(x)}$$

FORMULAS OF DIFFERENTIATION

	FORMULA	KEY POINT
Constant multiplication	$\{\alpha f(x)\}' = \alpha f'(x)$	The multiplicative constant can be factored out.
x^n (Power)	$(x^n)' = nx^{n-1}$	The exponent becomes the coefficient, reducing the degree by 1.
Sum	$\{f(x)+g(x)\}' = f'(x)+g'(x)$	The derivative of a sum is the sum of the derivatives.
Product	$\{f(x)g(x)\}' = f'(x)g(x)+f(x)g'(x)$	The sum of the products with each function differentiated in turn.
Quotient	$\left\{\frac{g(x)}{f(x)}\right\}' = \frac{g'(x)f(x)-g(x)f'(x)}{f(x)^2}$	The denominator is squared. The numerator is the difference between the products with only one function differentiated.
Composite functions	$\{g(f(x))\}' = g'(f(x))f'(x)$	The product of the derivative of the outer and that of the inner.
Inverse functions	$g'(y) = \frac{1}{f'(x)}$	The derivative of an inverse function is the reciprocal of the original.

EXERCISES

1. For natural number n, find the derivative $f'(x)$ of $f(x) = \frac{1}{x^n}$.
2. Calculate the extrema of $f(x) = x^3 - 12x$.
3. Find the derivative $f'(x)$ of $f(x) = (1 - x)^3$.
4. Calculate the maximum value of $g(x) = x^2(1 - x)^3$ in the interval $0 \leq x \leq 1$.

3

LET'S INTEGRATE A FUNCTION!

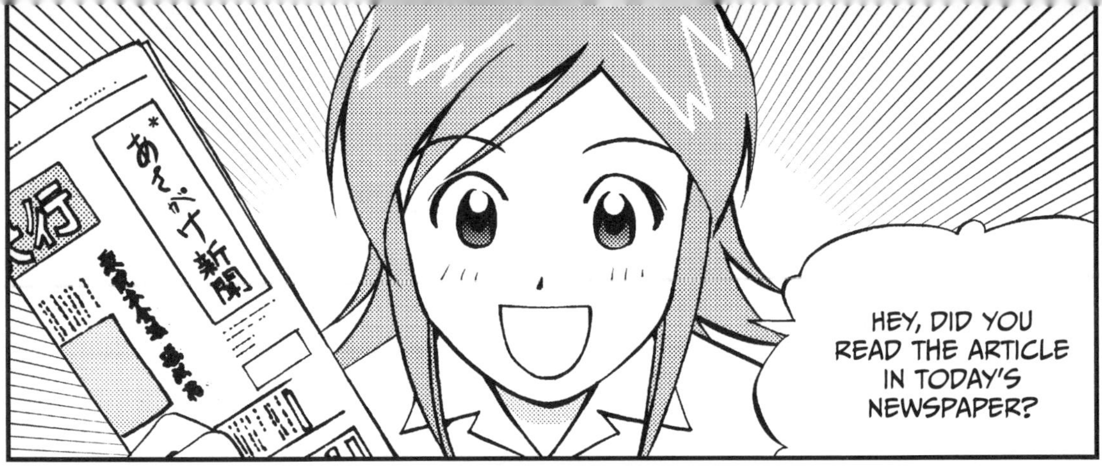
あさがけ新聞
HEY, DID YOU READ THE ARTICLE IN TODAY'S NEWSPAPER?

* THE ASAGAKE TIMES

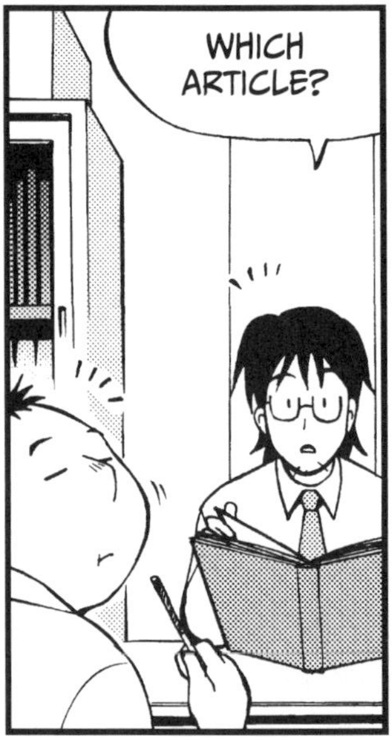
WHICH ARTICLE?

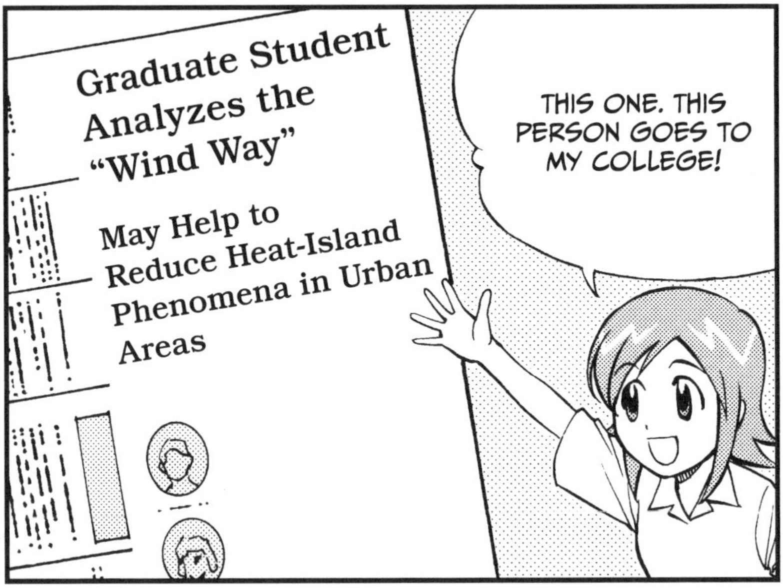
Graduate Student Analyzes the "Wind Way"
May Help to Reduce Heat-Island Phenomena in Urban Areas
THIS ONE. THIS PERSON GOES TO MY COLLEGE!

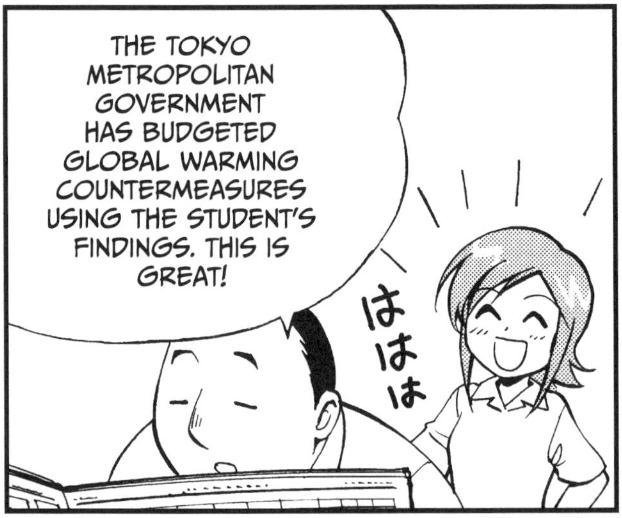
THE TOKYO METROPOLITAN GOVERNMENT HAS BUDGETED GLOBAL WARMING COUNTERMEASURES USING THE STUDENT'S FINDINGS. THIS IS GREAT!
ははは

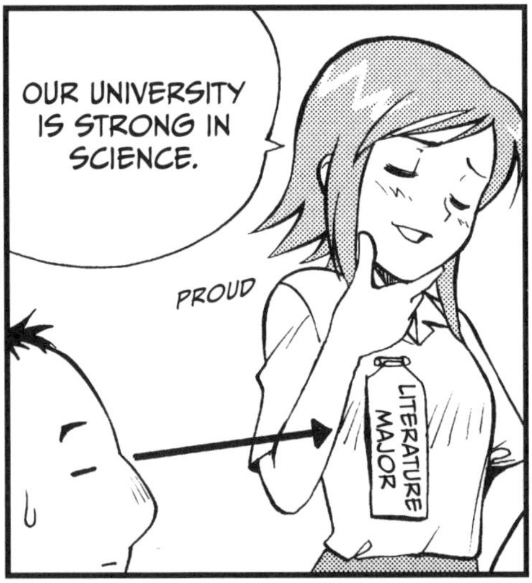
OUR UNIVERSITY IS STRONG IN SCIENCE.
PROUD
LITERATURE MAJOR

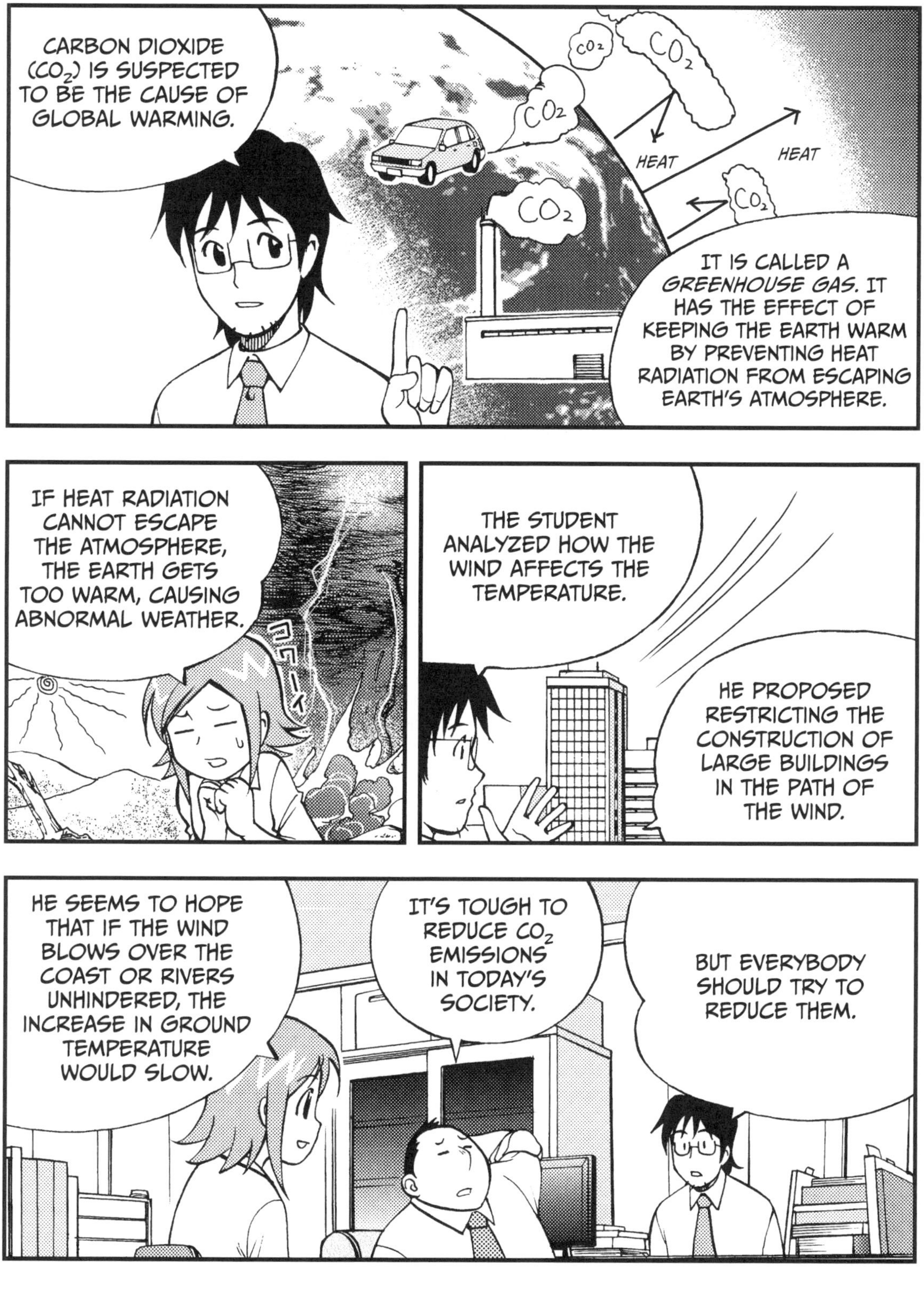
CARBON DIOXIDE (CO_2) IS SUSPECTED TO BE THE CAUSE OF GLOBAL WARMING.
CO_2
CO_2
CO_2
CO_2
HEAT
HEAT
CO_2
IT IS CALLED A *GREENHOUSE GAS*. IT HAS THE EFFECT OF KEEPING THE EARTH WARM BY PREVENTING HEAT RADIATION FROM ESCAPING EARTH'S ATMOSPHERE.
IF HEAT RADIATION CANNOT ESCAPE THE ATMOSPHERE, THE EARTH GETS TOO WARM, CAUSING ABNORMAL WEATHER.
THE STUDENT ANALYZED HOW THE WIND AFFECTS THE TEMPERATURE.
HE PROPOSED RESTRICTING THE CONSTRUCTION OF LARGE BUILDINGS IN THE PATH OF THE WIND.
HE SEEMS TO HOPE THAT IF THE WIND BLOWS OVER THE COAST OR RIVERS UNHINDERED, THE INCREASE IN GROUND TEMPERATURE WOULD SLOW.
IT'S TOUGH TO REDUCE CO_2 EMISSIONS IN TODAY'S SOCIETY.
BUT EVERYBODY SHOULD TRY TO REDUCE THEM.

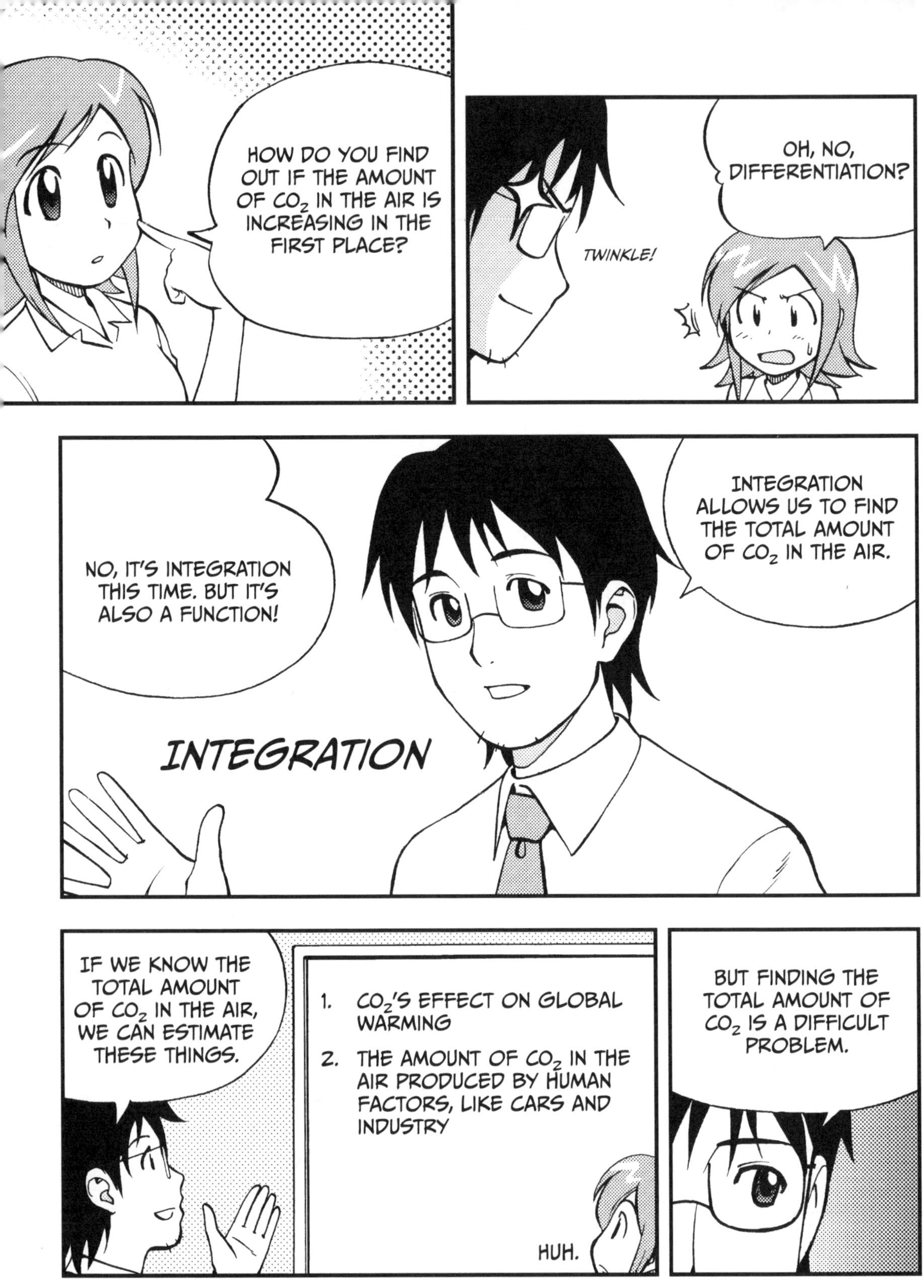
HOW DO YOU FIND OUT IF THE AMOUNT OF CO_2 IN THE AIR IS INCREASING IN THE FIRST PLACE?
OH, NO, DIFFERENTIATION?
TWINKLE!
NO, IT'S INTEGRATION THIS TIME. BUT IT'S ALSO A FUNCTION!
INTEGRATION
INTEGRATION ALLOWS US TO FIND THE TOTAL AMOUNT OF CO_2 IN THE AIR.
IF WE KNOW THE TOTAL AMOUNT OF CO_2 IN THE AIR, WE CAN ESTIMATE THESE THINGS.
1. CO_2'S EFFECT ON GLOBAL WARMING
2. THE AMOUNT OF CO_2 IN THE AIR PRODUCED BY HUMAN FACTORS, LIKE CARS AND INDUSTRY
HUH.
BUT FINDING THE TOTAL AMOUNT OF CO_2 IS A DIFFICULT PROBLEM.

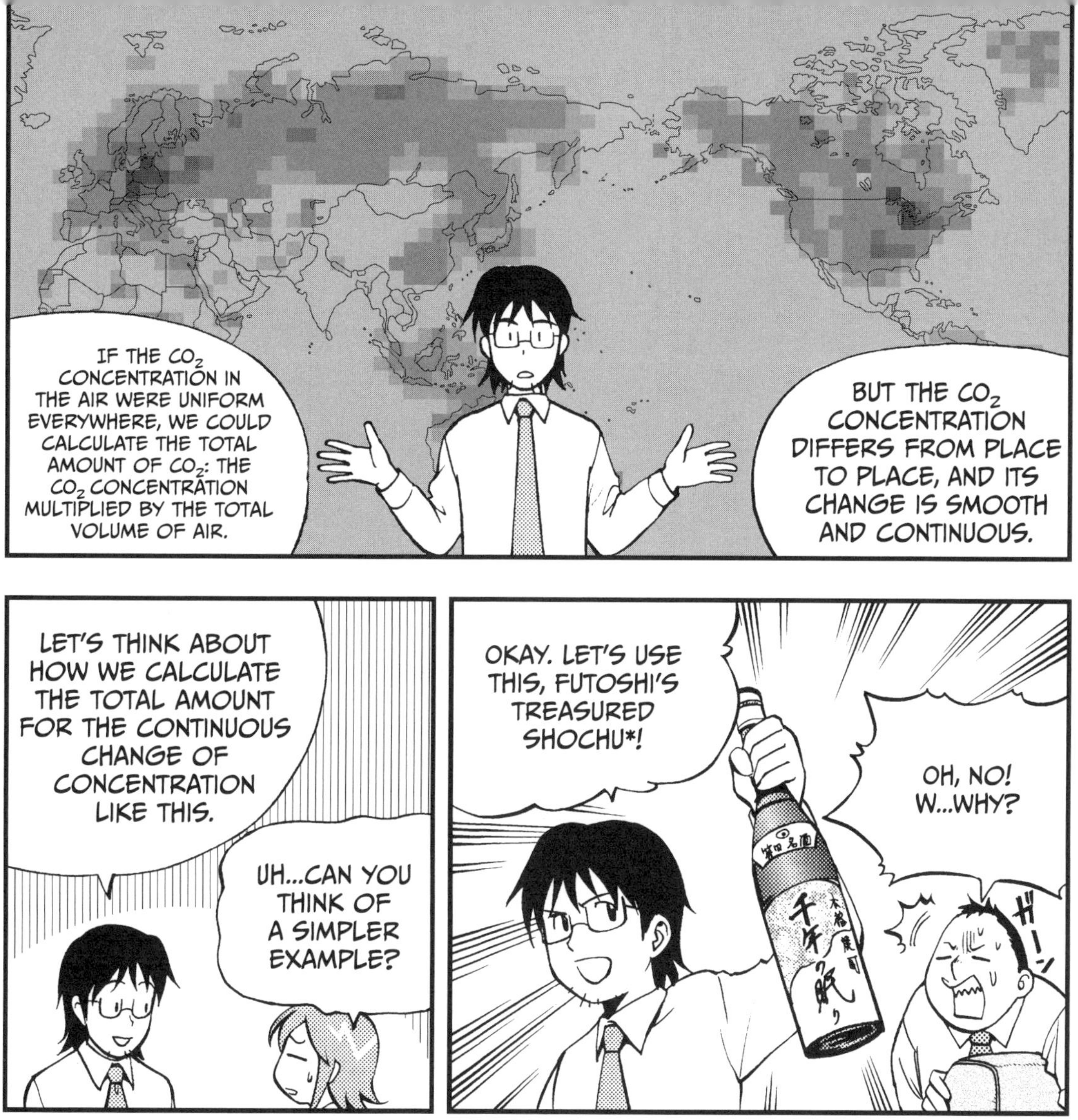
IF THE CO_2 CONCENTRATION IN THE AIR WERE UNIFORM EVERYWHERE, WE COULD CALCULATE THE TOTAL AMOUNT OF CO_2: THE CO_2 CONCENTRATION MULTIPLIED BY THE TOTAL VOLUME OF AIR.
BUT THE CO_2 CONCENTRATION DIFFERS FROM PLACE TO PLACE, AND ITS CHANGE IS SMOOTH AND CONTINUOUS.
LET'S THINK ABOUT HOW WE CALCULATE THE TOTAL AMOUNT FOR THE CONTINUOUS CHANGE OF CONCENTRATION LIKE THIS.
UH...CAN YOU THINK OF A SIMPLER EXAMPLE?
OKAY. LET'S USE THIS, FUTOSHI'S TREASURED SHOCHU*!
OH, NO! W...WHY?
ガーン

* A JAPANESE DISTILLED SPIRIT

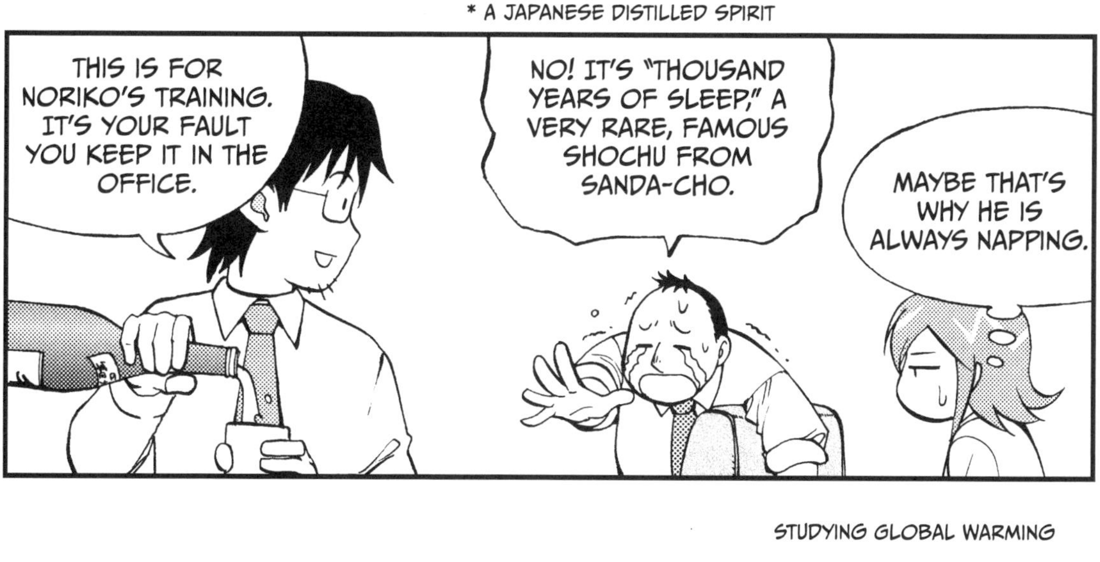
THIS IS FOR NORIKO'S TRAINING. IT'S YOUR FAULT YOU KEEP IT IN THE OFFICE.
NO! IT'S "THOUSAND YEARS OF SLEEP," A VERY RARE, FAMOUS SHOCHU FROM SANDA-CHO.
MAYBE THAT'S WHY HE IS ALWAYS NAPPING.

ILLUSTRATING THE FUNDAMENTAL THEOREM OF CALCULUS

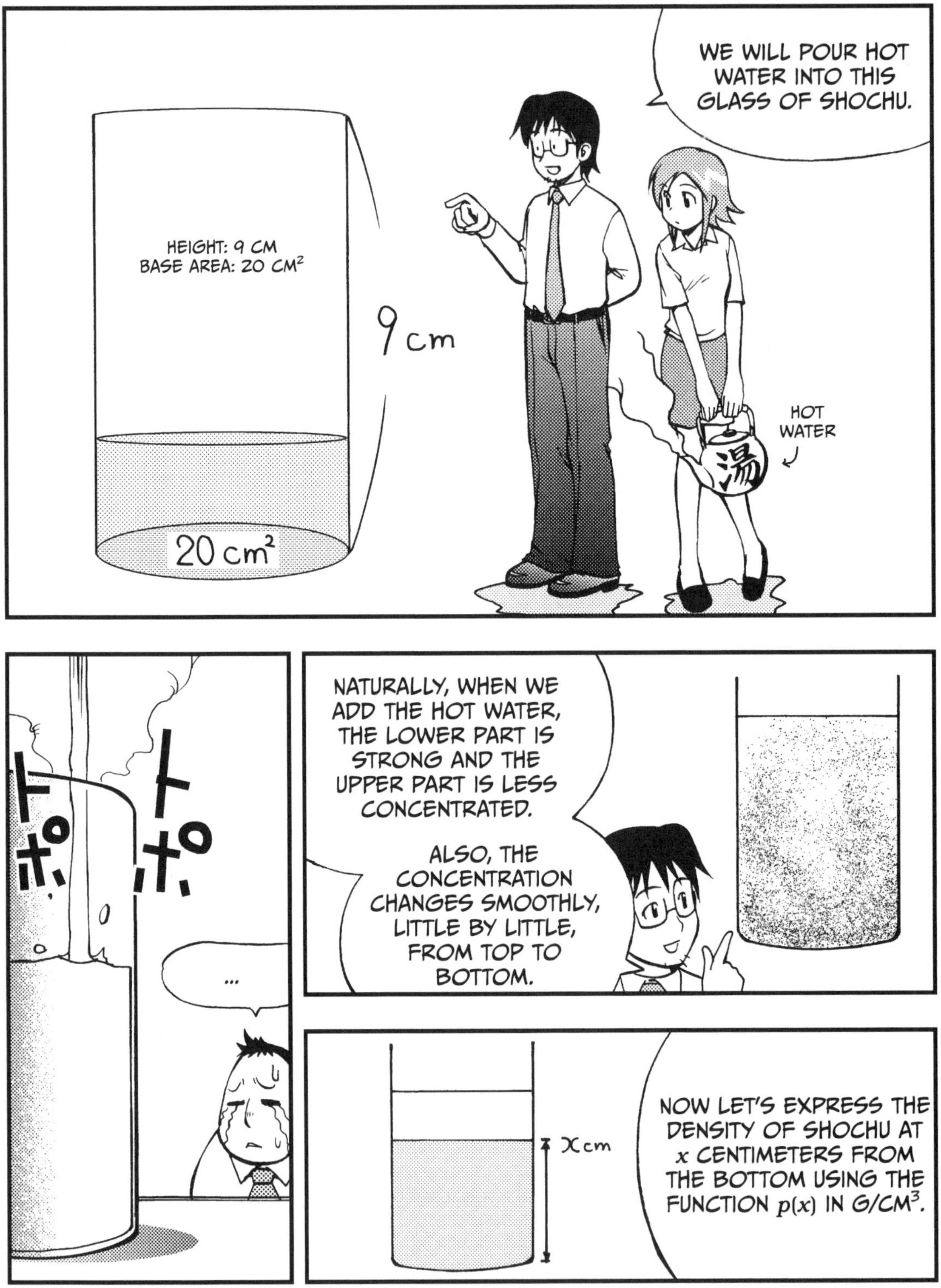

WE WILL POUR HOT WATER INTO THIS GLASS OF SHOCHU.
HEIGHT: 9 CM
BASE AREA: 20 CM^2
9 cm
20 cm^2
HOT WATER
湯
トポ
トポ
...
NATURALLY, WHEN WE ADD THE HOT WATER, THE LOWER PART IS STRONG AND THE UPPER PART IS LESS CONCENTRATED.
ALSO, THE CONCENTRATION CHANGES SMOOTHLY, LITTLE BY LITTLE, FROM TOP TO BOTTOM.
x cm
NOW LET'S EXPRESS THE DENSITY OF SHOCHU AT x CENTIMETERS FROM THE BOTTOM USING THE FUNCTION $p(x)$ IN G/CM^3.

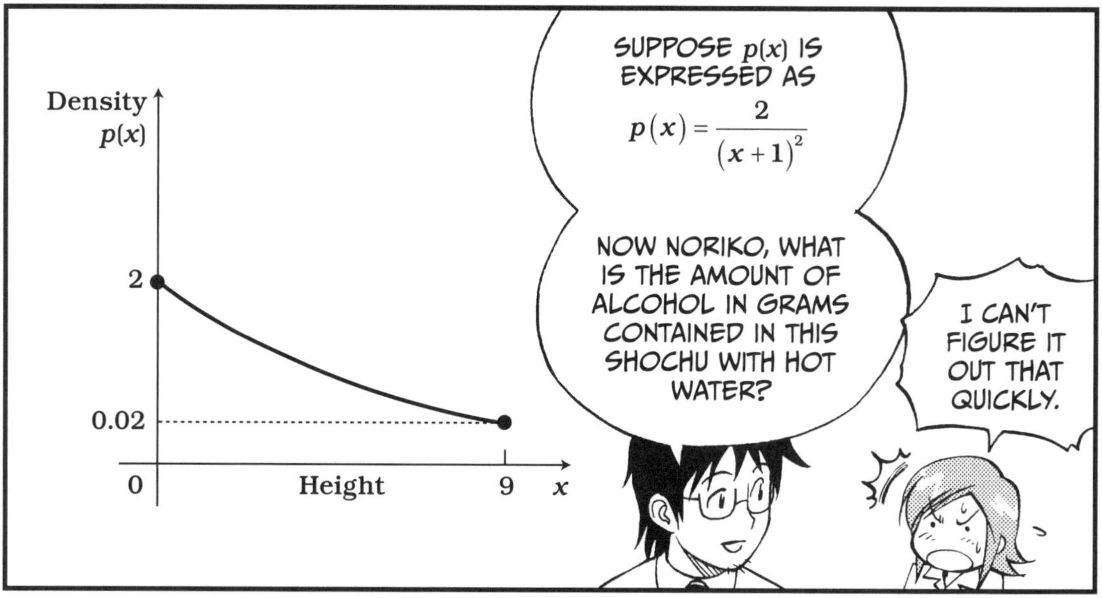
Density
$p(x)$
2
0.02
0
Height
9
x
SUPPOSE $p(x)$ IS EXPRESSED AS
$$p(x) = \frac{2}{(x+1)^2}$$
NOW NORIKO, WHAT IS THE AMOUNT OF ALCOHOL IN GRAMS CONTAINED IN THIS SHOCHU WITH HOT WATER?
I CAN'T FIGURE IT OUT THAT QUICKLY.

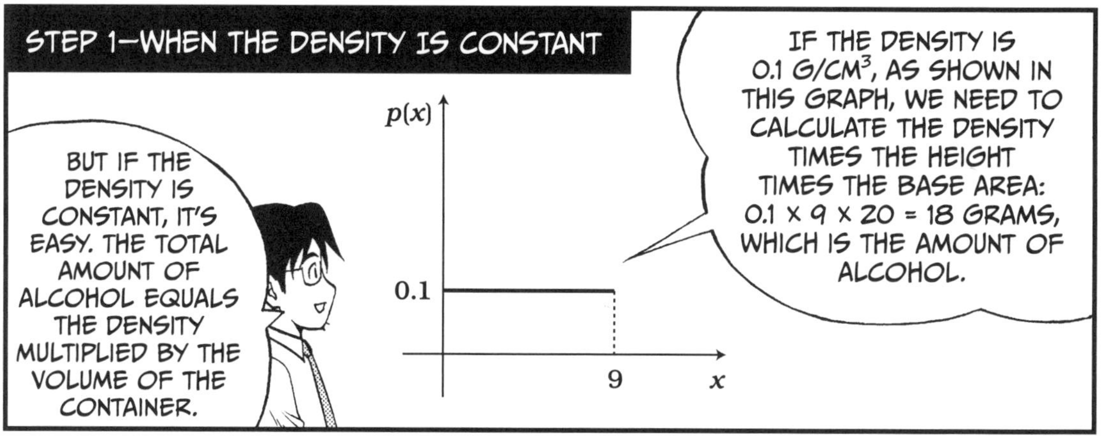
STEP 1—WHEN THE DENSITY IS CONSTANT
BUT IF THE DENSITY IS CONSTANT, IT'S EASY. THE TOTAL AMOUNT OF ALCOHOL EQUALS THE DENSITY MULTIPLIED BY THE VOLUME OF THE CONTAINER.
$p(x)$
0.1
9
x
IF THE DENSITY IS 0.1 G/CM^3, AS SHOWN IN THIS GRAPH, WE NEED TO CALCULATE THE DENSITY TIMES THE HEIGHT TIMES THE BASE AREA: 0.1 × 9 × 20 = 18 GRAMS, WHICH IS THE AMOUNT OF ALCOHOL.

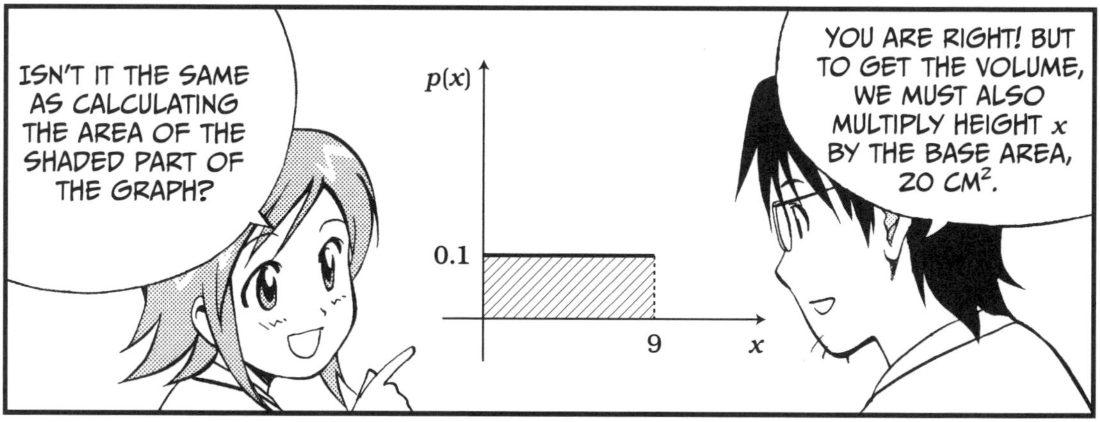
ISN'T IT THE SAME AS CALCULATING THE AREA OF THE SHADED PART OF THE GRAPH?
$p(x)$
0.1
9
x
YOU ARE RIGHT! BUT TO GET THE VOLUME, WE MUST ALSO MULTIPLY HEIGHT x BY THE BASE AREA, 20 CM^2.

STEP 2—WHEN THE DENSITY CHANGES STEPWISE

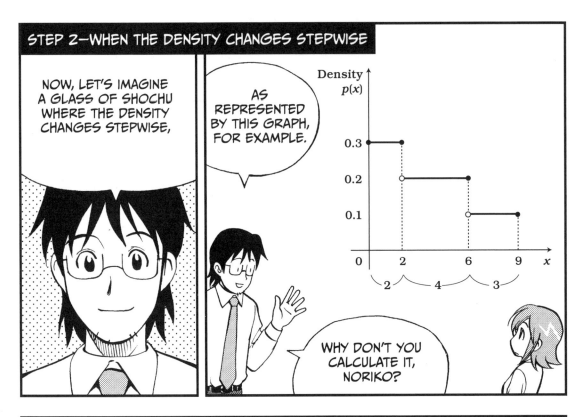

NOW, LET'S IMAGINE A GLASS OF SHOCHU WHERE THE DENSITY CHANGES STEPWISE,

AS REPRESENTED BY THIS GRAPH, FOR EXAMPLE.

WHY DON'T YOU CALCULATE IT, NORIKO?

WELL, SEPARATING THE GRAPH INTO THE STEPS...THE BASE AREA IS 20 CM²...

$$0.3 \times 2 \times 20 + 0.2 \times 4 \times 20 + 0.1 \times 3 \times 20$$

$$\left(\begin{array}{c} \text{Alcohol for} \\ \text{the portion of} \\ 0 \leq x \leq 2 \end{array} \right) \left(\begin{array}{c} \text{Alcohol for} \\ \text{the portion of} \\ 2 < x \leq 6 \end{array} \right) \left(\begin{array}{c} \text{Alcohol for} \\ \text{the portion of} \\ 6 < x \leq 9 \end{array} \right)$$

$$= (0.3 \times 2 + 0.2 \times 4 + 0.1 \times 3) \times 20 = 34$$

SO...

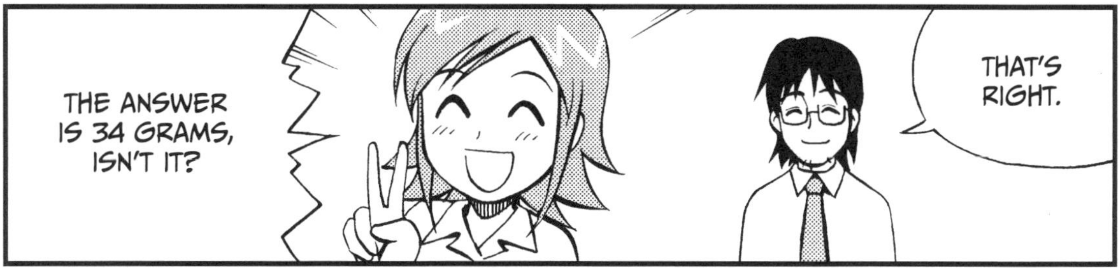
THE ANSWER IS 34 GRAMS, ISN'T IT?
THAT'S RIGHT.

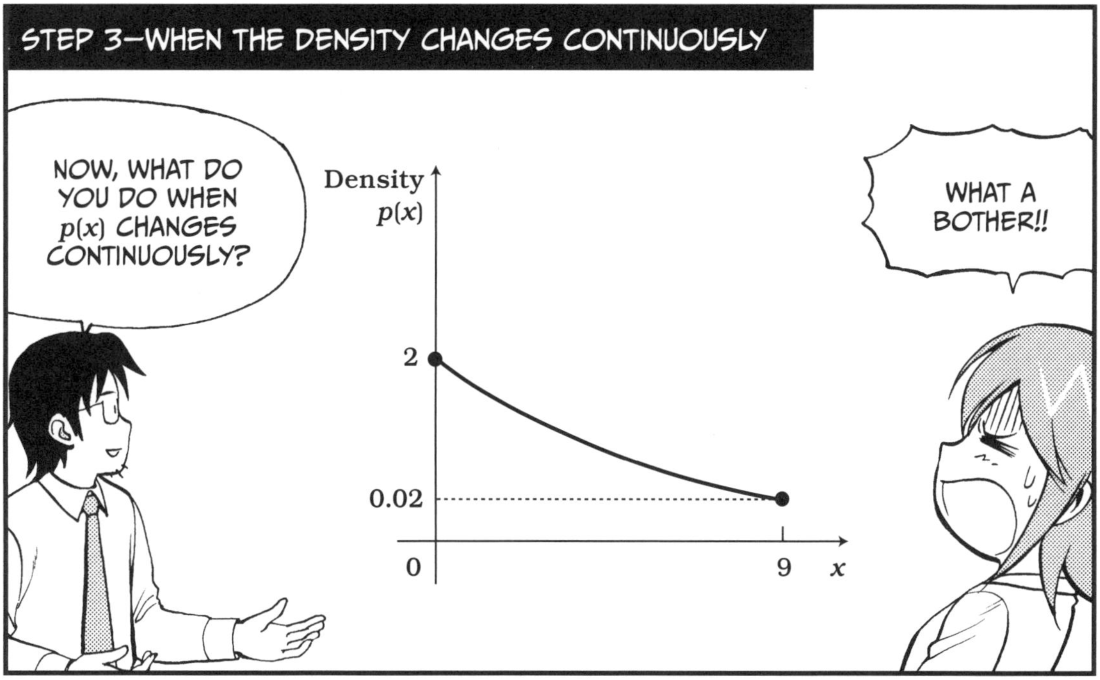
STEP 3—WHEN THE DENSITY CHANGES CONTINUOUSLY
NOW, WHAT DO YOU DO WHEN p(x) CHANGES CONTINUOUSLY?
Density
p(x)
2
0.02
0
9
x
WHAT A BOTHER!!

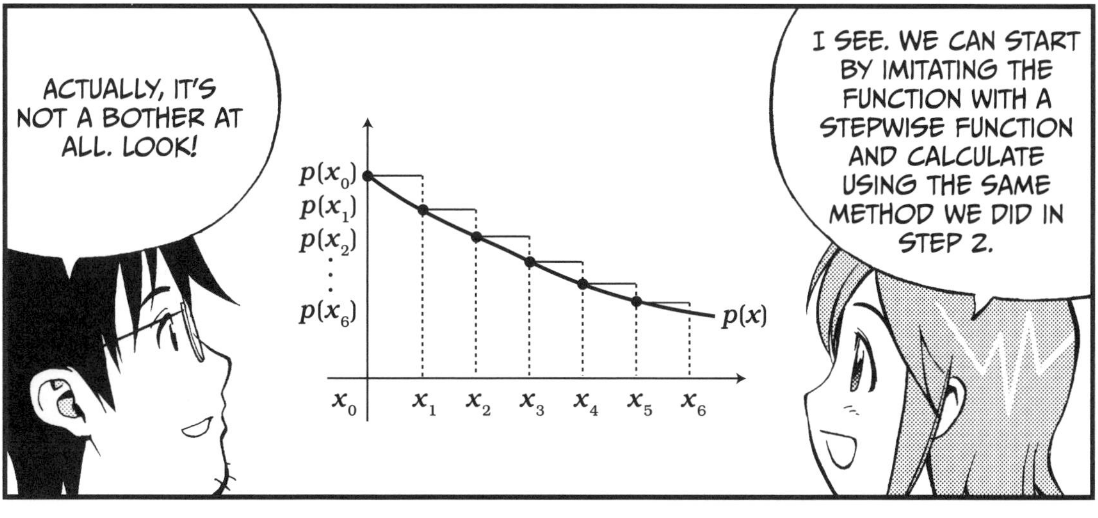
ACTUALLY, IT'S NOT A BOTHER AT ALL. LOOK!
p(x0)
p(x1)
p(x2)
p(x6)
p(x)
x0
x1
x2
x3
x4
x5
x6
I SEE. WE CAN START BY IMITATING THE FUNCTION WITH A STEPWISE FUNCTION AND CALCULATE USING THE SAME METHOD WE DID IN STEP 2.

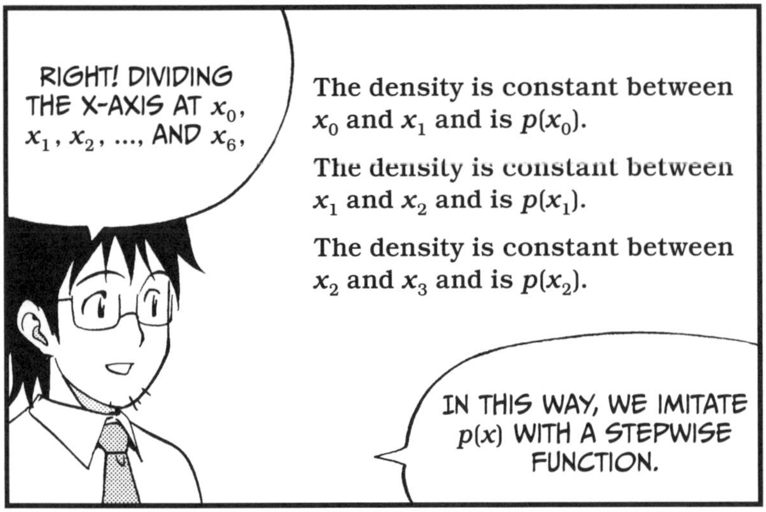
RIGHT! DIVIDING THE X-AXIS AT x_0, x_1, x_2, ..., AND x_6,
The density is constant between x_0 and x_1 and is $p(x_0)$.
The density is constant between x_1 and x_2 and is $p(x_1)$.
The density is constant between x_2 and x_3 and is $p(x_2)$.
IN THIS WAY, WE IMITATE $p(x)$ WITH A STEPWISE FUNCTION.

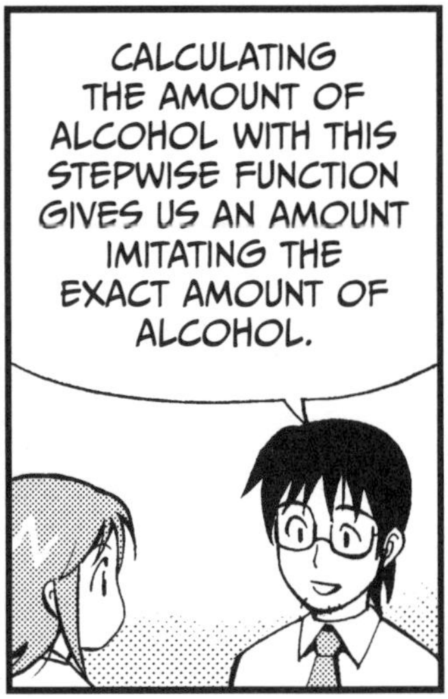
CALCULATING THE AMOUNT OF ALCOHOL WITH THIS STEPWISE FUNCTION GIVES US AN AMOUNT IMITATING THE EXACT AMOUNT OF ALCOHOL.

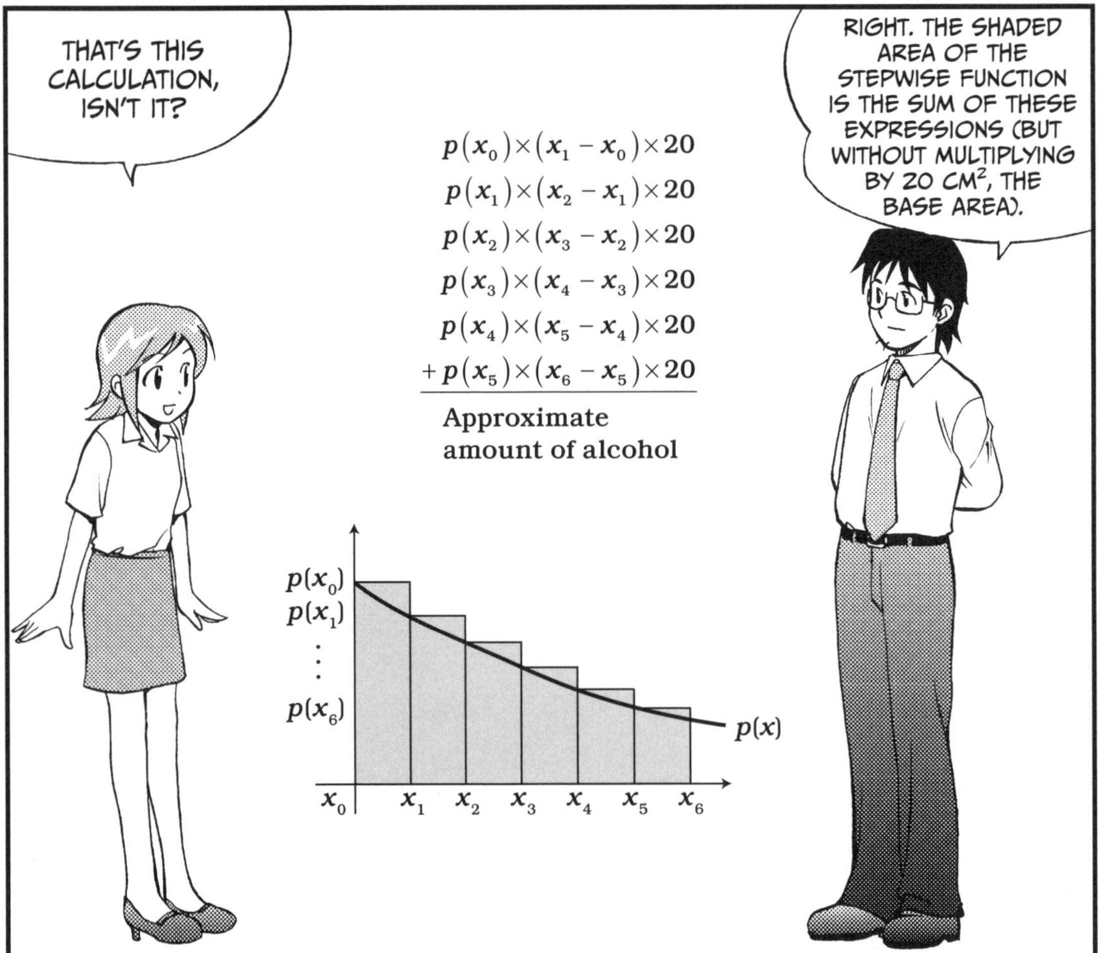
THAT'S THIS CALCULATION, ISN'T IT?
$p(x_0)\times(x_1 - x_0)\times 20$
$p(x_1)\times(x_2 - x_1)\times 20$
$p(x_2)\times(x_3 - x_2)\times 20$
$p(x_3)\times(x_4 - x_3)\times 20$
$p(x_4)\times(x_5 - x_4)\times 20$
$+\,p(x_5)\times(x_6 - x_5)\times 20$
Approximate amount of alcohol
$p(x_0)$
$p(x_1)$
$p(x_6)$
$p(x)$
x_0 x_1 x_2 x_3 x_4 x_5 x_6
RIGHT. THE SHADED AREA OF THE STEPWISE FUNCTION IS THE SUM OF THESE EXPRESSIONS (BUT WITHOUT MULTIPLYING BY 20 CM^2, THE BASE AREA).

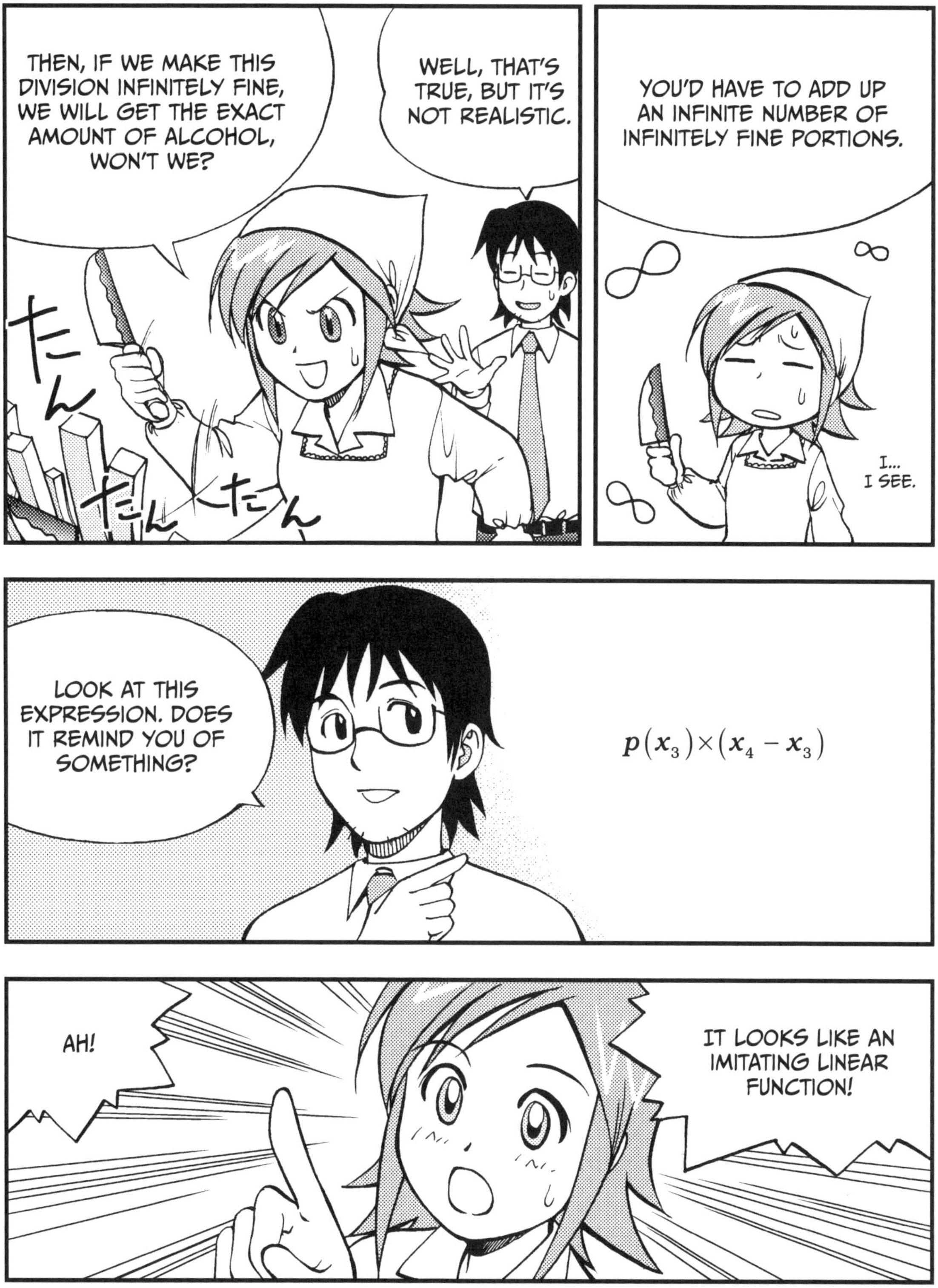
THEN, IF WE MAKE THIS DIVISION INFINITELY FINE, WE WILL GET THE EXACT AMOUNT OF ALCOHOL, WON'T WE?
WELL, THAT'S TRUE, BUT IT'S NOT REALISTIC.
たん
たん
たん
YOU'D HAVE TO ADD UP AN INFINITE NUMBER OF INFINITELY FINE PORTIONS.
∞
∞
∞
I... I SEE.
LOOK AT THIS EXPRESSION. DOES IT REMIND YOU OF SOMETHING?
$p(x_3) \times (x_4 - x_3)$
AH!
IT LOOKS LIKE AN IMITATING LINEAR FUNCTION!

STEP 4—REVIEW OF THE IMITATING LINEAR FUNCTION

When the derivative of $f(x)$ is given by $f'(x)$, we had $f(x) \approx f'(a)\ (x - a) + f(a)$ near $x = a$.

Transposing $f(a)$, we get

$$❶ \quad f(x) - f(a) \approx f'(a)(x - a)$$

or (Difference in f) ≈ (Derivative of f) × (Difference in x)

If we assume that the interval between two consecutive values of x_0, x_1, x_2, x_3, ..., x_6 is small enough, x_1 is close to x_0, x_2 is close to x_1, and so on.

Now, let's introduce a new function, $q(x)$, whose derivative is $p(x)$. This means $q'(x) = p(x)$.

Using ❶ for this $q(x)$, we get

(Difference in q) ≈ (Derivative of q) × (Difference in x)

$$q(x_1) - q(x_0) \approx p(x_0)(x_1 - x_0)$$

$$q(x_2) - q(x_1) \approx p(x_1)(x_2 - x_1)$$

The sum of the right sides of these expressions is the same as the sum of the left sides.

Some terms in the expressions for the sum cancel each other out.

$$\begin{aligned}
&q(x_1) - q(x_0) \approx p(x_0)(x_1 - x_0)\\
&q(x_2) - q(x_1) \approx p(x_1)(x_2 - x_1)\\
&q(x_3) - q(x_2) \approx p(x_2)(x_3 - x_2)\\
&q(x_4) - q(x_3) \approx p(x_3)(x_4 - x_3)\\
&q(x_5) - q(x_4) \approx p(x_4)(x_5 - x_4)\\
+&q(x_6) - q(x_5) \approx p(x_5)(x_6 - x_5)\\
\hline
&q(x_6) - q(x_0) \approx \text{The sum}
\end{aligned}$$

Substituting $x_6 = 9$ and $x_0 = 0$, we get

The approximate amount of alcohol = the sum × 20

$$\{q(x_6) - q(x_0)\} \times 20$$

$$\{q(9) - q(0)\} \times 20$$

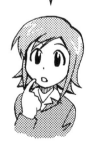

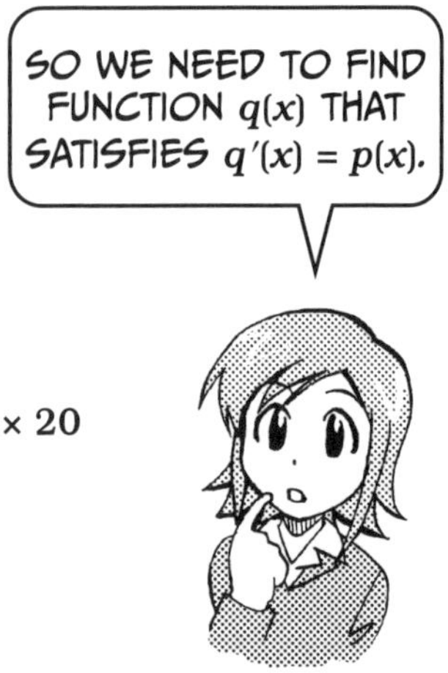

SO WE NEED TO FIND FUNCTION q(x) THAT SATISFIES q′(x) = p(x).

STEP 5—APPROXIMATION → EXACT VALUE

WE HAVE JUST OBTAINED THE FOLLOWING RELATIONSHIP OF EXPRESSIONS SHOWN IN THE DIAGRAM.

The approximate amount of alcohol ($\div$ 20) given by the stepwise function:

$$p(x_0)(x_1 - x_0) + p(x_1)(x_2 - x_1) + \ldots$$

❷
$\approx$
$$q(9) - q(0)$$
(Constant)

❶ $\approx$

The exact amount of alcohol ($\div$ 20)

BUT IF WE INCREASE THE NUMBER OF POINTS x_0, x_1, x_2, x_3, AND SO ON, UNTIL IT BECOMES INFINITE,

WE CAN SAY THAT RELATIONSHIP ❶ CHANGES FROM "APPROXIMATION" TO "EQUALITY."

BUT, SINCE THE SUM OF THE EXPRESSIONS HAVE BEEN IMITATING THE CONSTANT VALUE $q(9) - q(0)$,

The sum of $p(x_i)(x_{i+1} - x_i)$ for an infinite number of x_i

$=$

$$q(9) - q(0)$$

$\|$ $\|$

The exact amount of alcohol ($\div$ 20)

WE GET THE RELATIONSHIP SHOWN HERE.*

* WE WILL OBTAIN THIS RELATIONSHIP MORE RIGOROUSLY ON PAGE 94.

ILLUSTRATING THE FUNDAMENTAL THEOREM OF CALCULUS 89

STEP 6—$p(x)$ IS THE DERIVATIVE OF $q(x)$

NOW NORIKO, THE NEXT EXPRESSION WE WILL LOOK AT IS THIS.

If we suppose $q(x) = -\dfrac{2}{x+1}$, then $q'(x) = -\dfrac{2}{(x+1)^2} = p(x)$

In other words, $p(x)$ is the derivative of $q(x)$.
$q(x)$ is called the *antiderivative* of $p(x)$.

SO, THIS $q(x)$ IS THE FUNCTION WE WANTED.

The amount of alcohol

$$= \{q(9) - q(0)\} \times 20$$

$$= \left\{ -\frac{2}{9+1} - \left(-\frac{2}{0+1} \right) \right\} \times 20$$

$$= 36 \text{ grams}$$

THE AMOUNT OF ALCOHOL IN A GLASS OF SHOCHU WITH HOT WATER IS GENERALLY 24.3 GRAMS.

SO, WE HAVE A VERY STRONG DRINK HERE.

SINCE THE SUM OF INFINITE TERMS WE HAVE BEEN DOING

REQUIRES A LOT OF TIME TO WRITE DOWN, I WILL SHOW YOU ITS SYMBOL.

USING THE FUNDAMENTAL THEOREM OF CALCULUS

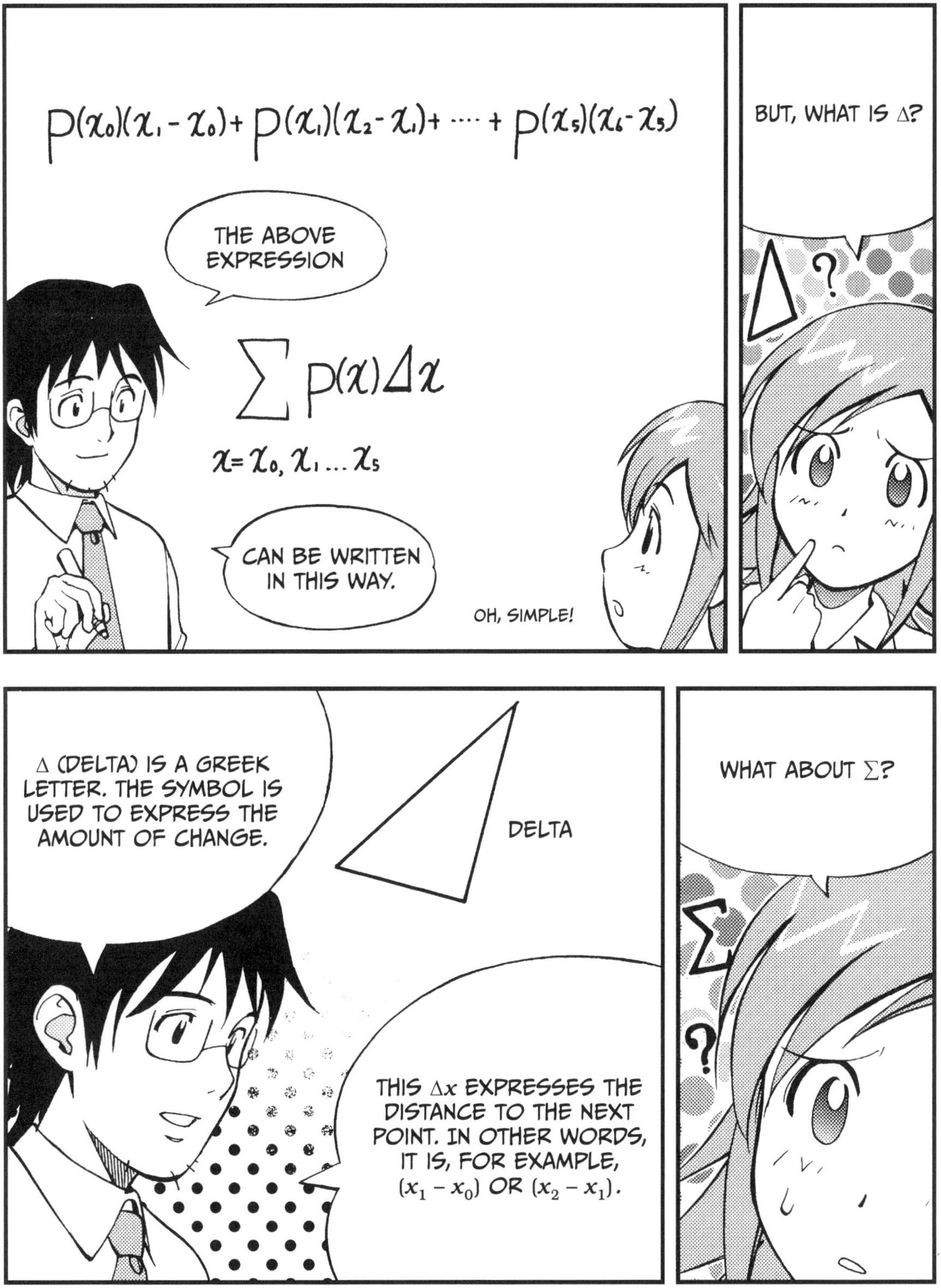

$p(x_0)(x_1 - x_0) + p(x_1)(x_2 - x_1) + \cdots + p(x_5)(x_6 - x_5)$
THE ABOVE EXPRESSION
$\sum p(x)\Delta x$
$x = x_0, x_1 \ldots x_5$
CAN BE WRITTEN IN THIS WAY.
OH, SIMPLE!
BUT, WHAT IS Δ?
Δ?
Δ (DELTA) IS A GREEK LETTER. THE SYMBOL IS USED TO EXPRESS THE AMOUNT OF CHANGE.
DELTA
THIS Δx EXPRESSES THE DISTANCE TO THE NEXT POINT. IN OTHER WORDS, IT IS, FOR EXAMPLE, $(x_1 - x_0)$ OR $(x_2 - x_1)$.
WHAT ABOUT Σ?
Σ?

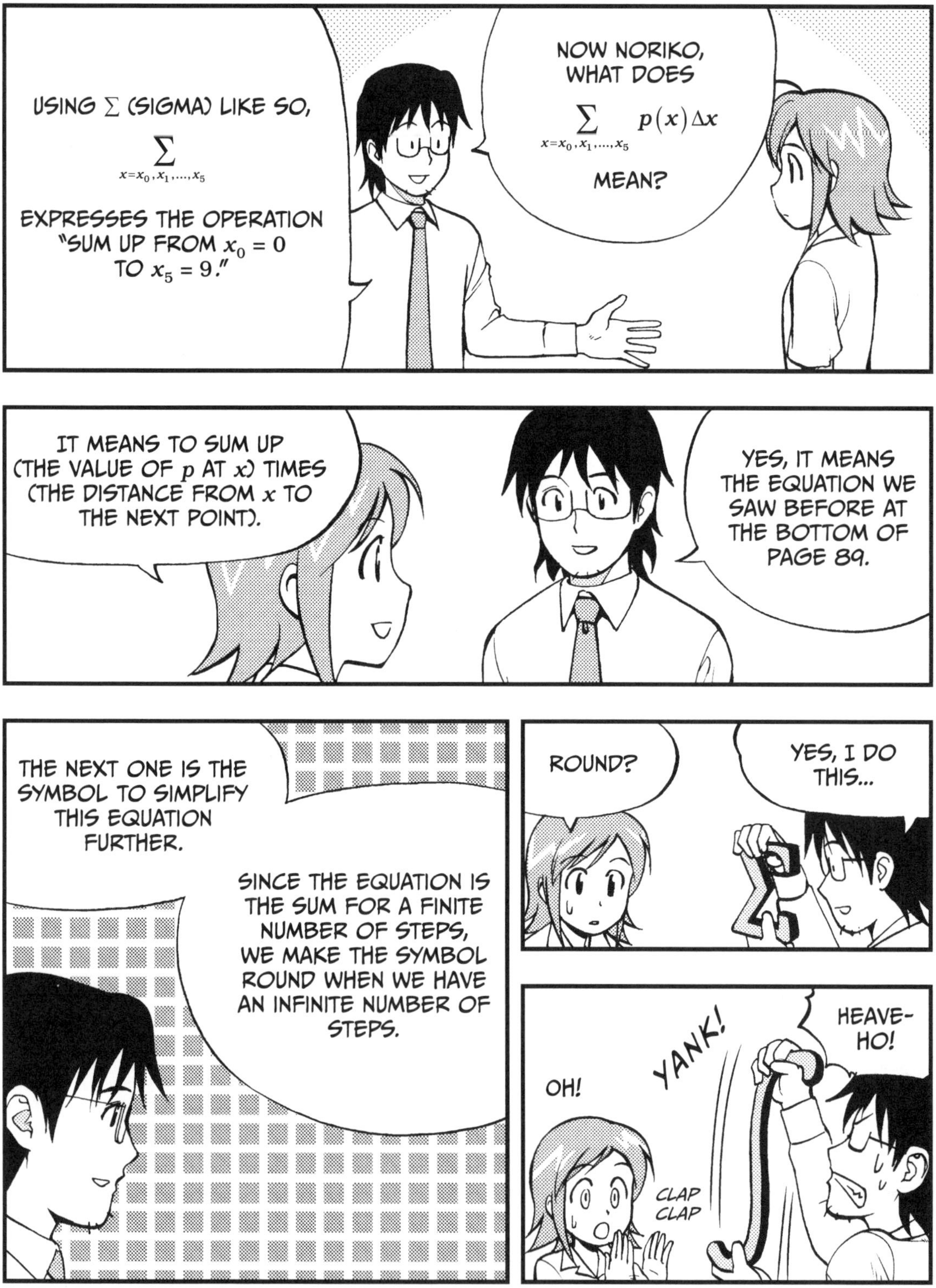
USING $\sum$ (SIGMA) LIKE SO, $\sum_{x=x_0, x_1, \ldots, x_5}$ EXPRESSES THE OPERATION "SUM UP FROM $x_0 = 0$ TO $x_5 = 9$."
NOW NORIKO, WHAT DOES $\sum_{x=x_0, x_1, \ldots, x_5} p(x)\Delta x$ MEAN?
IT MEANS TO SUM UP (THE VALUE OF p AT x) TIMES (THE DISTANCE FROM x TO THE NEXT POINT).
YES, IT MEANS THE EQUATION WE SAW BEFORE AT THE BOTTOM OF PAGE 89.
THE NEXT ONE IS THE SYMBOL TO SIMPLIFY THIS EQUATION FURTHER.
SINCE THE EQUATION IS THE SUM FOR A FINITE NUMBER OF STEPS, WE MAKE THE SYMBOL ROUND WHEN WE HAVE AN INFINITE NUMBER OF STEPS.
ROUND?
YES, I DO THIS...
OH!
YANK!
HEAVE-HO!
CLAP CLAP

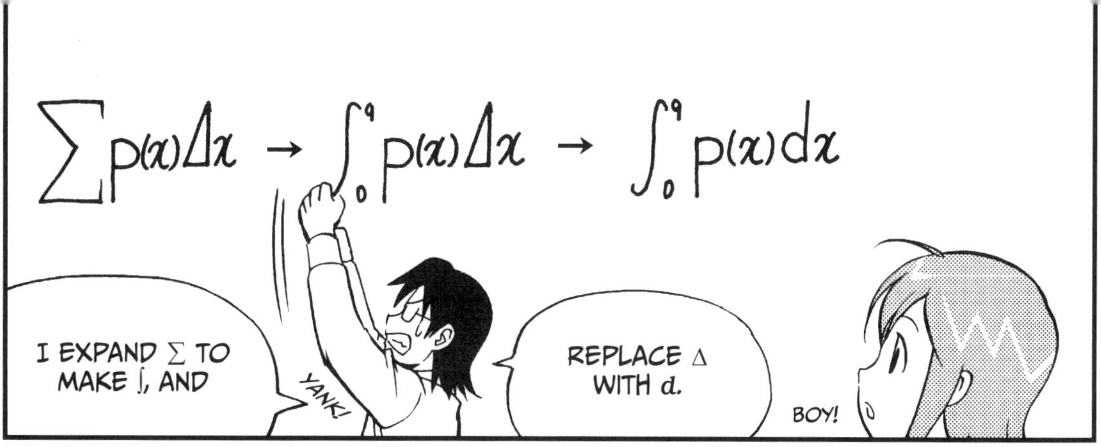
$\sum p(x)\Delta x \rightarrow \int_0^9 p(x)\Delta x \rightarrow \int_0^9 p(x)dx$
I EXPAND Σ TO MAKE ∫, AND
YANK!
REPLACE Δ WITH d.
BOY!

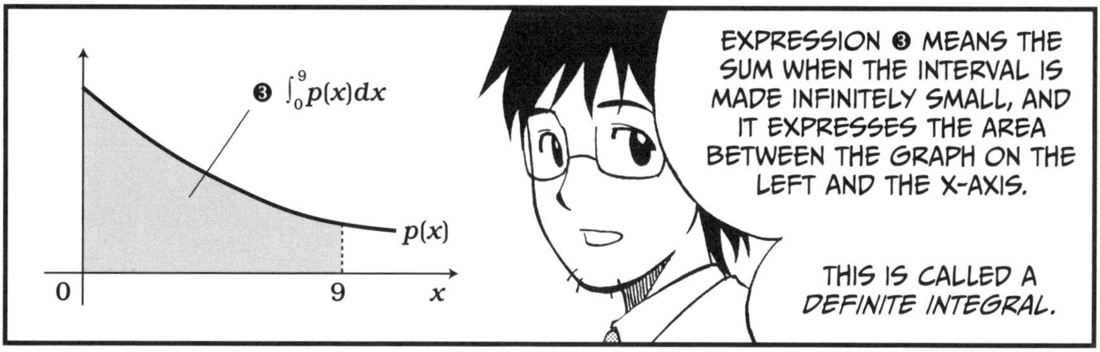
❸ $\int_0^9 p(x)dx$
p(x)
0
9
x
EXPRESSION ❸ MEANS THE SUM WHEN THE INTERVAL IS MADE INFINITELY SMALL, AND IT EXPRESSES THE AREA BETWEEN THE GRAPH ON THE LEFT AND THE X-AXIS.
THIS IS CALLED A *DEFINITE INTEGRAL.*

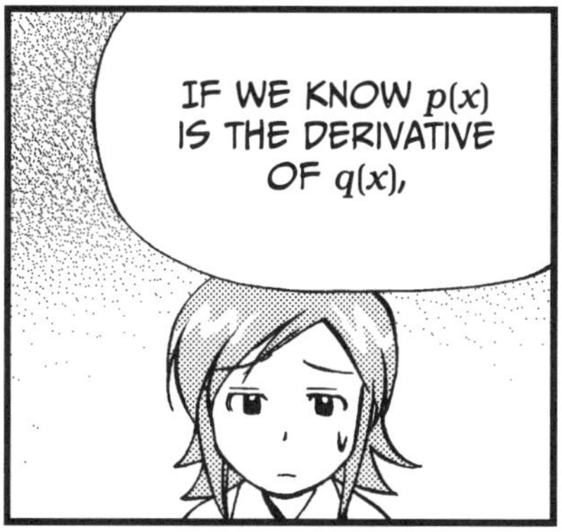
IF WE KNOW p(x) IS THE DERIVATIVE OF q(x),

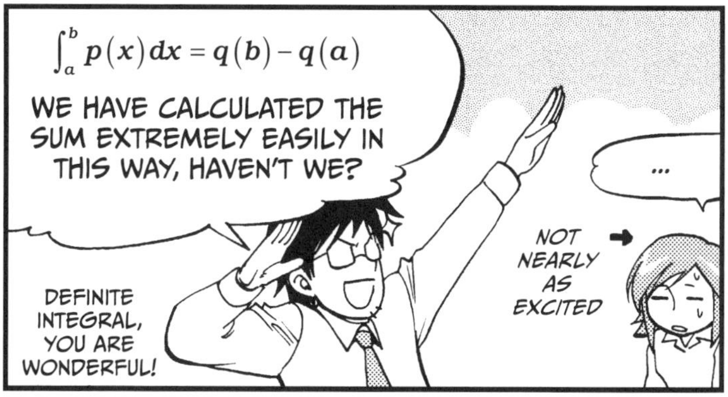
$\int_a^b p(x)dx = q(b) - q(a)$
WE HAVE CALCULATED THE SUM EXTREMELY EASILY IN THIS WAY, HAVEN'T WE?
DEFINITE INTEGRAL, YOU ARE WONDERFUL!
NOT NEARLY AS EXCITED
...

SUMMARY

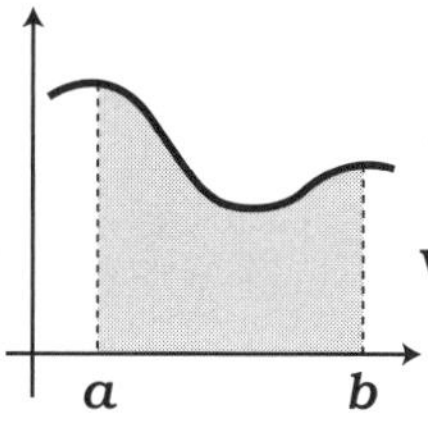
a
b

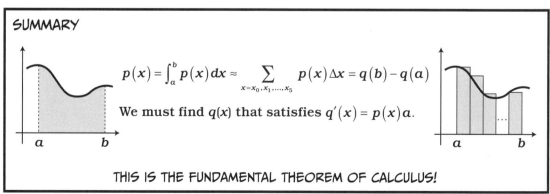

$$p(x) = \int_a^b p(x)dx \approx \sum_{x=x_0,x_1,\ldots,x_5} p(x)\Delta x = q(b) - q(a)$$

We must find $q(x)$ that satisfies $q'(x) = p(x)a$.

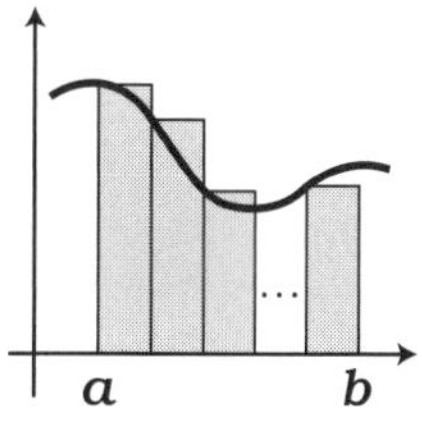
a
b

THIS IS THE FUNDAMENTAL THEOREM OF CALCULUS!

A STRICT EXPLANATION OF STEP 5

In the explanation given before (page 89), we used, as the basic expression, $q(x_1) - q(x_0) \approx p(x_0)(x_1 - x_0)$, a "crude" expression which roughly imitates the exact expression. For those who think this is a sloppy explanation, we will explain more carefully here. Using the mean value theorem, we can reproduce the same result.

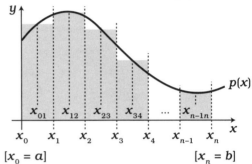

    We first find $q(x)$ that satisfies $q'(x) = p(x)$.

    We place points $x_0 (= a)$, x_1, x_2, x_3, ..., $x_n (= b)$ on the x-axis.

    We then find point x_{01} that exists between x_0 and x_1 and satisfies $q(x_1) - q(x_0) \approx q'(x_{01})(x_1 - x_0)$.

    The existence of such a point is guaranteed by the mean value theorem. Similarly, we find x_{12} between x_1 and x_2 and get

$$q(x_2) - q(x_1) \approx q'(x_{12})(x_2 - x_1)$$

Areas of these steps

    Repeating this operation, we get

$$
\begin{aligned}
q(x_1) - q(x_0) &= q'(x_{01})(x_1 - x_0) &= p(x_{01})(x_1 - x_0) \\
q(x_2) - q(x_1) &= q'(x_{12})(x_2 - x_1) &= p(x_{12})(x_2 - x_1) \\
q(x_3) - q(x_2) &= q'(x_{23})(x_3 - x_2) &= p(x_{23})(x_3 - x_2) \\
&\quad\cdots &\quad\cdots
\end{aligned}
$$

Summing up

$$+ \; q(x_n) - q(x_{n-1}) = q'(x_{n-1n})(x_n - x_{n-1}) = p(x_{n-1n})(x_n - x_{n-1})$$

$$q(x_n) - q(x_0) \quad \longleftarrow \boxed{\text{Always equal}} \longrightarrow \quad \text{Approximate area}$$

$$\downarrow \qquad\qquad\qquad\qquad\qquad\qquad \downarrow \text{Infinitely fine sections}$$

$$q(b) - q(a) \quad \longleftarrow \boxed{\text{Equal}} \longrightarrow \quad \text{Exact area}$$

This corresponds to the diagram in step 5.

USING INTEGRAL FORMULAS

FORMULA 3-1: THE INTEGRAL FORMULAS

❶ $$\int_a^b f(x)\,dx + \int_b^c f(x)\,dx = \int_a^c f(x)\,dx$$

The intervals of definite integrals of the same function can be joined.

❷ $$\int_a^b \{f(x) + g(x)\}\,dx = \int_a^b f(x)\,dx + \int_a^b g(x)\,dx$$

A definite integral of a sum can be divided into the sum of definite integrals.

❸ $$\int_a^b \alpha f(x)\,dx = \alpha \int_a^b f(x)\,dx$$

The multiplicative constant within a definite integral can be moved outside the integral.

Expressions ❶ through ❸ can be understood intuitively if we draw their figures.

❶

+
=
a
b
c
a
b
c
a
b
c

❷

Area for g
Area for f
=
f(x)
+
g(x)
a
b
a
b
a
b

❸

αf(x)
f(x)
Area is multiplied by α.

I SEE.

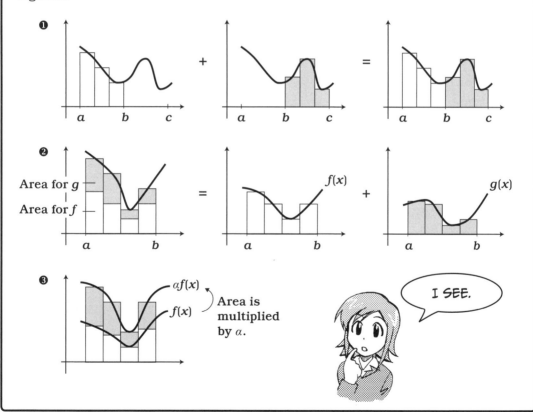

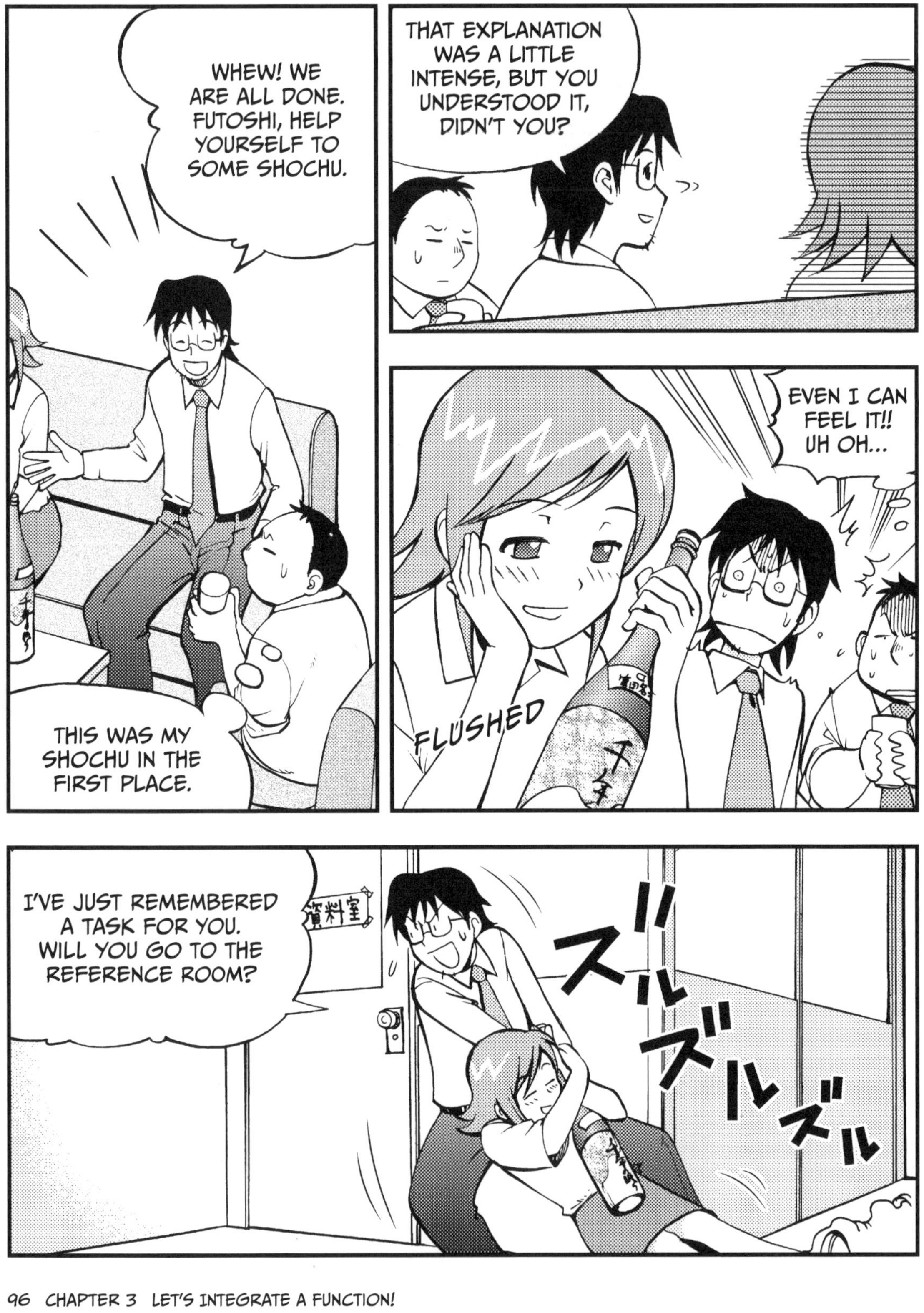
WHEW! WE ARE ALL DONE. FUTOSHI, HELP YOURSELF TO SOME SHOCHU.
THIS WAS MY SHOCHU IN THE FIRST PLACE.
THAT EXPLANATION WAS A LITTLE INTENSE, BUT YOU UNDERSTOOD IT, DIDN'T YOU?
EVEN I CAN FEEL IT!! UH OH...
FLUSHED
I'VE JUST REMEMBERED A TASK FOR YOU. WILL YOU GO TO THE REFERENCE ROOM?
資料室
ズルズルズルズル

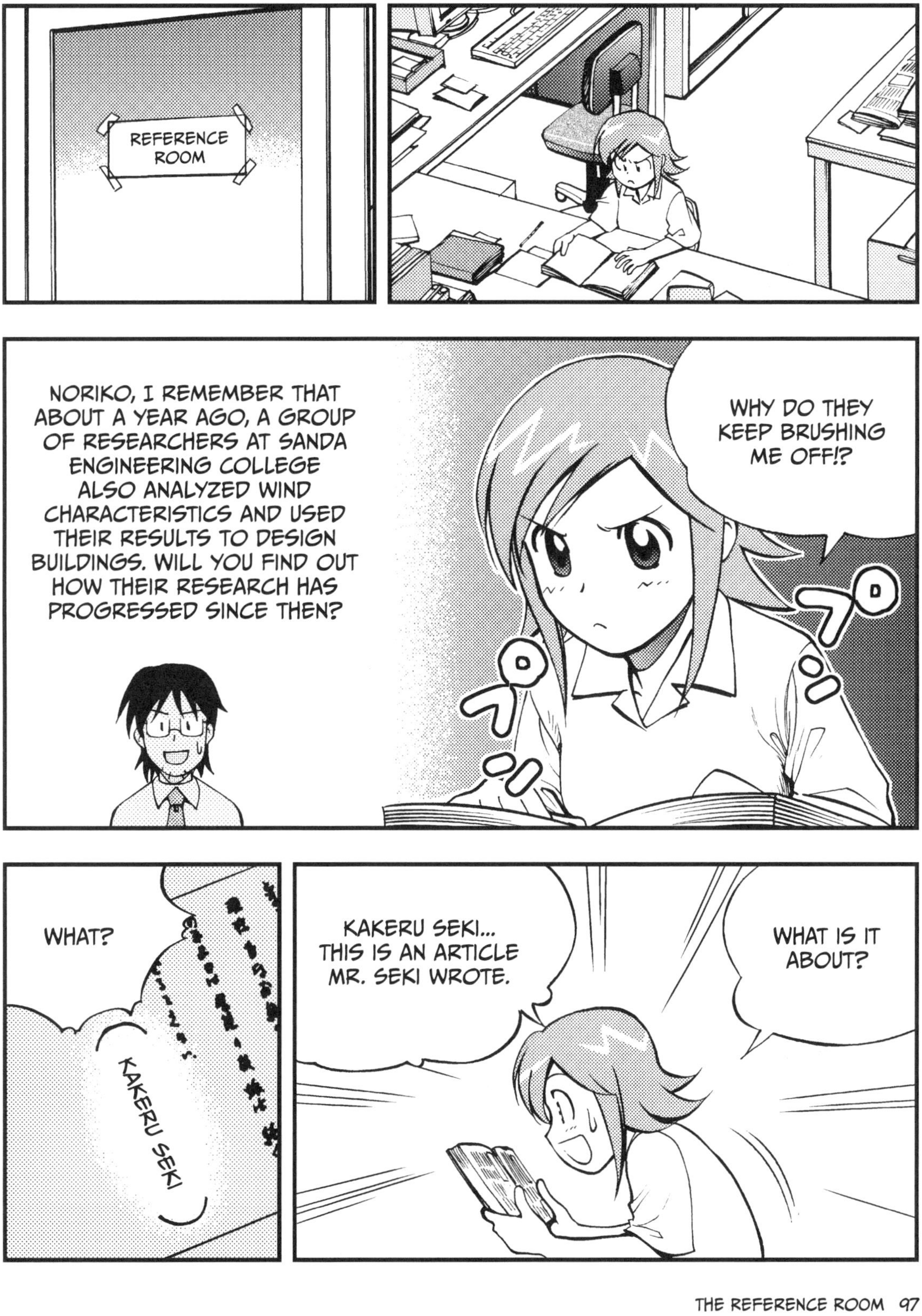
REFERENCE ROOM
NORIKO, I REMEMBER THAT ABOUT A YEAR AGO, A GROUP OF RESEARCHERS AT SANDA ENGINEERING COLLEGE ALSO ANALYZED WIND CHARACTERISTICS AND USED THEIR RESULTS TO DESIGN BUILDINGS. WILL YOU FIND OUT HOW THEIR RESEARCH HAS PROGRESSED SINCE THEN?
WHY DO THEY KEEP BRUSHING ME OFF!?
WHAT?
KAKERU SEKI
KAKERU SEKI... THIS IS AN ARTICLE MR. SEKI WROTE.
WHAT IS IT ABOUT?

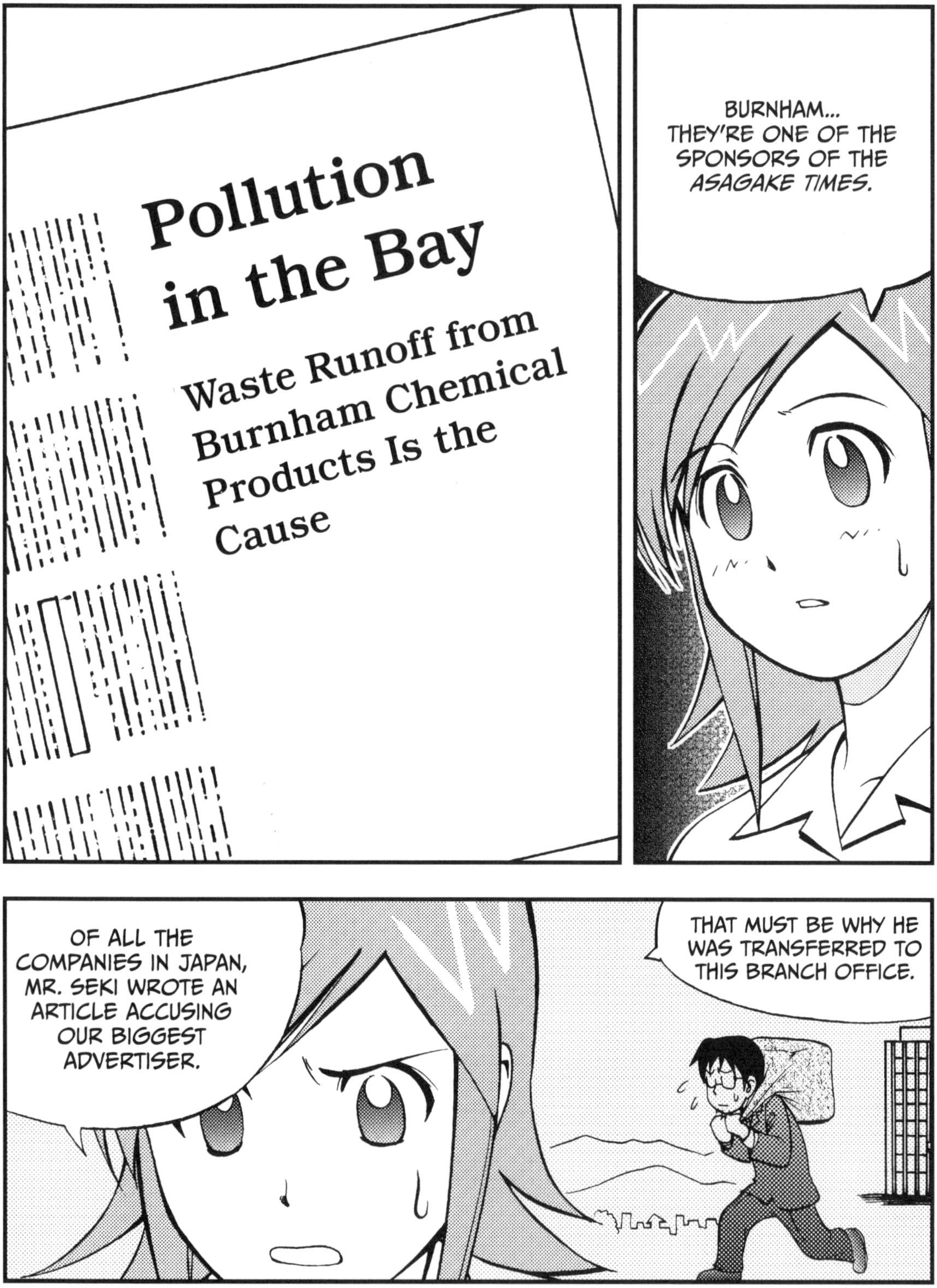
Pollution in the Bay
Waste Runoff from Burnham Chemical Products Is the Cause
BURNHAM... THEY'RE ONE OF THE SPONSORS OF THE *ASAGAKE TIMES.*
OF ALL THE COMPANIES IN JAPAN, MR. SEKI WROTE AN ARTICLE ACCUSING OUR BIGGEST ADVERTISER.
THAT MUST BE WHY HE WAS TRANSFERRED TO THIS BRANCH OFFICE.

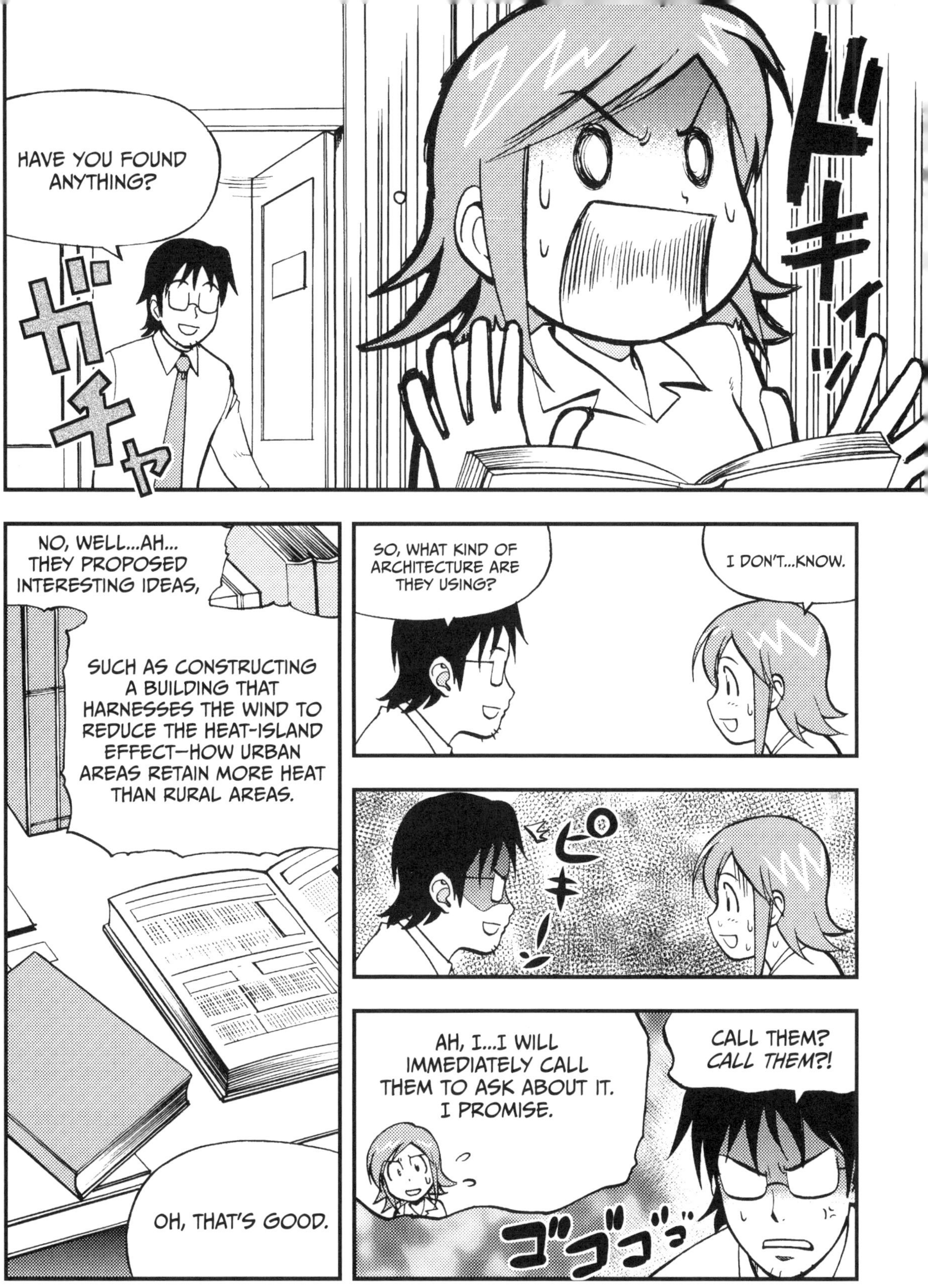
HAVE YOU FOUND ANYTHING?
ガチャ
NO, WELL...AH... THEY PROPOSED INTERESTING IDEAS,
SUCH AS CONSTRUCTING A BUILDING THAT HARNESSES THE WIND TO REDUCE THE HEAT-ISLAND EFFECT—HOW URBAN AREAS RETAIN MORE HEAT THAN RURAL AREAS.
OH, THAT'S GOOD.
SO, WHAT KIND OF ARCHITECTURE ARE THEY USING?
I DON'T...KNOW.
ピキ!
AH, I...I WILL IMMEDIATELY CALL THEM TO ASK ABOUT IT. I PROMISE.
CALL THEM? *CALL THEM?!*

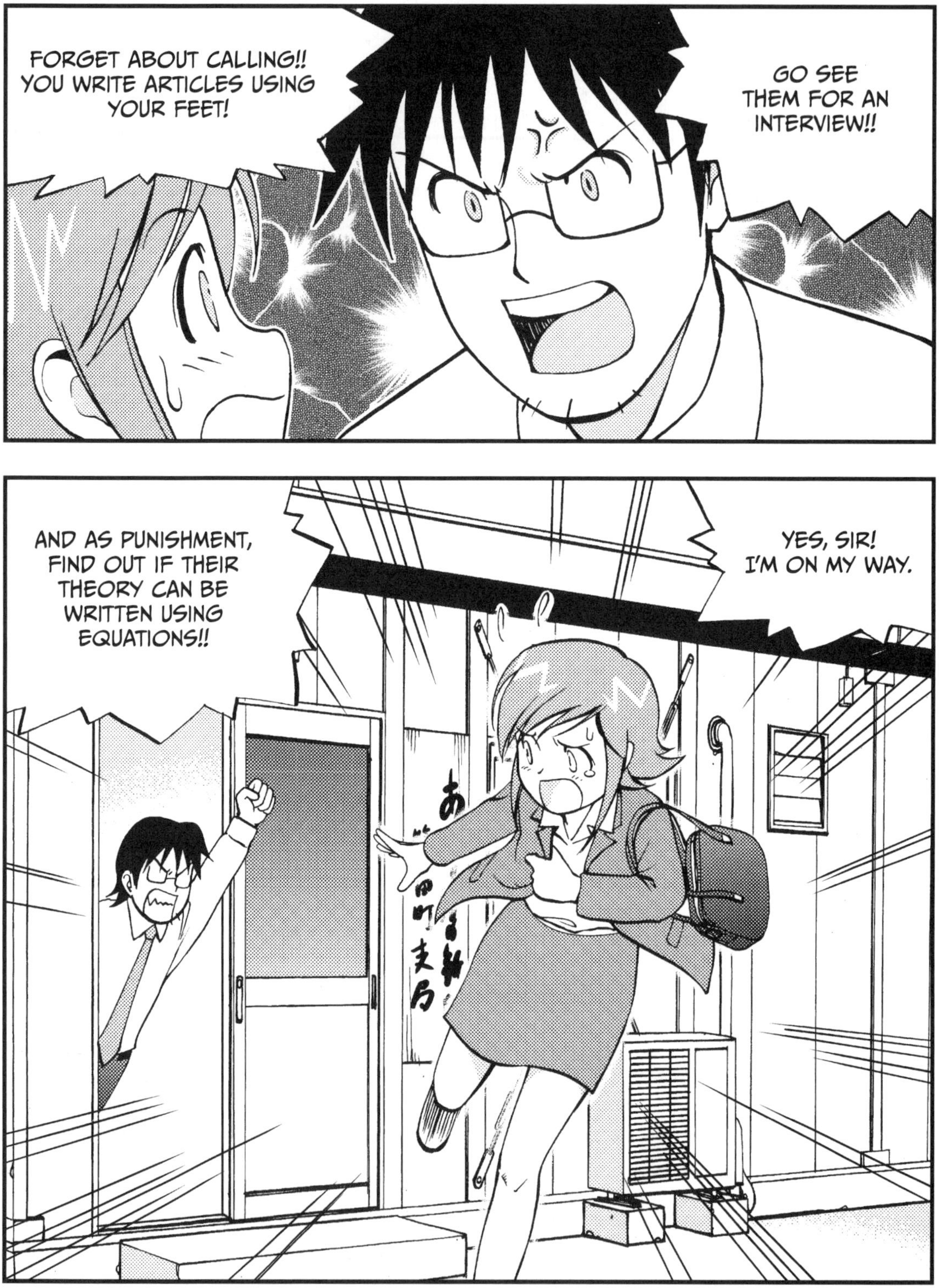
FORGET ABOUT CALLING!! YOU WRITE ARTICLES USING YOUR FEET!
GO SEE THEM FOR AN INTERVIEW!!
AND AS PUNISHMENT, FIND OUT IF THEIR THEORY CAN BE WRITTEN USING EQUATIONS!!
YES, SIR! I'M ON MY WAY.

APPLYING THE FUNDAMENTAL THEOREM

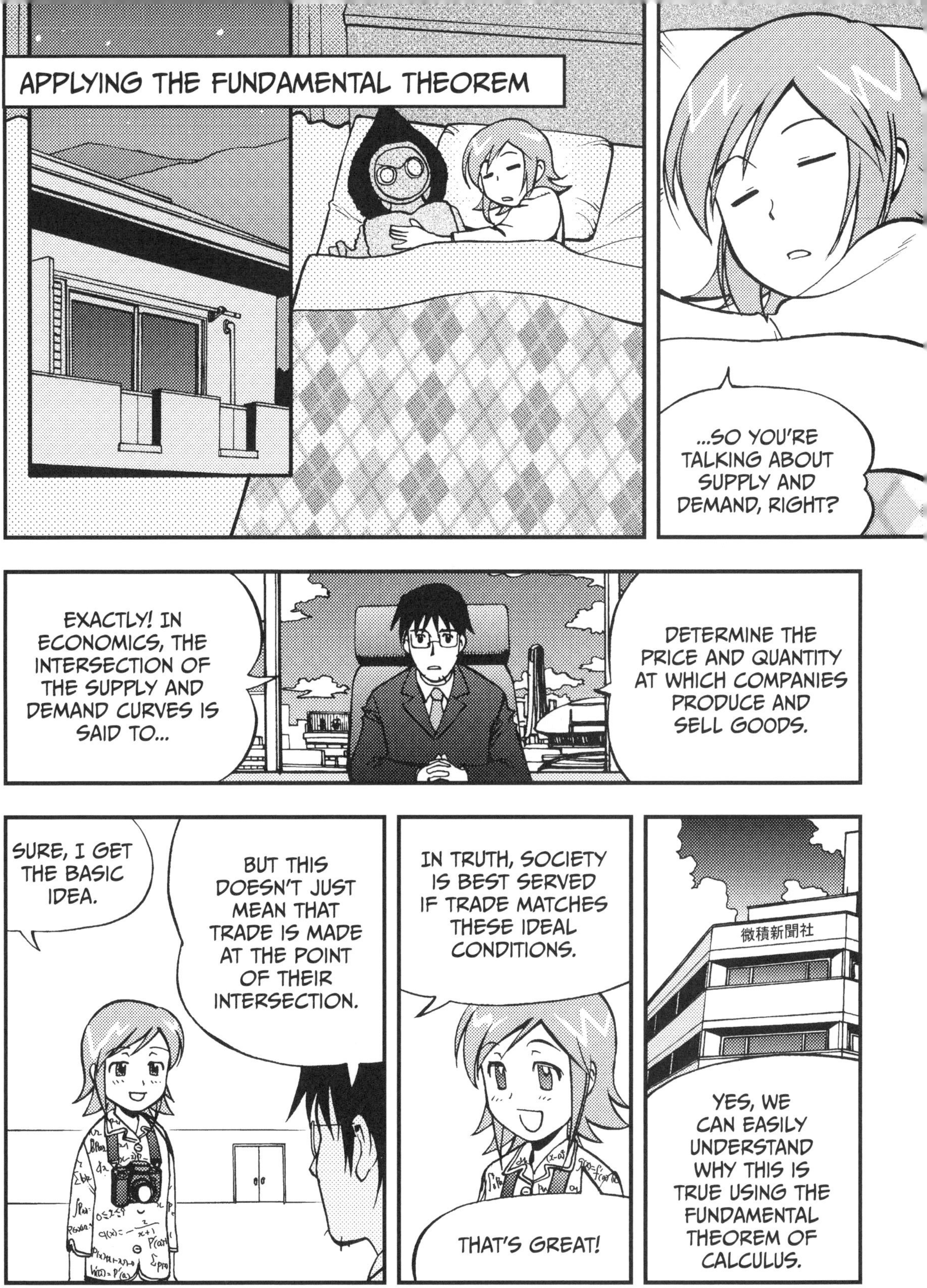

...SO YOU'RE TALKING ABOUT SUPPLY AND DEMAND, RIGHT?
EXACTLY! IN ECONOMICS, THE INTERSECTION OF THE SUPPLY AND DEMAND CURVES IS SAID TO...
DETERMINE THE PRICE AND QUANTITY AT WHICH COMPANIES PRODUCE AND SELL GOODS.
SURE, I GET THE BASIC IDEA.
BUT THIS DOESN'T JUST MEAN THAT TRADE IS MADE AT THE POINT OF THEIR INTERSECTION.
IN TRUTH, SOCIETY IS BEST SERVED IF TRADE MATCHES THESE IDEAL CONDITIONS.
THAT'S GREAT!
微積新聞社
YES, WE CAN EASILY UNDERSTAND WHY THIS IS TRUE USING THE FUNDAMENTAL THEOREM OF CALCULUS.

SUPPLY CURVE

FIRST, LET'S CONSIDER HOW COMPANIES MAXIMIZE PROFIT IN A PERFECTLY COMPETITIVE MARKET. WE'LL TRY TO DERIVE A SUPPLY CURVE FIRST.

The profit $P(x)$ when x units of a commodity are produced is given by the following function:

$$\underbrace{\text{(Profit)}}_{P(x)} = \underbrace{\text{(Price)}}_{p} \times \underbrace{\text{(Production Quantity)}}_{x} - \underbrace{\text{(Cost)}}_{C(x)} = px - C(x)$$

where $C(x)$ is the cost of production.

Let's assume the x value that maximizes the profit $P(x)$ is the quantity of production s.

A company wants to maximize its profits. Recall that to find a function's extrema, we take the derivative and set it to zero. This means that the company's maximum profit occurs when

$$P'(s) = p - C'(s) = 0$$

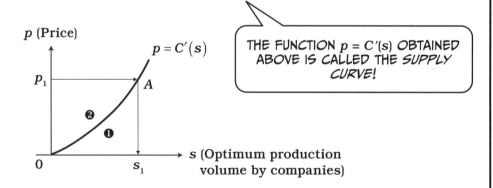

p (Price)
p = C′(s)
p1
A
❷
❶
0
s1
s (Optimum production volume by companies)
THE FUNCTION p = C′(s) OBTAINED ABOVE IS CALLED THE SUPPLY CURVE!

Price p_1 corresponds to point A on the function, which leads us to optimum production volume s_1.

The rectangle bounded by these four points (p_1, A, s_1, and the origin) equals the price multiplied by the production quantity. This should be the companies' gross profits, before subtracting their costs of production. But look, the area ❶ of this graph corresponds to the companies' production costs, and we can obtain it using an integral.

$$\int_0^{s_1} C'(s)\,ds = C(s_1) - C(0) = C(s_1) = \text{Costs}$$

We used the Fundamental Theorem here.

To simplify, we assume C(0) = 0.

This means we can easily find the companies' net profit, which is represented by area ❷ in the graph, or the area of the rectangle minus area ❶.

DEMAND CURVE

Next, let's consider the maximum benefit for consumers.

When consumers purchase x units of a commodity, the benefit $B(x)$ for them is given by the equation:

$$B(x) = \text{Total Value of Consumption} - (\text{Price} \times \text{Quantity}) = u(x) - px$$

where $u(x)$ is a function describing the value of the commodity for all consumers.

Consumers will purchase the most of this commodity when $B(x)$ is maximized.

If we set the consumption value to t when the derivative of $B(x) = 0$, we get the following equation:*

$$B'(t) = u'(t) - p = 0$$

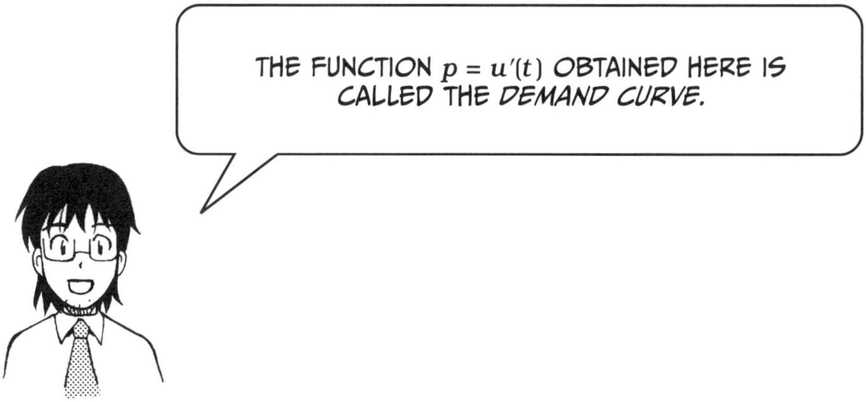
THE FUNCTION $p = u'(t)$ OBTAINED HERE IS CALLED THE *DEMAND CURVE.*

* Again, you can see we're looking for extrema (where $B'(t) = 0$), as consumers want to maximize their benefits.

p (Price)
❹
B
p_1
$p = u'(t)$
❸
t (Optimum consumption)
0
t_1

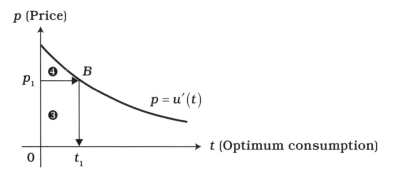

So let's consider the area of the rectangle labeled ❸, above, which corresponds to the price multiplied by the product consumption. In other words, this is the total amount consumers pay for a product.

The total area of ❸ and ❹ can be obtained using integration.

$$\int_0^{t_1} u'(t)\,dt = u(t_1) - u(0) = u(t_1) = \text{Total value of consumption}$$

To simplify, we assume $u(0) = 0$.

If you simply subtract the value of the rectangle ❸ from the integral from 0 to t_1, you can find the area of ❹, the benefit to consumers.

THE BENEFIT FOR THE CONSUMERS ❹ IS THE TOTAL VALUE OF CONSUMPTION MINUS THE AMOUNT THEY PAID ❸, RIGHT?
YES, THAT'S IT. NOW LET'S LOOK AT THE SUPPLY AND DEMAND CURVES COMBINED TOGETHER.

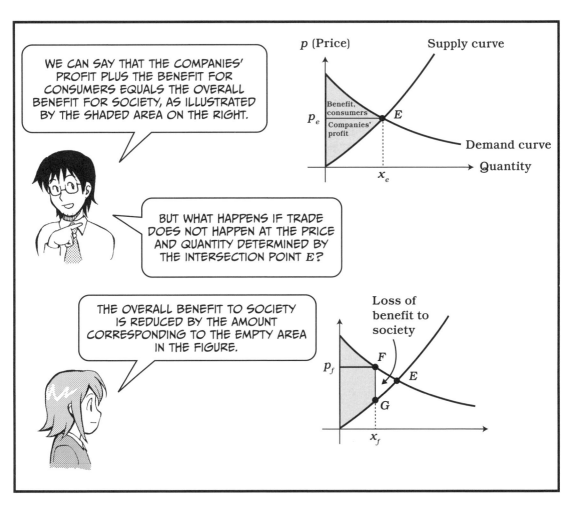

WE CAN SAY THAT THE COMPANIES' PROFIT PLUS THE BENEFIT FOR CONSUMERS EQUALS THE OVERALL BENEFIT FOR SOCIETY, AS ILLUSTRATED BY THE SHADED AREA ON THE RIGHT.

p (Price)

Supply curve

Benefit, consumers

p_e

Companies' profit

E

Demand curve

Quantity

x_e

BUT WHAT HAPPENS IF TRADE DOES NOT HAPPEN AT THE PRICE AND QUANTITY DETERMINED BY THE INTERSECTION POINT E?

THE OVERALL BENEFIT TO SOCIETY IS REDUCED BY THE AMOUNT CORRESPONDING TO THE EMPTY AREA IN THE FIGURE.

Loss of benefit to society

p_f

F

E

G

x_f

DO YOU GET IT?

YES, I WILL REPORT MY STORIES USING CALCULUS, TOO.

I ALSO THINK VELOCITY AND FALLING BODIES ARE GOOD TOPICS TO WRITE ABOUT.

I'M GOING TO LOOK INTO THEM!

¥50 **THE CALCULUS NEWS-GAZETTE** Vol. 1

The Integral of Velocity Proven to Be Distance!

The integral of velocity = difference in position = distance traveled

If we understand this formula, it's said that we can correctly calculate the distance traveled for objects whose velocity changes constantly. But is that true? Our promising freshman reporter Noriko Hikima closes in on the truth of this matter in her hard-hitting report.

y
y_3
y_2
y_1
x_1 x_2 x_3
x

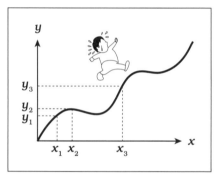

Figure 1: This graph represents Futoshi's distance traveled over time. He moves to point y_1, y_2, y_3... as time progresses to x_1, x_2, x_3...

Sanda-Cho—Some readers will recall our earlier example describing Futoshi walking on a moving walkway. Others have likely deliberately blocked his sweaty image from their minds. But you almost certainly remember that the derivative of the distance is the speed.

❶ $y = F(x)$

❷ $\int_a^b v(x)\,dx = F(b) - F(a)$

Equation ❶ expresses the position of the monstrous, sweating Futoshi. In other words, after x seconds he has lumbered a total distance of y.

Integral of Velocity = Difference in Position

The derivative $F'(x)$ of expression ❶ is the "instantaneous velocity" at x seconds. If we rewrite $F'(x)$ as $v(x)$, using v for *velocity*, the Fundamental Theorem of Calculus can be used to obtain equation ❷! Look at the graph of $v(x)$ in Figure 2-A—Futoshi's velocity over time. The shaded part of the graph is equal to the integral—equation ❷.

But also look at Figure 2-B, which shows the distance Futoshi has traveled over time. If we look at Figures 2-A and 2-B side by side, we see that the integral of the velocity is equal to the difference in position (or distance)! Notice how the two graphs match—when Futoshi's velocity is positive, his distance increases, and vice versa.

A Velocity
B Distance
y
Area = Difference
+
−
+
x_0 x_1 x_3
x
y
y = F(x)
y_3
y_1
x_1 x_3
x

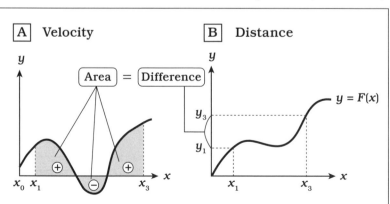

Figure 2

Free Fall from Tokyo Tower

How Many Seconds to the Ground?

It's easy to take things for granted—consider gravity. If you drop an object from your hand, it naturally falls to the ground. We can say that this is a motion that changes every second—it is *accelerating* due to the Earth's gravitational pull. This motion can be easily described using calculus.

But let's consider a bigger drop—all the way from the top of Tokyo Tower—and find out, "How many seconds does it take an object to reach the ground?" Pay no attention to Futoshi's remark, "Why don't you go to the top of Tokyo Tower with a stopwatch and find out for yourself?"

The increase in velocity when an object is in free fall is called *gravitational acceleration*, or 9.8 m/s^2. In other words, this means that an object's velocity increases by 9.8 m/s every second. Why is this the rate of acceleration? Well, let's just assume the scientists are right for today.

Expression ❶ gives the distance the object falls in T seconds. Since the integral of the velocity is the difference in position (or the distance the object travels), equation ❷ can be derived. Look at Figure 3—we've calculated the area by taking half of the product of the x and y values—in this case, ½ × 9.8t × t. And we know that the height of Tokyo Tower is 333 m. The square root of (333 / 4.9) equals about 8.2, so an object takes about 8.2 seconds to reach the ground. (We've neglected air resistance here for convenience.)

❶ $F(T) - F(0) = \int_0^T v(x)\,dx = \int_0^T 9.8(x)\,dx$

❷ $4.9T^2 - 4.9 \times 0^2 = 4.9T^2$

$$333 = 4.9T^2 \Rightarrow T = \sqrt{\frac{333}{4.9}} = 8.2 \text{ seconds}$$

Velocity
v(x) = 9.8x
Area of the velocity
9.8t × t × ½ = 4.9t²
9.8t
Time
t
Distance
4.9t²
t
Distance fallen
Time

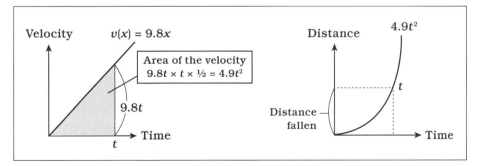

Figure 3

The Die Is Cast!!!

The Fundamental Theorem of Calculus Applies to Dice, Too

You probably remember playing games with dice as a child. Since ancient times, these hexahedrons have been rolled around the world, not only in games, but also for fortune telling and gambling.

Mathematically, you can say that dice are the world's smallest random-number generator. Dice are wonderful. Now we'll cast them for calculus! A die can show a 1, 2, 3, 4, 5, or 6—the probability of any one number is 1 in 6. This can be shown with a histogram (Figure 4), with their numbers on the x-axis and the probability on the y-axis.

This can be expressed by equation ❶, or $f(x)$ = Probability of rolling x. This becomes equation ❷ when we try to predict a single result—for example, a roll of 4.

f(x)
1/6
1 2 3 4 5 6

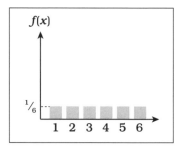

Figure 4: Density function

F(x)
1
1/6
1 2 3 4 5 6

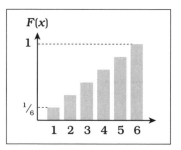

Figure 5: Distribution function

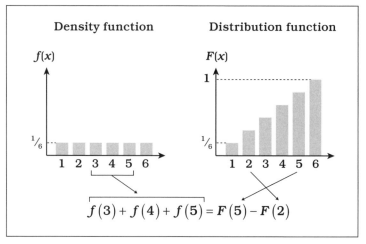

Density function
Distribution function
f(x)
F(x)
1
1/6
1 2 3 4 5 6
f(3) + f(4) + f(5) = F(5) − F(2)

Figure 6: Derivative of distribution function $F(x)$ = density function $f(x)$

❶ $f(x)$ = Probability of rolling x

❷ $f(4) = \frac{1}{6}$ = Probability of rolling 4

Now let's take a look at Figure 5, which describes a distribution function. First, start at 1 on the x-axis. Since no number less than 1 exists on a die, the probability in this region is 0. At $x = 1$, the graph jumps to 1/6, because the probability of rolling a number less than or equal to 1 is 1 in 6. You can also see that the probability of rolling a number equal to or greater than 1 and less than 2 is 1/6 as well. This should make intuitive sense. At 2, the probability jumps up to 2/6, which means the probability for rolling a number equal to or less than 2 is 2/6. Since this probability remains until right below 3, the probability of numbers less than 3 is 2/6.

❸ $\int_a^b f(x)dx = F(b) - F(a)$

= Probability of rolling x where $a \le x \le b$

In the same way, we can find that the probability of rolling a 6 or any number smaller than 6 (that is, any number on the die) is 1. After all, a die cannot stand on one of its corners. Now let's look at the probability of rolling numbers greater than 2 and equal to or less than 5. The equation in Figure 6 explains this relationship.

If we look at equation ❸, we see that it describes what we know—"A definite integral of a differentiated function = The difference in the original function." This is nothing but the Fundamental Theorem of Calculus! How wonderful dice are.

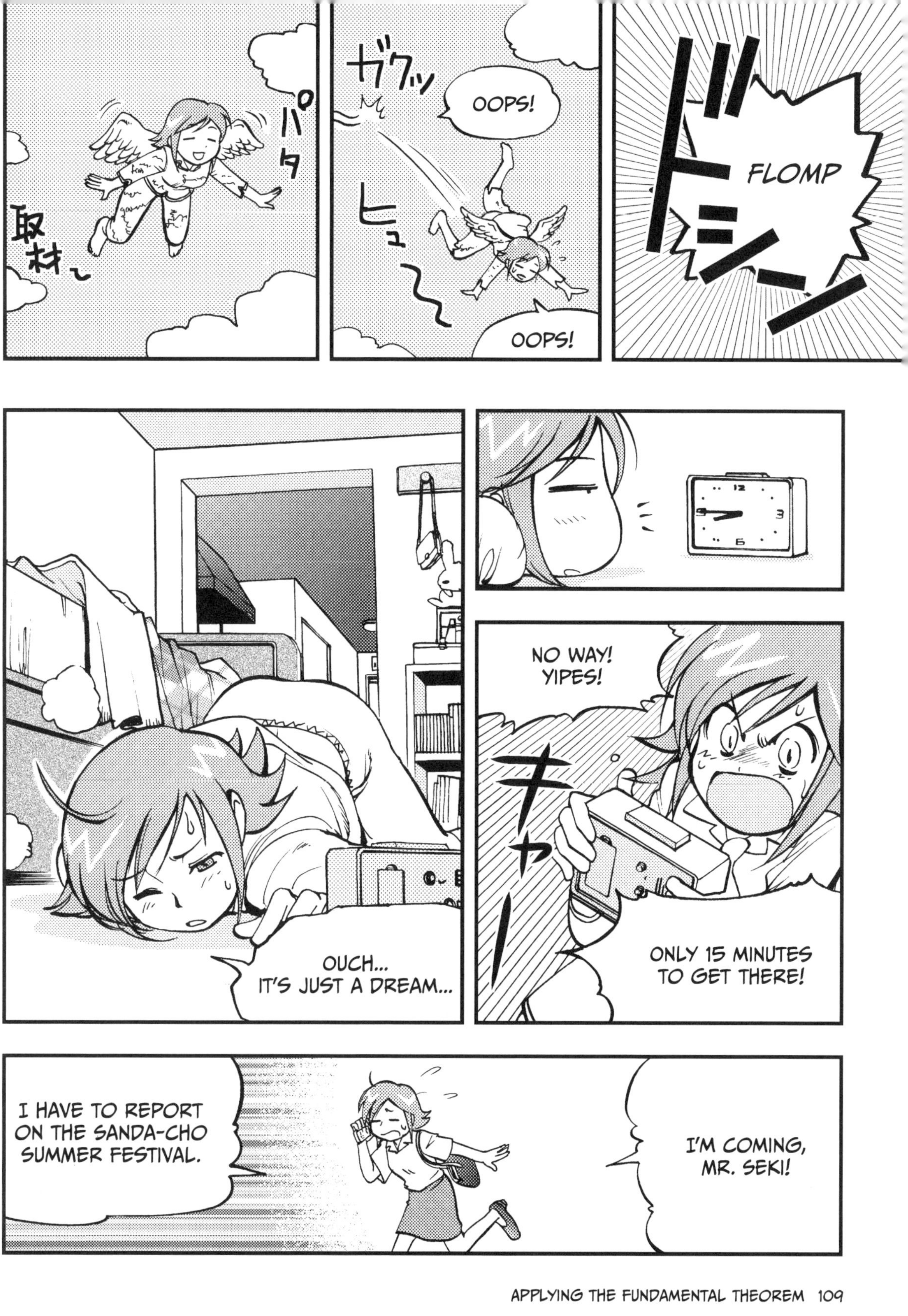
パタ
取材～
ガクッ
OOPS!
ヒュ～
OOPS!
ドシーン
FLOMP
OUCH... IT'S JUST A DREAM...
NO WAY! YIPES!
キャー
ONLY 15 MINUTES TO GET THERE!
I HAVE TO REPORT ON THE SANDA-CHO SUMMER FESTIVAL.
I'M COMING, MR. SEKI!

REVIEW OF THE FUNDAMENTAL THEOREM OF CALCULUS

When the derivative of $F(x)$ is $f(x)$, that is, if $f(x) = F'(x)$

$$\int_a^b f(x)\,dx = F(b) - F(a)$$

This can also be written as

$$\int_a^b F'(x)\,dx = F(b) - F(a)$$

These expressions mean the following.

> (Differentiated function) dx
> = Difference of the original function between b and a

It also means graphically that

$$\left(\begin{array}{c} \text{Area surrounded by the differentiated function} \\ \text{and the x-axis, between } x = a \text{ and } x = b \end{array} \right) = \left(\begin{array}{c} \text{Change in the original} \\ \text{function from } a \text{ to } b \end{array} \right)$$

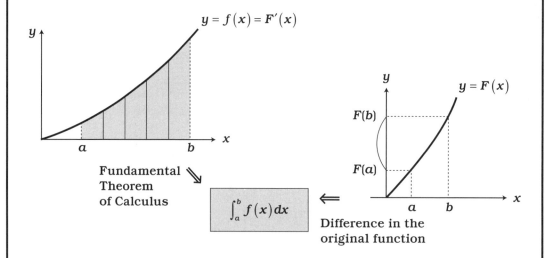

Fundamental
Theorem
of Calculus ⟱

$$\int_a^b f(x)\,dx$$

⟸ Difference in the
original function

FORMULA OF THE SUBSTITUTION RULE OF INTEGRATION

When a function of y is substituted for variable x as $x = g(y)$, how do we express

$$S = \int_a^b f(x)\,dx$$

a definite integral with respect to x, as a definite integral with respect to y?

First, we express the definite integral in terms of a stepwise function approximately as

$$S \approx \sum_{k=0,1,2,\ldots,n-1} f(x_k)(x_{k+1} - x_k) \quad (x_0 = a, x_n = b)$$

Transforming variable x as $x = g(y)$, we set

$$y_0 = \alpha, y_1, y_2, \ldots, y_n = \beta$$

so that

$$a = g(\alpha), x_1 = g(y_1), x_2 = g(y_2), \ldots, b = g(\beta)$$

Note here that using an approximate linear function of

$$x_{k+1} - x_k = g(y_{k+1}) - g(y_k) \approx g'(y_k)(y_{k+1} - y_k)$$

Substituting these expressions in S, we get

$$S \approx \sum_{k=0,1,2,\ldots,n-1} f(x_k)(x_{k+1} - x_k) \approx \sum_{k=0,1,2,\ldots,n-1} f(g(y_k))g'(y_k)(y_{k+1} - y_k)$$

The last expression is an approximation of

$$\int_\alpha^\beta f(g(y))g'(y)\,dy$$

Therefore, by making the divisions infinitely small, we obtain the following formula.

FORMULA 3-2: THE SUBSTITUTION RULE OF INTEGRATION

$$\int_a^b f(x)\,dx = \int_\alpha^\beta f(g(y))g'(y)\,dy$$

EXAMPLE:

Calculate:

$$\int_0^1 10(2x+1)^4\,dx$$

We first substitute the variable so that $y = 2x + 1$, or $x = g(y) = \dfrac{y-1}{2}$.

Since $y = 2x + 1$, if we take the derivative of both sides, we get $dy = 2dx$. Then we get $dx = \dfrac{1}{2}dy$.

Since we now integrate with respect to y, the new interval of integration is obtained from $0 = g(1)$ and $1 = g(3)$ to be $1 - 3$.[*]

$$\int_0^1 10(2x+1)^4\,dx = \int_1^3 10y^4\frac{1}{2}\,dy = \int_1^3 5y^4\,dy = 3^5 - 1^5 = 242$$

THE POWER RULE OF INTEGRATION

In the example above we remembered that $5y^4$ is the derivative of y^5 to finish the problem. Since we know that if $F(x) = x^n$, then $F'(x) = f(x) = nx^{(n+1)}$, we should be able to find a general rule for finding $F(x)$ when $f(x) = x^n$.

We know that $F(x)$ should have $x^{(n+1)}$ in it, but what about that coefficient? We don't have a coefficient in our derivative, so we'll need to start with one. When we take the derivative, the coefficient will be $(n + 1)$, so it follows that $1 / (n + 1)$ will cancel it out. That means that the general rule for finding the antiderivative $F(x)$ of $f(x) = x^n$ is

$$F(x) = \frac{1}{n+1} \times x^{(n+1)} = x^{\frac{n+1}{n+1}}$$

---

[*] In other words, when $x = 0$, $y = 1$, and when $x = 1$, $y = 3$. We then use that as the range of our definite integral.

EXERCISES

1. Calculate the definite integrals given below.

   ❶ $\int_1^3 3x^2\,dx$

   ❷ $\int_2^4 \dfrac{x^3+1}{x^2}\,dx$

   ❸ $\int_0^5 x+\left(1+x^2\right)^7\,dx + \int_0^5 x-\left(1+x^2\right)^7\,dx$

2. Answer the following questions.

   A. Write an expression of the definite integral which calculates the area surrounded by the graph of $y = f(x) = x^2 - 3x$ and the x-axis.

   B. Calculate the area given by this expression.

4
LET'S LEARN INTEGRATION TECHNIQUES!

USING TRIGONOMETRIC FUNCTIONS

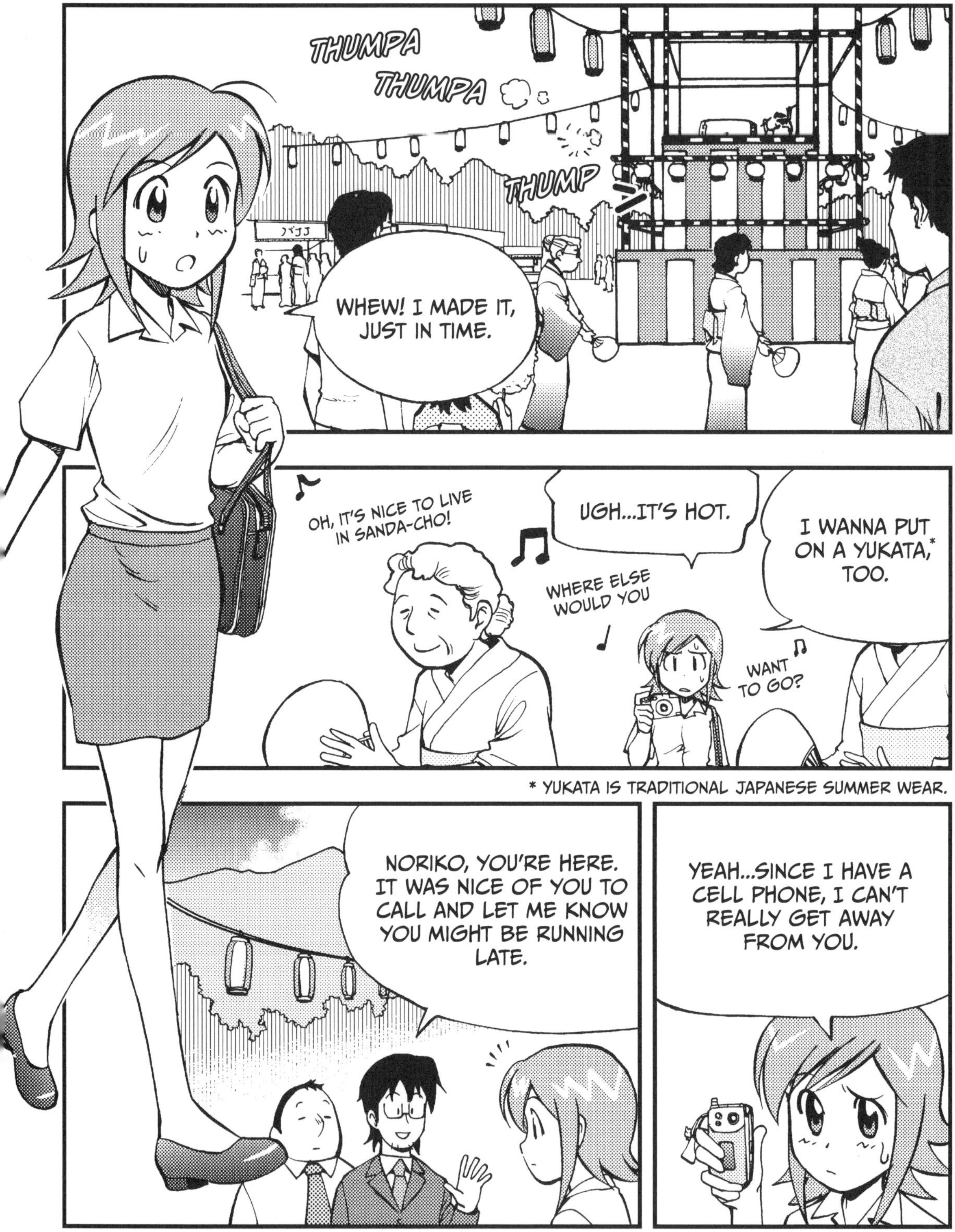
THUMPA
THUMPA
THUMP
WHEW! I MADE IT, JUST IN TIME.
OH, IT'S NICE TO LIVE IN SANDA-CHO!
UGH...IT'S HOT.
WHERE ELSE WOULD YOU
WANT TO GO?
I WANNA PUT ON A YUKATA,* TOO.
NORIKO, YOU'RE HERE. IT WAS NICE OF YOU TO CALL AND LET ME KNOW YOU MIGHT BE RUNNING LATE.
YEAH...SINCE I HAVE A CELL PHONE, I CAN'T REALLY GET AWAY FROM YOU.

* YUKATA IS TRADITIONAL JAPANESE SUMMER WEAR.

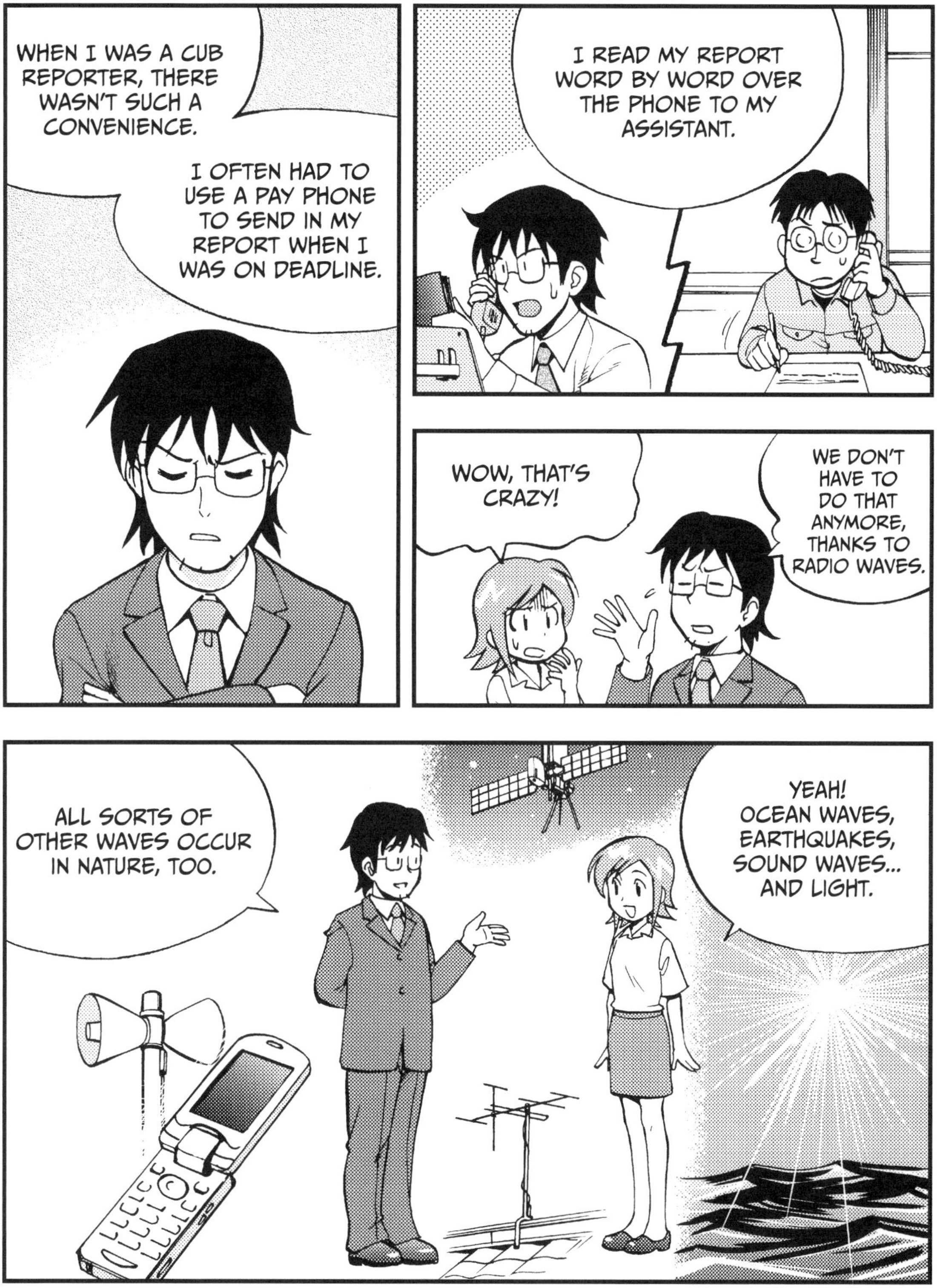
WHEN I WAS A CUB REPORTER, THERE WASN'T SUCH A CONVENIENCE.
I OFTEN HAD TO USE A PAY PHONE TO SEND IN MY REPORT WHEN I WAS ON DEADLINE.
I READ MY REPORT WORD BY WORD OVER THE PHONE TO MY ASSISTANT.
WOW, THAT'S CRAZY!
WE DON'T HAVE TO DO THAT ANYMORE, THANKS TO RADIO WAVES.
ALL SORTS OF OTHER WAVES OCCUR IN NATURE, TOO.
YEAH! OCEAN WAVES, EARTHQUAKES, SOUND WAVES... AND LIGHT.

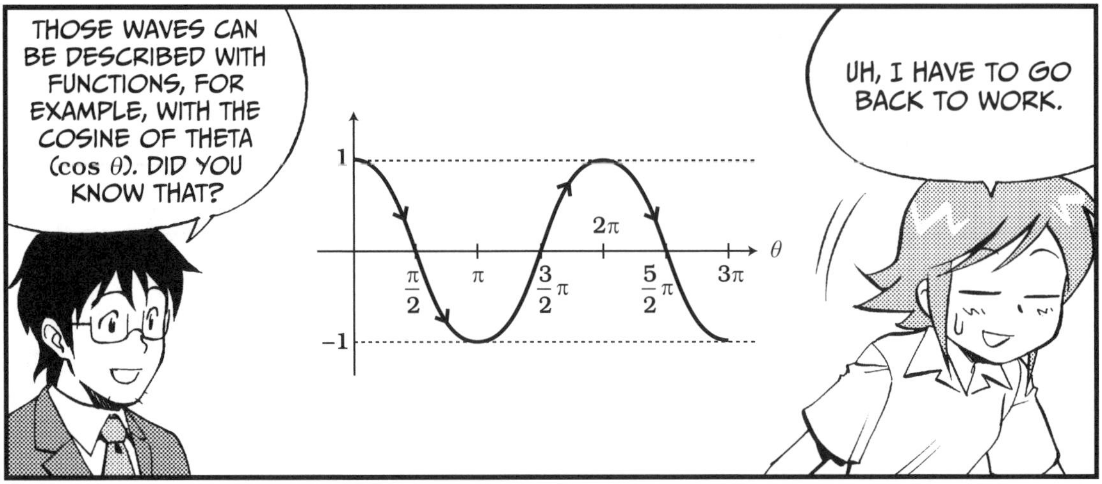
THOSE WAVES CAN BE DESCRIBED WITH FUNCTIONS, FOR EXAMPLE, WITH THE COSINE OF THETA (cos θ). DID YOU KNOW THAT?
1
−1
$\frac{\pi}{2}$
π
$\frac{3}{2}\pi$
2π
$\frac{5}{2}\pi$
3π
θ
UH, I HAVE TO GO BACK TO WORK.

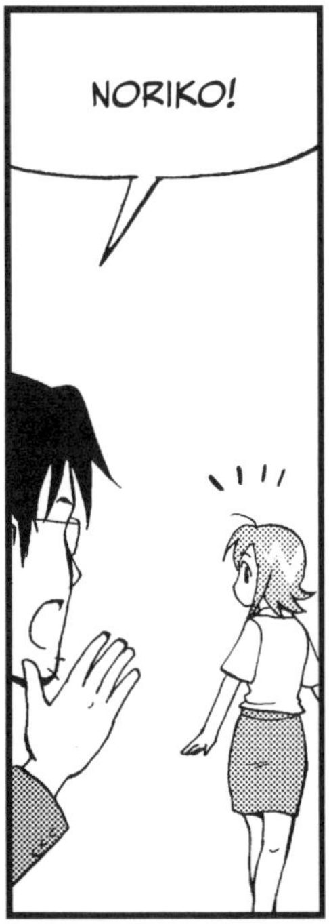
NORIKO!

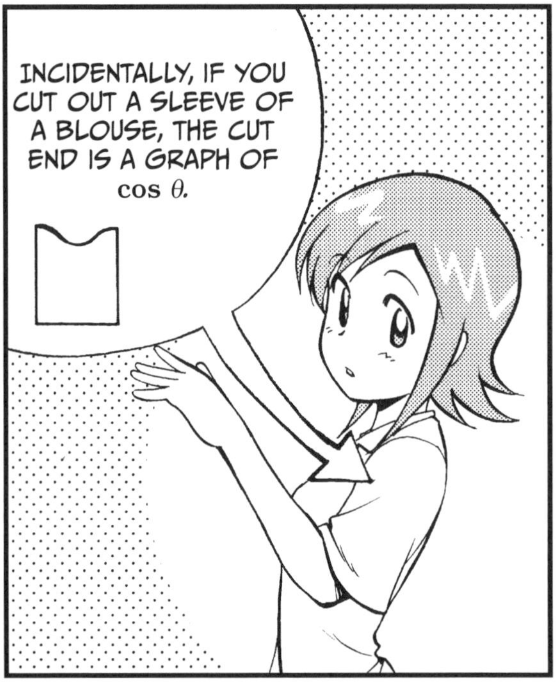
INCIDENTALLY, IF YOU CUT OUT A SLEEVE OF A BLOUSE, THE CUT END IS A GRAPH OF cos θ.

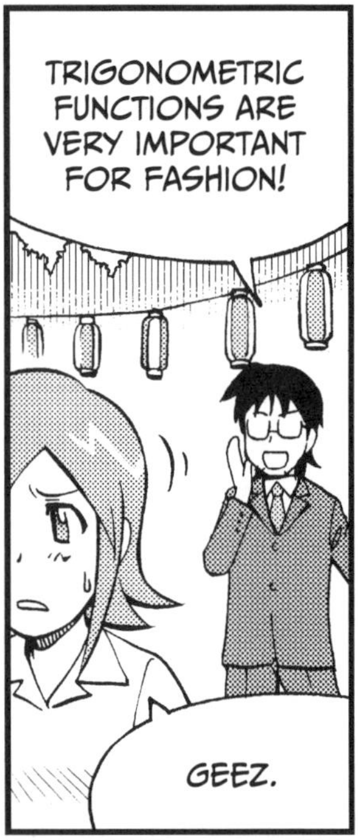
TRIGONOMETRIC FUNCTIONS ARE VERY IMPORTANT FOR FASHION!
GEEZ.

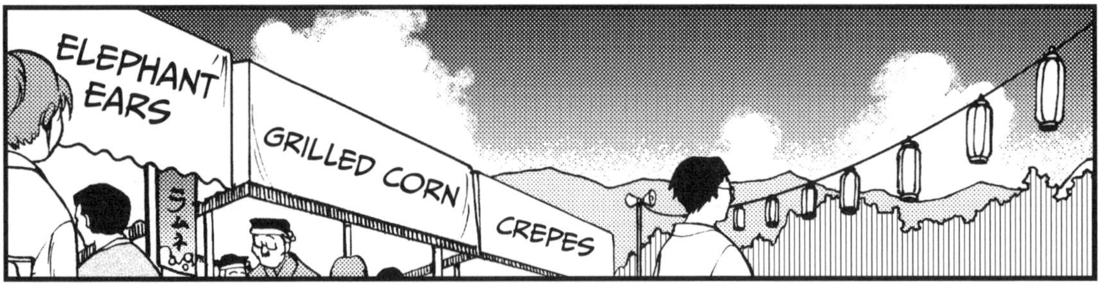
ELEPHANT EARS
GRILLED CORN
CREPES

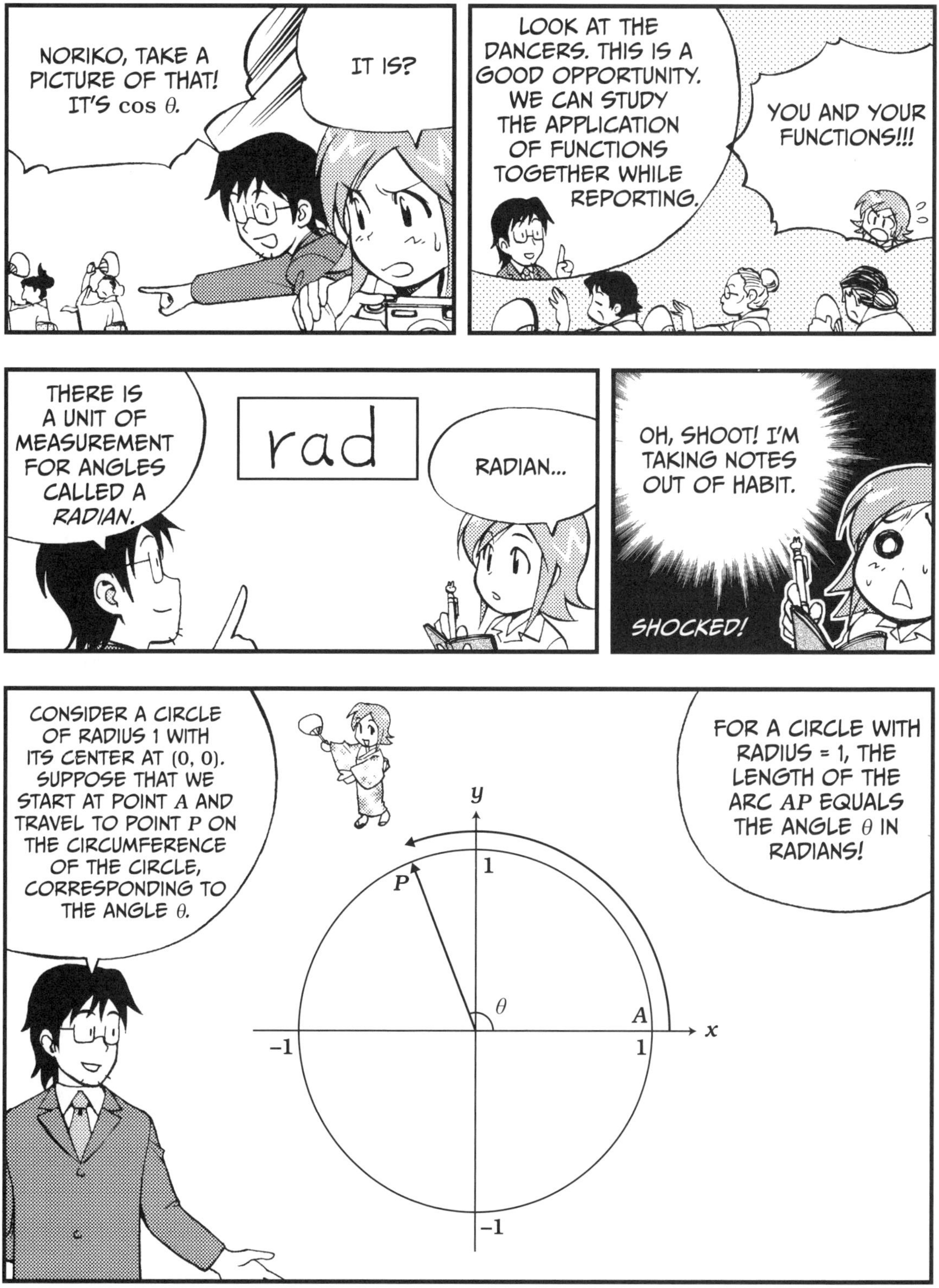
NORIKO, TAKE A PICTURE OF THAT! IT'S $\cos\theta$.
IT IS?
LOOK AT THE DANCERS. THIS IS A GOOD OPPORTUNITY. WE CAN STUDY THE APPLICATION OF FUNCTIONS TOGETHER WHILE REPORTING.
YOU AND YOUR FUNCTIONS!!!
THERE IS A UNIT OF MEASUREMENT FOR ANGLES CALLED A *RADIAN*.
rad
RADIAN...
OH, SHOOT! I'M TAKING NOTES OUT OF HABIT.
SHOCKED!
CONSIDER A CIRCLE OF RADIUS 1 WITH ITS CENTER AT (0, 0). SUPPOSE THAT WE START AT POINT A AND TRAVEL TO POINT P ON THE CIRCUMFERENCE OF THE CIRCLE, CORRESPONDING TO THE ANGLE θ.
y
x
1
-1
1
-1
P
A
θ
FOR A CIRCLE WITH RADIUS = 1, THE LENGTH OF THE ARC AP EQUALS THE ANGLE θ IN RADIANS!

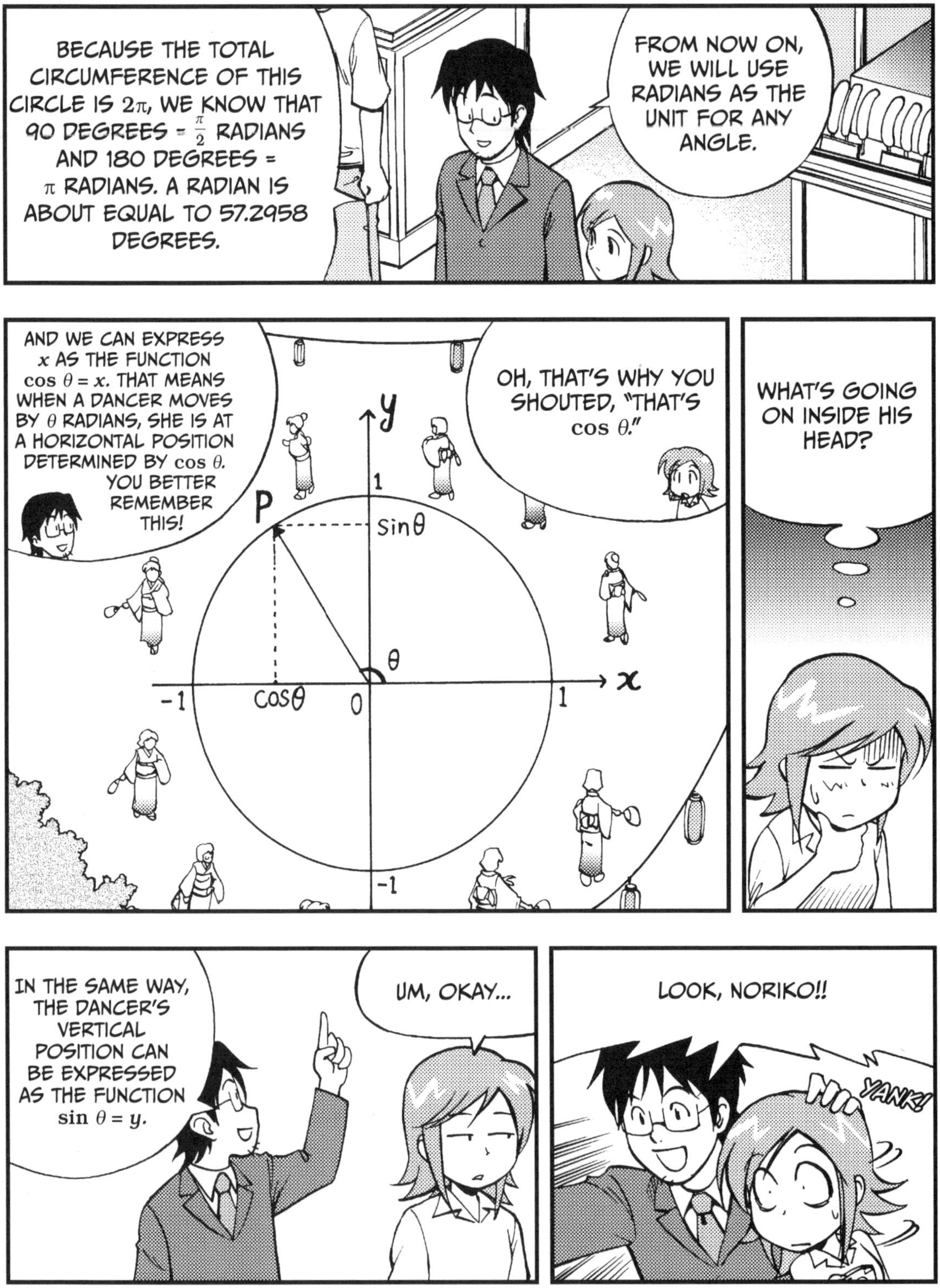
BECAUSE THE TOTAL CIRCUMFERENCE OF THIS CIRCLE IS 2π, WE KNOW THAT 90 DEGREES = $\frac{\pi}{2}$ RADIANS AND 180 DEGREES = π RADIANS. A RADIAN IS ABOUT EQUAL TO 57.2958 DEGREES.
FROM NOW ON, WE WILL USE RADIANS AS THE UNIT FOR ANY ANGLE.
AND WE CAN EXPRESS x AS THE FUNCTION $\cos\theta = x$. THAT MEANS WHEN A DANCER MOVES BY θ RADIANS, SHE IS AT A HORIZONTAL POSITION DETERMINED BY $\cos\theta$. YOU BETTER REMEMBER THIS!
OH, THAT'S WHY YOU SHOUTED, "THAT'S $\cos\theta$."
y
1
P
$\sin\theta$
θ
x
−1
$\cos\theta$
0
1
−1
WHAT'S GOING ON INSIDE HIS HEAD?
IN THE SAME WAY, THE DANCER'S VERTICAL POSITION CAN BE EXPRESSED AS THE FUNCTION $\sin\theta = y$.
UM, OKAY...
LOOK, NORIKO!!
YANK!

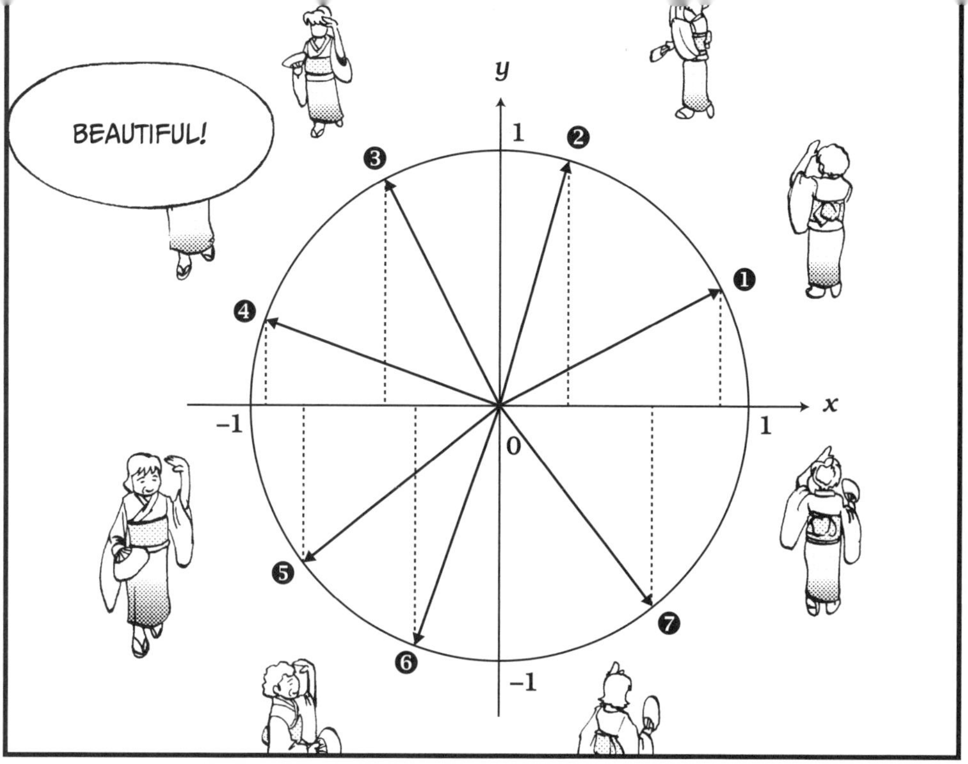
BEAUTIFUL!
y
1
x
−1
1
0
−1
❶
❷
❸
❹
❺
❻
❼

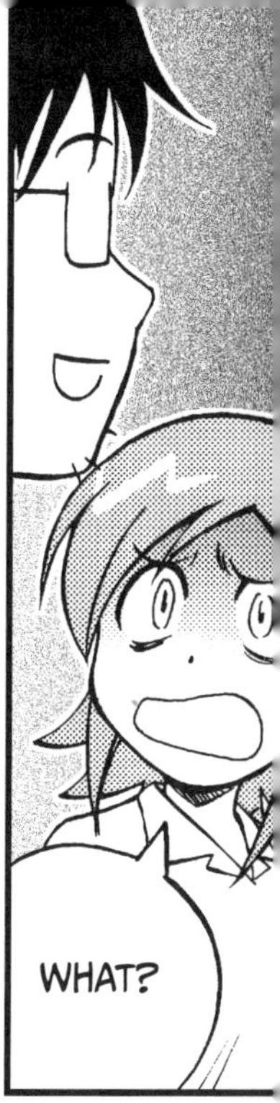
WHAT?

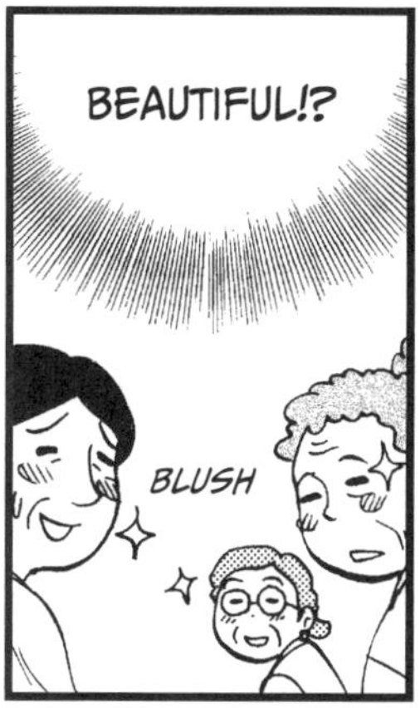
BEAUTIFUL!?
BLUSH

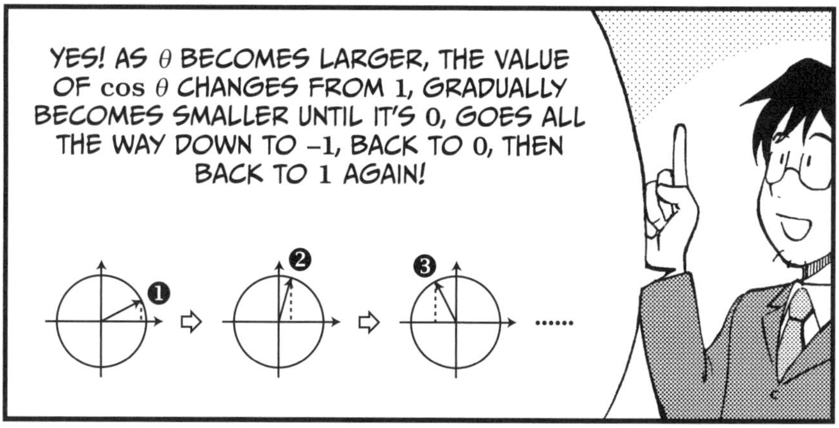
YES! AS θ BECOMES LARGER, THE VALUE OF cos θ CHANGES FROM 1, GRADUALLY BECOMES SMALLER UNTIL IT'S 0, GOES ALL THE WAY DOWN TO −1, BACK TO 0, THEN BACK TO 1 AGAIN!
❶
❷
❸
......

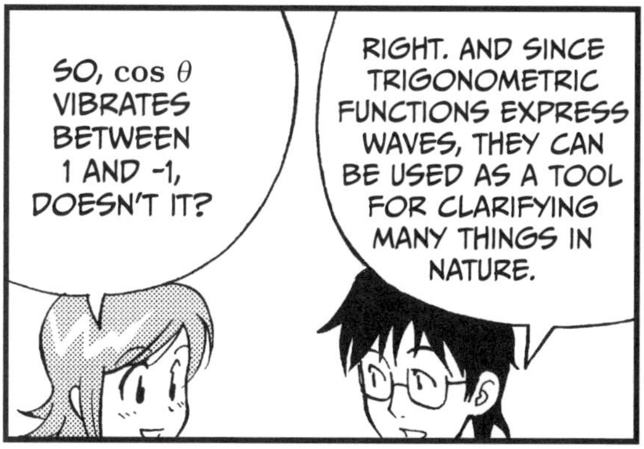
SO, cos θ VIBRATES BETWEEN 1 AND -1, DOESN'T IT?
RIGHT. AND SINCE TRIGONOMETRIC FUNCTIONS EXPRESS WAVES, THEY CAN BE USED AS A TOOL FOR CLARIFYING MANY THINGS IN NATURE.

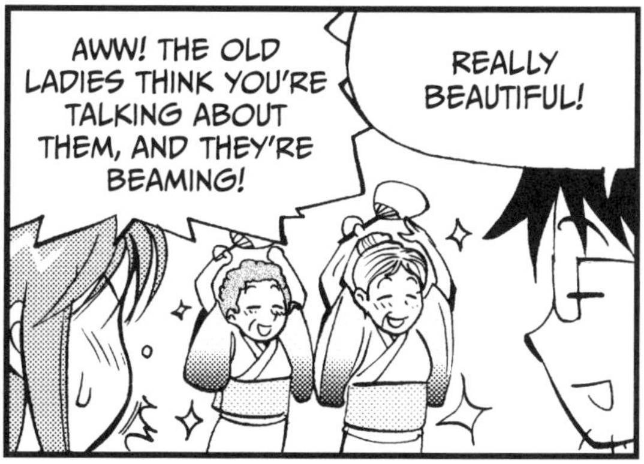
AWW! THE OLD LADIES THINK YOU'RE TALKING ABOUT THEM, AND THEY'RE BEAMING!
REALLY BEAUTIFUL!

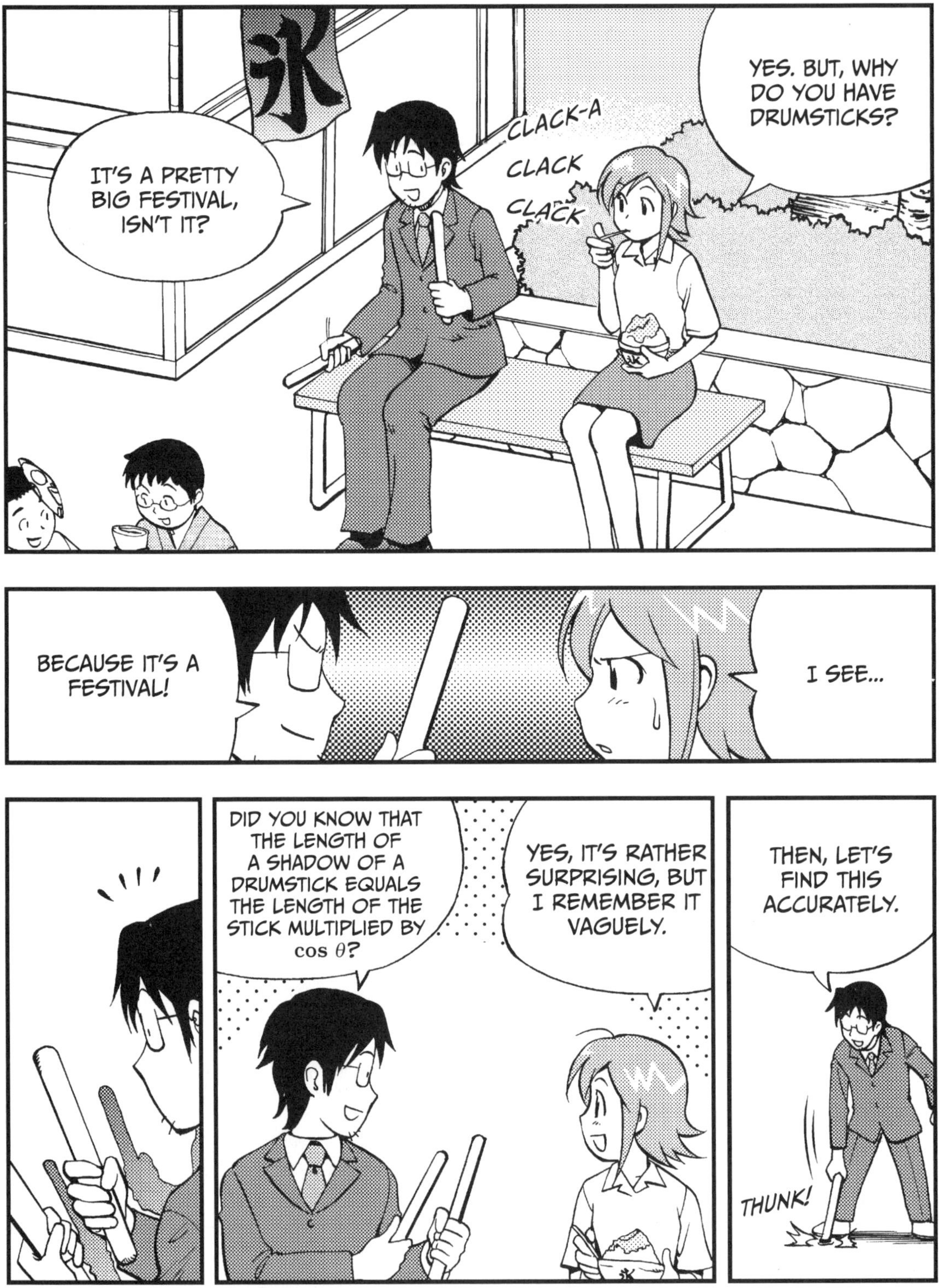
IT'S A PRETTY BIG FESTIVAL, ISN'T IT?
CLACK-A
CLACK
CLACK
YES. BUT, WHY DO YOU HAVE DRUMSTICKS?
BECAUSE IT'S A FESTIVAL!
I SEE...
DID YOU KNOW THAT THE LENGTH OF A SHADOW OF A DRUMSTICK EQUALS THE LENGTH OF THE STICK MULTIPLIED BY cos θ?
YES, IT'S RATHER SURPRISING, BUT I REMEMBER IT VAGUELY.
THEN, LET'S FIND THIS ACCURATELY.
THUNK!

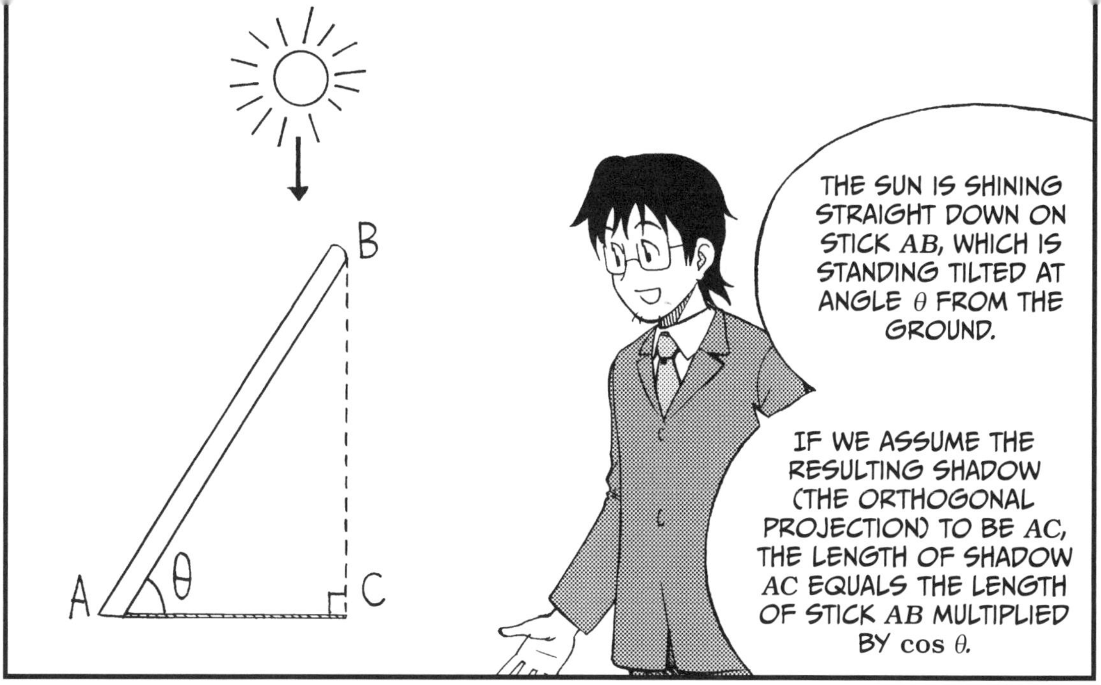
B
A
θ
C
THE SUN IS SHINING STRAIGHT DOWN ON STICK AB, WHICH IS STANDING TILTED AT ANGLE θ FROM THE GROUND.
IF WE ASSUME THE RESULTING SHADOW (THE ORTHOGONAL PROJECTION) TO BE AC, THE LENGTH OF SHADOW AC EQUALS THE LENGTH OF STICK AB MULTIPLIED BY $\cos\theta$.

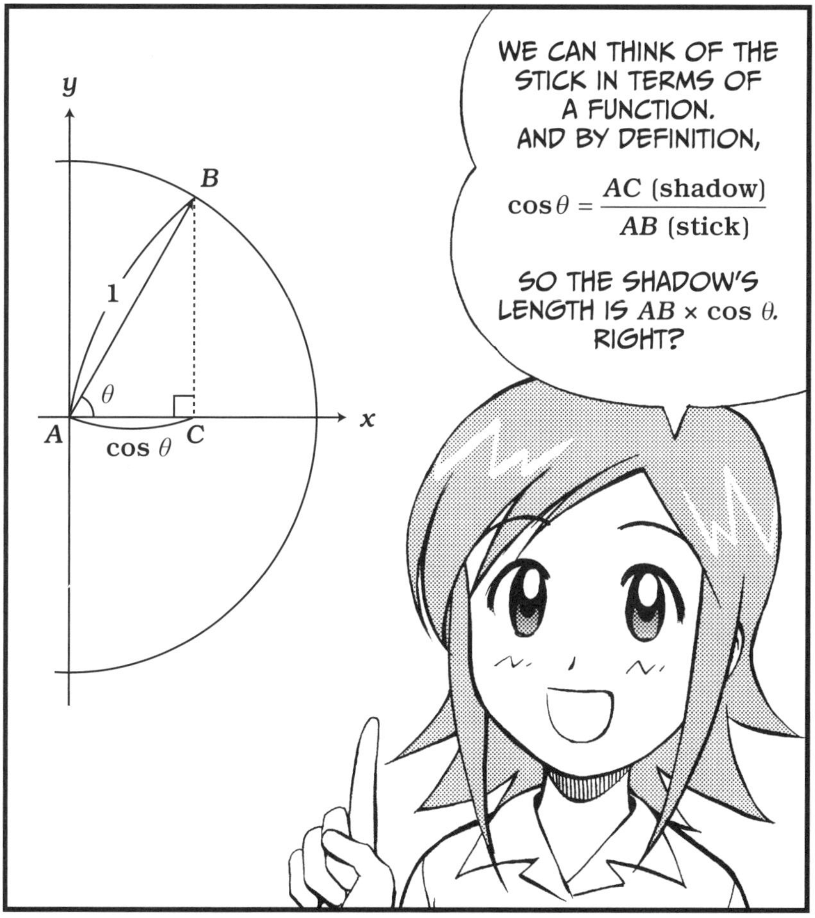
y
B
1
θ
A
cos θ
C
x
WE CAN THINK OF THE STICK IN TERMS OF A FUNCTION. AND BY DEFINITION,
$\cos\theta = \frac{AC\ (\text{shadow})}{AB\ (\text{stick})}$
SO THE SHADOW'S LENGTH IS $AB \times \cos\theta$. RIGHT?

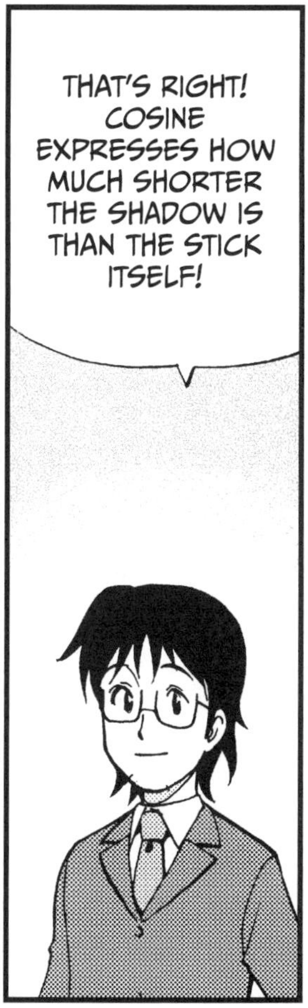
THAT'S RIGHT! COSINE EXPRESSES HOW MUCH SHORTER THE SHADOW IS THAN THE STICK ITSELF!

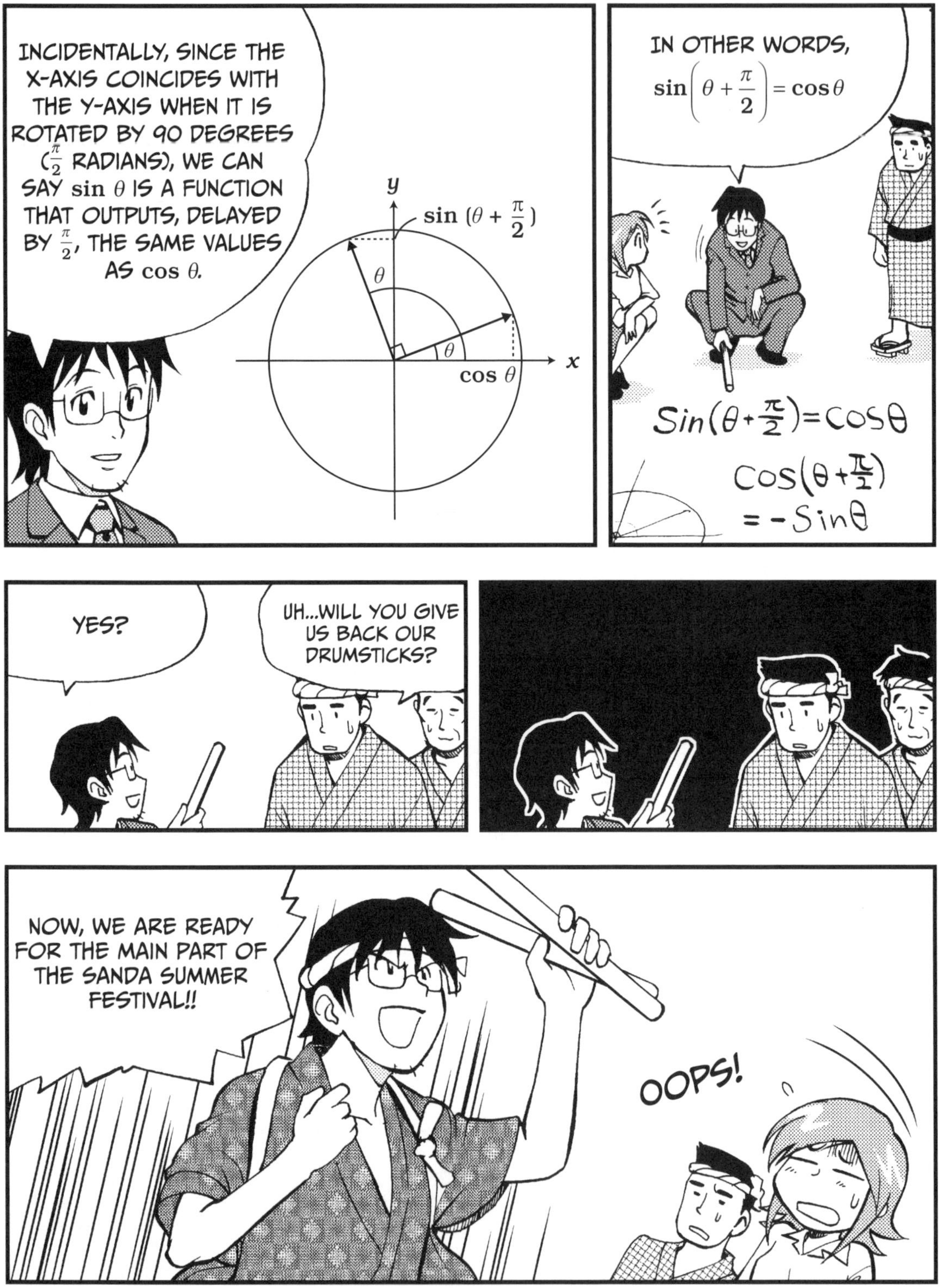
INCIDENTALLY, SINCE THE X-AXIS COINCIDES WITH THE Y-AXIS WHEN IT IS ROTATED BY 90 DEGREES ($\frac{\pi}{2}$ RADIANS), WE CAN SAY $\sin\theta$ IS A FUNCTION THAT OUTPUTS, DELAYED BY $\frac{\pi}{2}$, THE SAME VALUES AS $\cos\theta$.
y
$\sin(\theta + \frac{\pi}{2})$
θ
θ
x
$\cos\theta$
IN OTHER WORDS, $\sin\left(\theta + \frac{\pi}{2}\right) = \cos\theta$
$\sin(\theta+\frac{\pi}{2})=\cos\theta$
$\cos(\theta+\frac{\pi}{2}) = -\sin\theta$
YES?
UH...WILL YOU GIVE US BACK OUR DRUMSTICKS?
NOW, WE ARE READY FOR THE MAIN PART OF THE SANDA SUMMER FESTIVAL!!
OOPS!

USING INTEGRALS WITH TRIGONOMETRIC FUNCTIONS

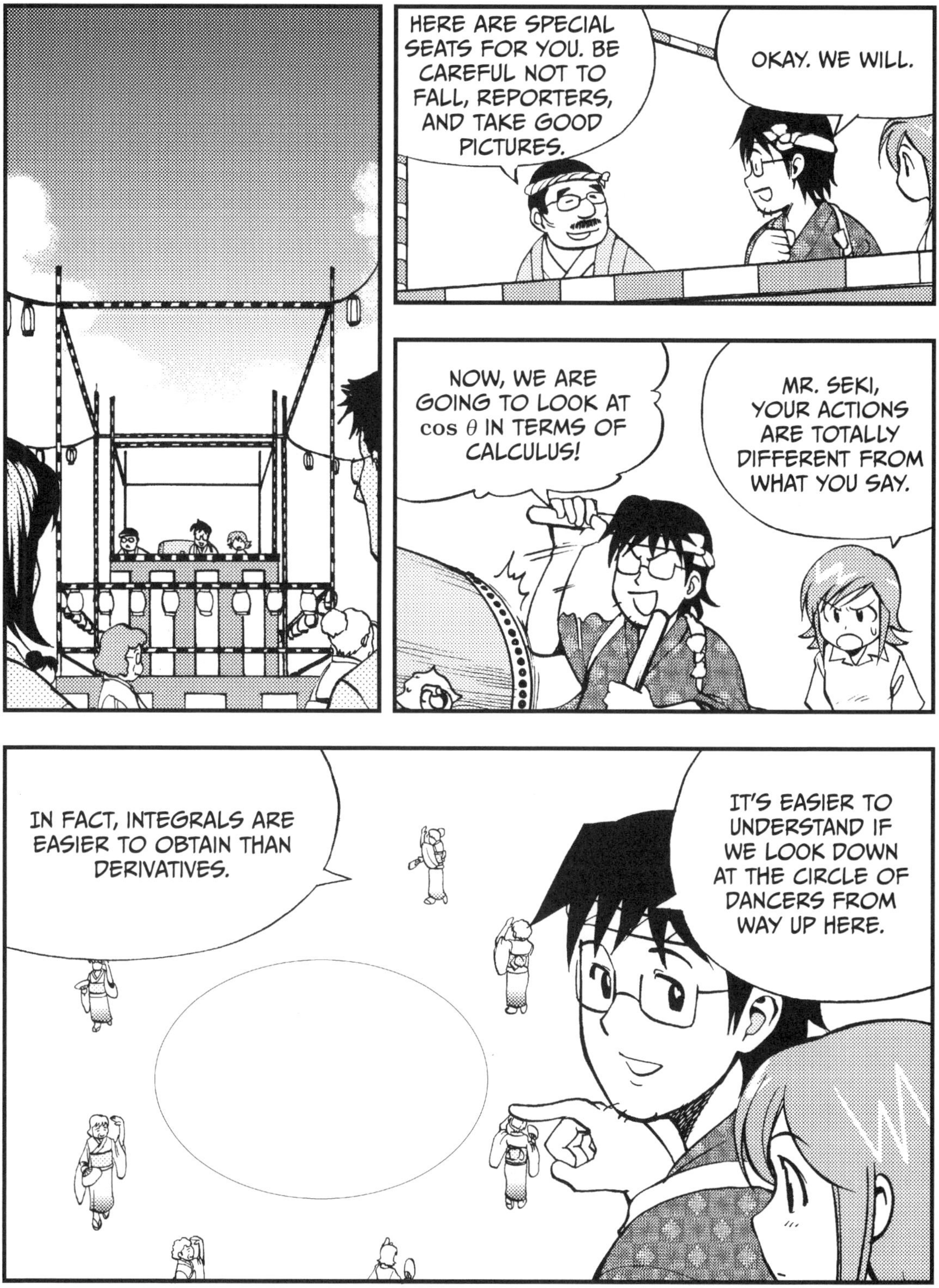
HERE ARE SPECIAL SEATS FOR YOU. BE CAREFUL NOT TO FALL, REPORTERS, AND TAKE GOOD PICTURES.
OKAY. WE WILL.
NOW, WE ARE GOING TO LOOK AT $\cos\theta$ IN TERMS OF CALCULUS!
MR. SEKI, YOUR ACTIONS ARE TOTALLY DIFFERENT FROM WHAT YOU SAY.
IN FACT, INTEGRALS ARE EASIER TO OBTAIN THAN DERIVATIVES.
IT'S EASIER TO UNDERSTAND IF WE LOOK DOWN AT THE CIRCLE OF DANCERS FROM WAY UP HERE.

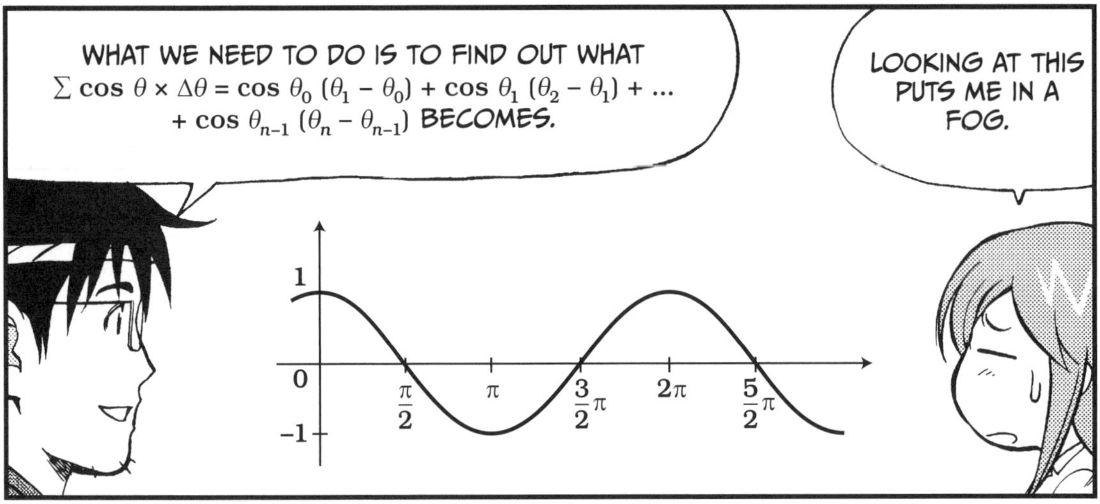
WHAT WE NEED TO DO IS TO FIND OUT WHAT $\sum \cos\theta \times \Delta\theta = \cos\theta_0(\theta_1 - \theta_0) + \cos\theta_1(\theta_2 - \theta_1) + \ldots + \cos\theta_{n-1}(\theta_n - \theta_{n-1})$ BECOMES.
LOOKING AT THIS PUTS ME IN A FOG.
1
0
−1
$\frac{\pi}{2}$
π
$\frac{3}{2}\pi$
2π
$\frac{5}{2}\pi$

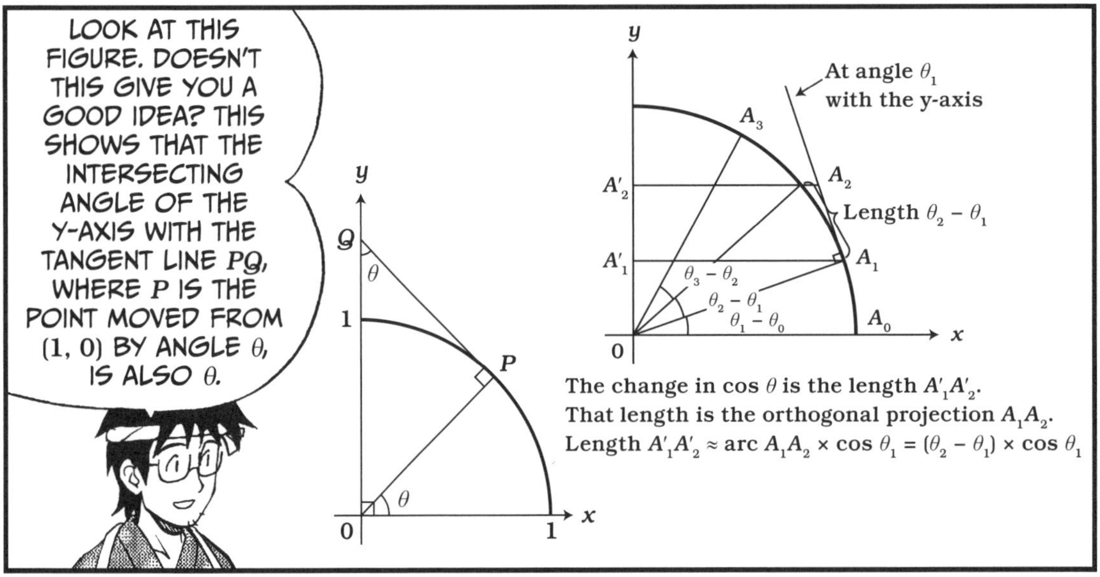
LOOK AT THIS FIGURE. DOESN'T THIS GIVE YOU A GOOD IDEA? THIS SHOWS THAT THE INTERSECTING ANGLE OF THE Y-AXIS WITH THE TANGENT LINE PQ, WHERE P IS THE POINT MOVED FROM (1, 0) BY ANGLE θ, IS ALSO θ.
y
Q
θ
1
P
θ
0
1
x
y
At angle θ_1 with the y-axis
A_3
A'_2
A_2
Length $\theta_2 - \theta_1$
A'_1
A_1
$\theta_3 - \theta_2$
$\theta_2 - \theta_1$
$\theta_1 - \theta_0$
A_0
x
0
The change in $\cos\theta$ is the length $A'_1A'_2$.
That length is the orthogonal projection A_1A_2.
Length $A'_1A'_2 \approx$ arc $A_1A_2 \times \cos\theta_1 = (\theta_2 - \theta_1) \times \cos\theta_1$

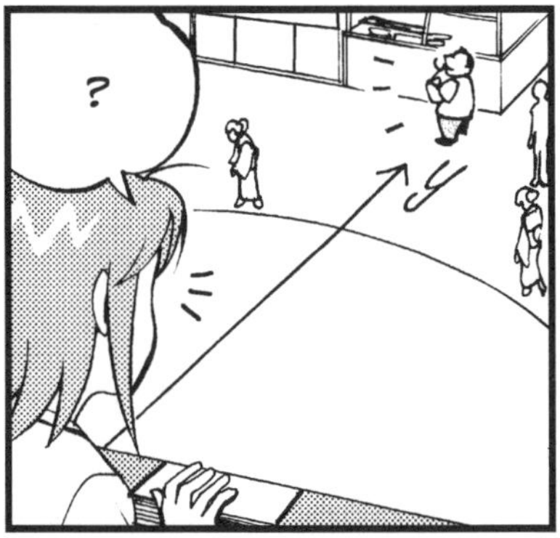
?

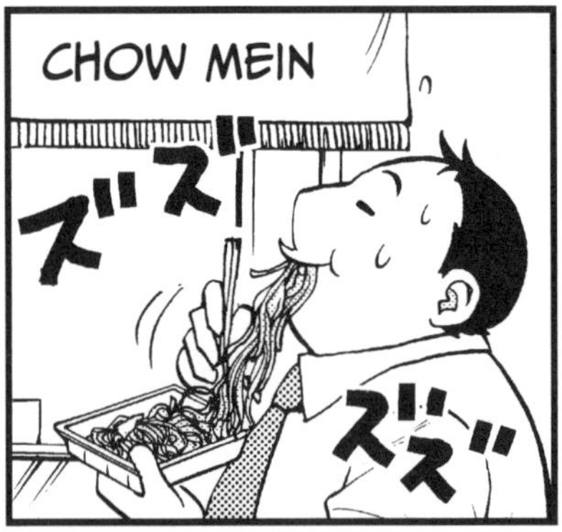
CHOW MEIN
ズズ
ズズ

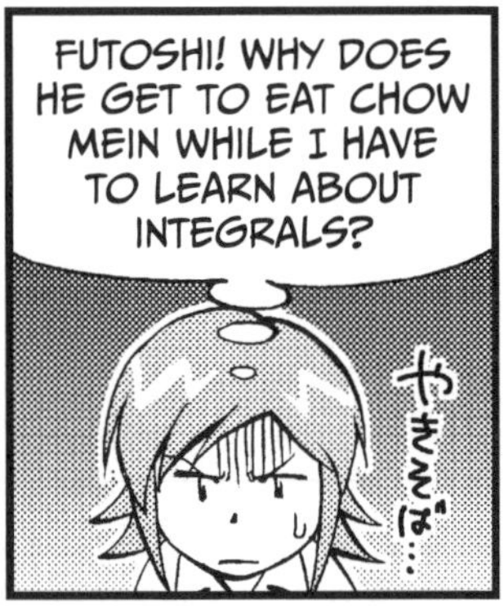
FUTOSHI! WHY DOES HE GET TO EAT CHOW MEIN WHILE I HAVE TO LEARN ABOUT INTEGRALS?
やきそば…

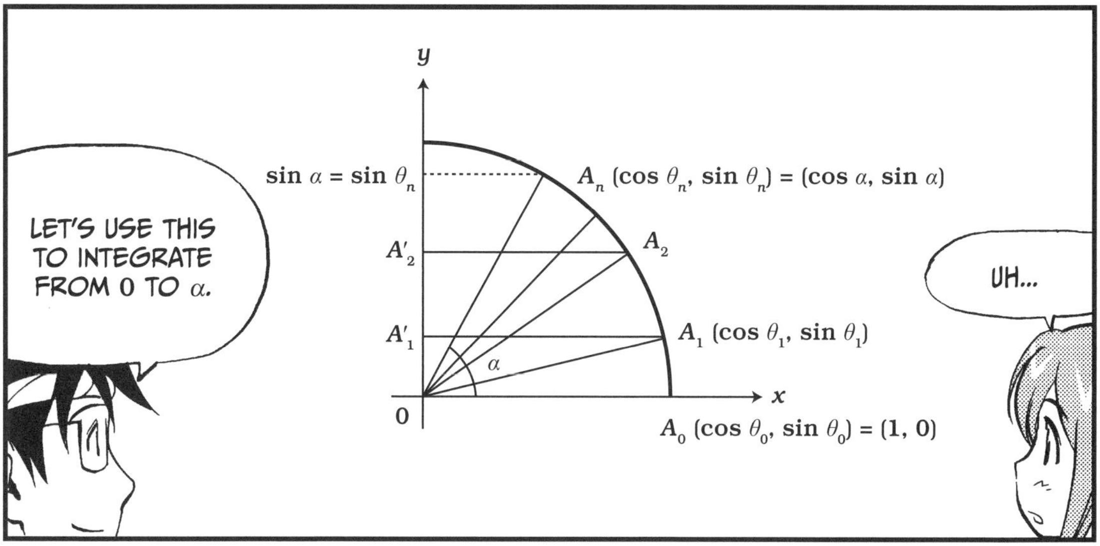
LET'S USE THIS TO INTEGRATE FROM 0 TO α.
y
$\sin \alpha = \sin \theta_n$
A_n $(\cos \theta_n, \sin \theta_n) = (\cos \alpha, \sin \alpha)$
A'_2
A_2
A'_1
A_1 $(\cos \theta_1, \sin \theta_1)$
α
x
0
A_0 $(\cos \theta_0, \sin \theta_0) = (1, 0)$
UH...

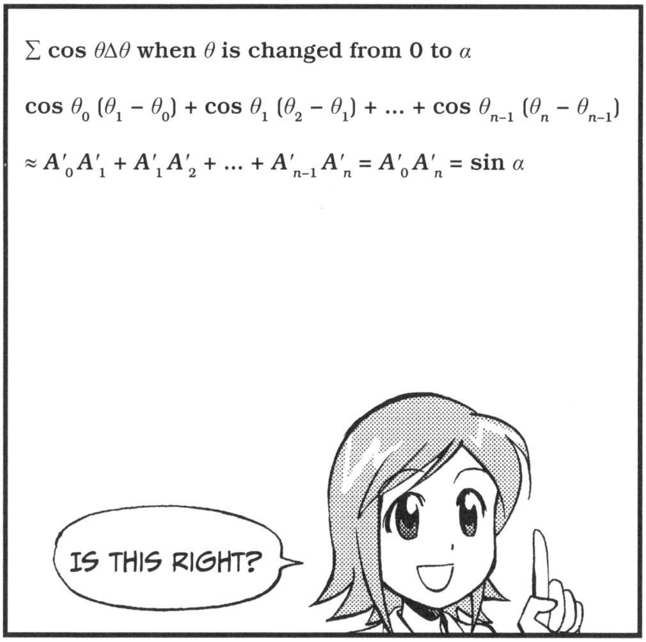
$\sum \cos \theta \Delta \theta$ when θ is changed from 0 to α
$\cos \theta_0 (\theta_1 - \theta_0) + \cos \theta_1 (\theta_2 - \theta_1) + \ldots + \cos \theta_{n-1} (\theta_n - \theta_{n-1})$
$\approx A'_0A'_1 + A'_1A'_2 + \ldots + A'_{n-1}A'_n = A'_0A'_n = \sin \alpha$
IS THIS RIGHT?

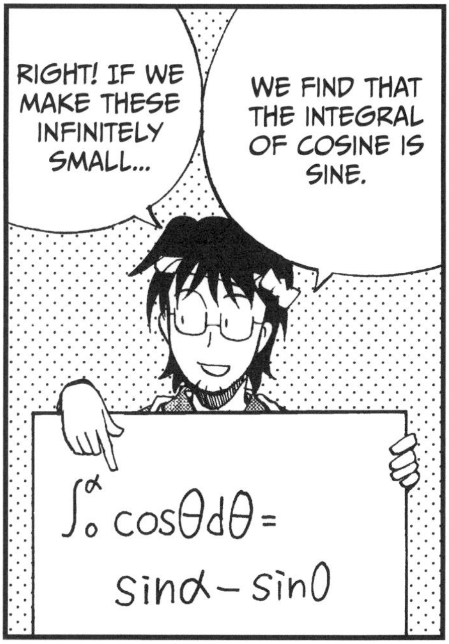
RIGHT! IF WE MAKE THESE INFINITELY SMALL...
WE FIND THAT THE INTEGRAL OF COSINE IS SINE.
$\int_0^\alpha \cos\theta d\theta = \sin\alpha - \sin 0$

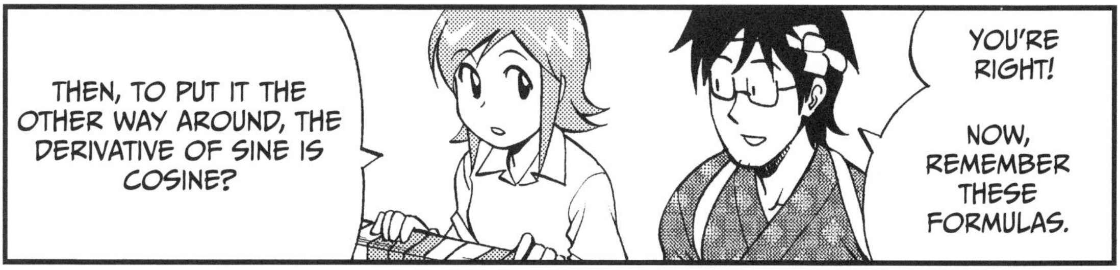
THEN, TO PUT IT THE OTHER WAY AROUND, THE DERIVATIVE OF SINE IS COSINE?
YOU'RE RIGHT!
NOW, REMEMBER THESE FORMULAS.

FORMULA 4-1: THE DIFFERENTIATION AND INTEGRATION OF TRIGONOMETRIC FUNCTIONS

Since ❶ $\int_0^{\alpha} \cos\theta d\theta = \sin\alpha - \sin 0$, we know that sine must be cosine's derivative.

❷ $(\sin\theta)' = \cos\theta$

Now, substitute $\theta + \frac{\pi}{2}$ for θ in ❷. We get $\left\{\sin\left(\theta + \frac{\pi}{2}\right)\right\}' = \cos\left(\theta + \frac{\pi}{2}\right)$. Using the equations from page 124, we then know that

❸ $(\cos\theta)' = -\sin\theta$

We find that differentiating or integrating sine gives cosine and vice versa.

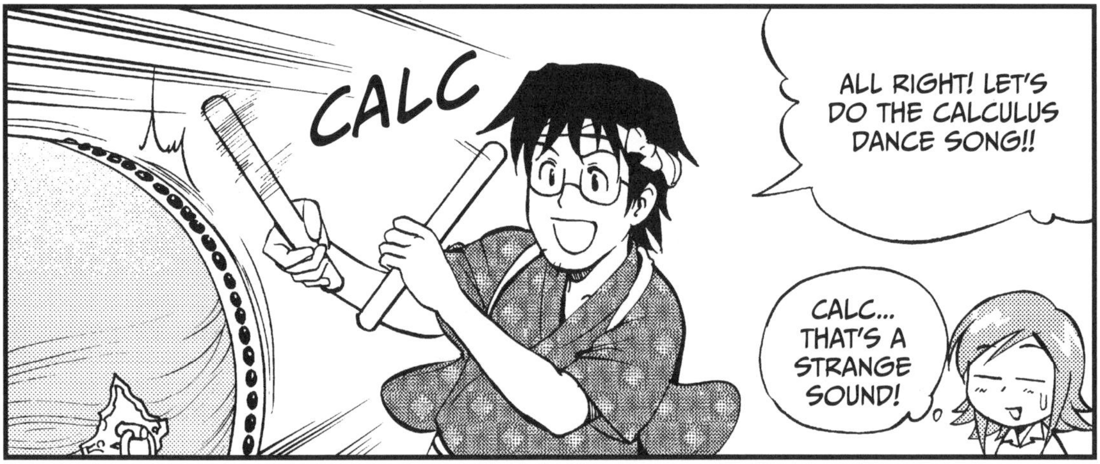
CALC
ALL RIGHT! LET'S DO THE CALCULUS DANCE SONG!!
CALC... THAT'S A STRANGE SOUND!

CALCULUS DANCE SONG
TRIGONOMETRIC VERSION

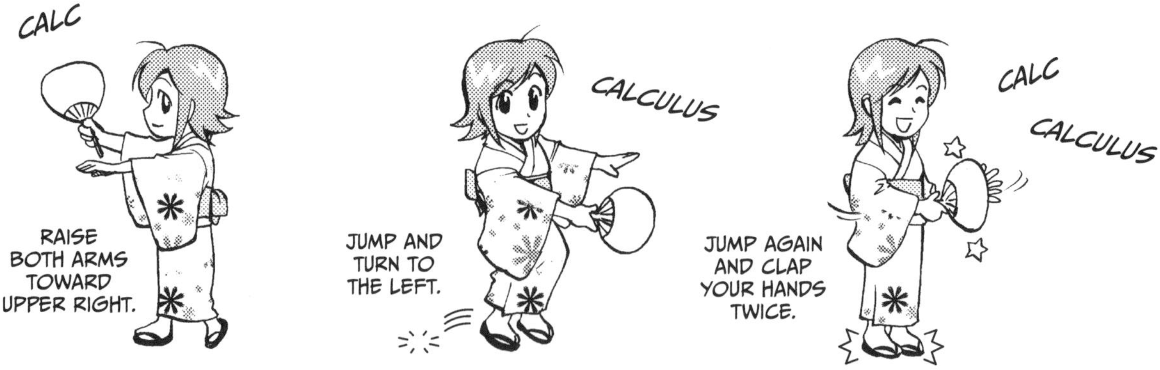
CALC
RAISE BOTH ARMS TOWARD UPPER RIGHT.
CALCULUS
JUMP AND TURN TO THE LEFT.
CALC
CALCULUS
JUMP AGAIN AND CLAP YOUR HANDS TWICE.

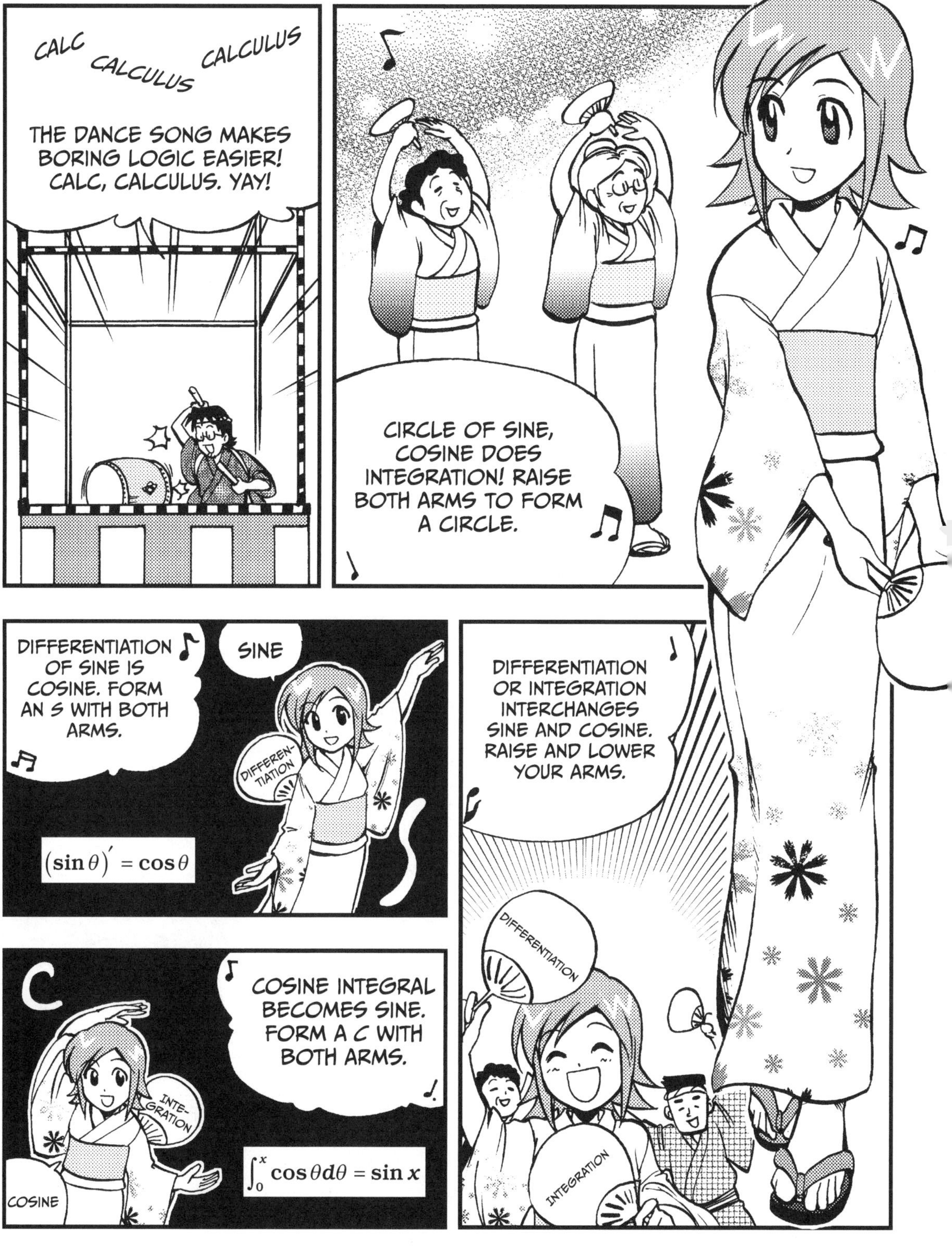
CALC
CALCULUS
CALCULUS
THE DANCE SONG MAKES BORING LOGIC EASIER! CALC, CALCULUS. YAY!
CIRCLE OF SINE, COSINE DOES INTEGRATION! RAISE BOTH ARMS TO FORM A CIRCLE.
DIFFERENTIATION OF SINE IS COSINE. FORM AN S WITH BOTH ARMS.
SINE
DIFFEREN-TIATION
$(\sin\theta)' = \cos\theta$
DIFFERENTIATION OR INTEGRATION INTERCHANGES SINE AND COSINE. RAISE AND LOWER YOUR ARMS.
DIFFERENTIATION
INTEGRATION
C
COSINE INTEGRAL BECOMES SINE. FORM A C WITH BOTH ARMS.
INTE-GRATION
COSINE
$\int_0^x \cos\theta d\theta = \sin x$

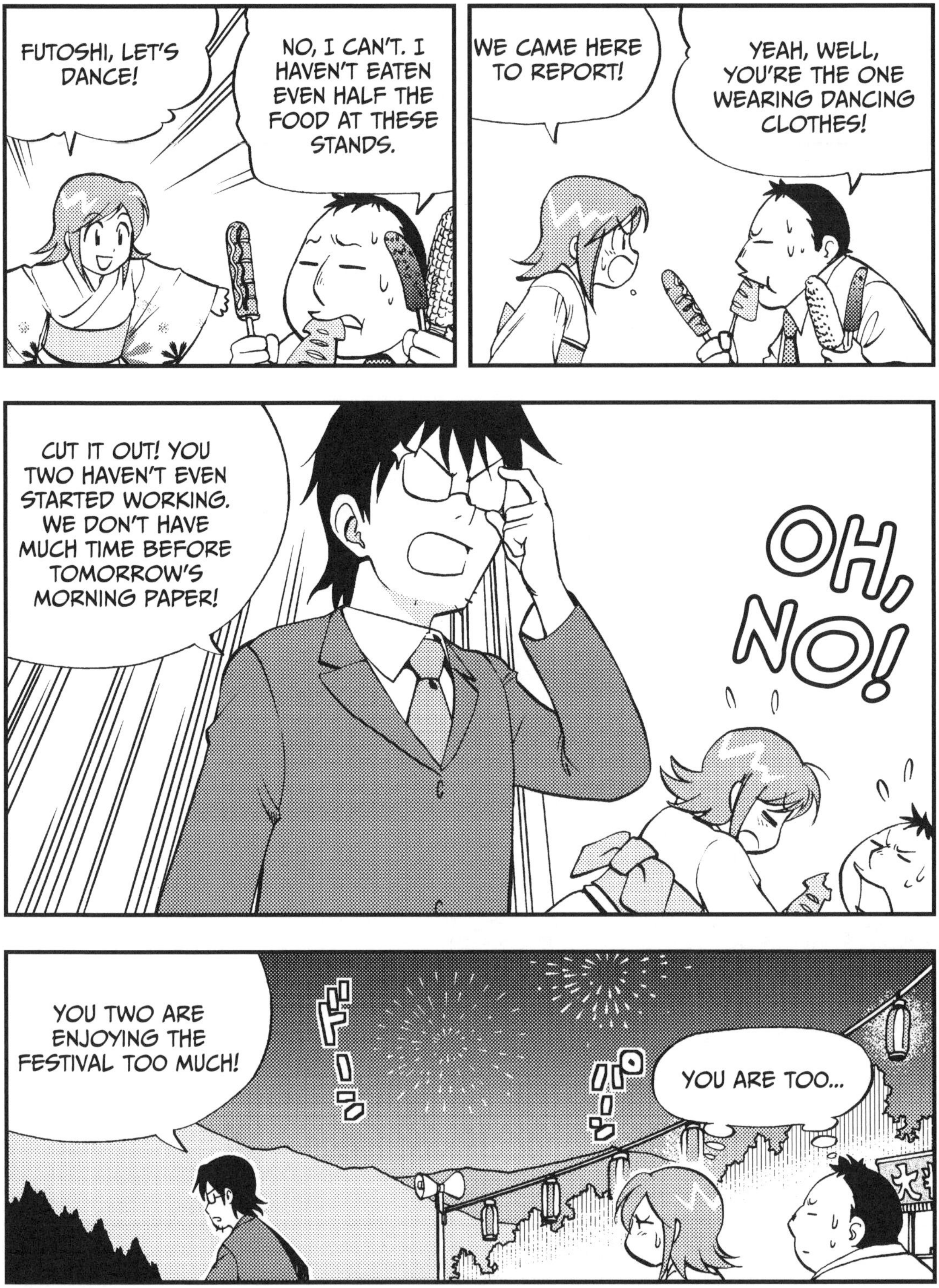
FUTOSHI, LET'S DANCE!
NO, I CAN'T. I HAVEN'T EATEN EVEN HALF THE FOOD AT THESE STANDS.
WE CAME HERE TO REPORT!
YEAH, WELL, YOU'RE THE ONE WEARING DANCING CLOTHES!
CUT IT OUT! YOU TWO HAVEN'T EVEN STARTED WORKING. WE DON'T HAVE MUCH TIME BEFORE TOMORROW'S MORNING PAPER!
OH, NO!
YOU TWO ARE ENJOYING THE FESTIVAL TOO MUCH!
ドーン
パーン
YOU ARE TOO...

USING EXPONENTIAL AND LOGARITHMIC FUNCTIONS

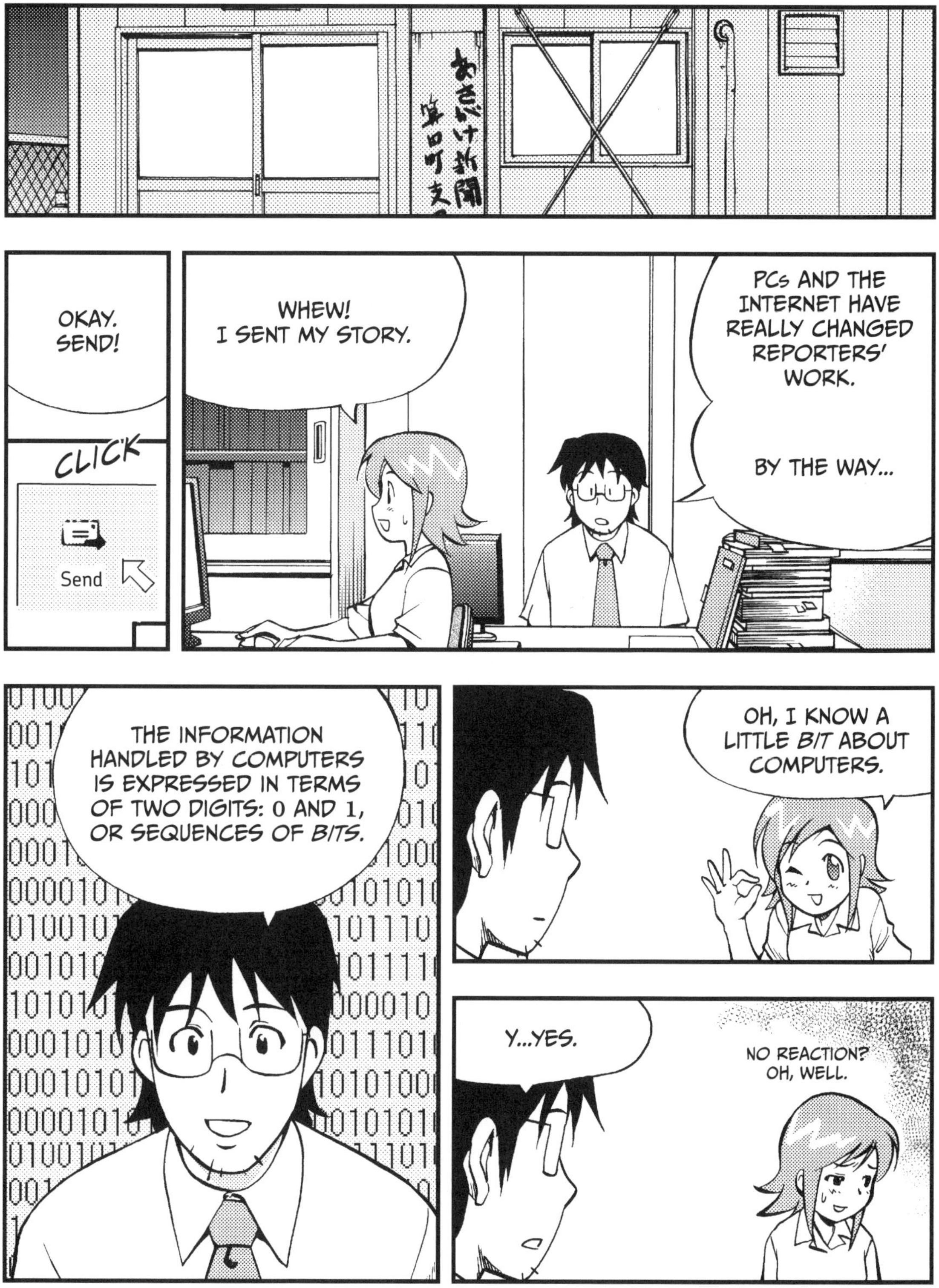
OKAY. SEND!
CLICK
Send
WHEW! I SENT MY STORY.
PCs AND THE INTERNET HAVE REALLY CHANGED REPORTERS' WORK.
BY THE WAY...
THE INFORMATION HANDLED BY COMPUTERS IS EXPRESSED IN TERMS OF TWO DIGITS: 0 AND 1, OR SEQUENCES OF *BITS*.
OH, I KNOW A LITTLE *BIT* ABOUT COMPUTERS.
Y...YES.
NO REACTION? OH, WELL.

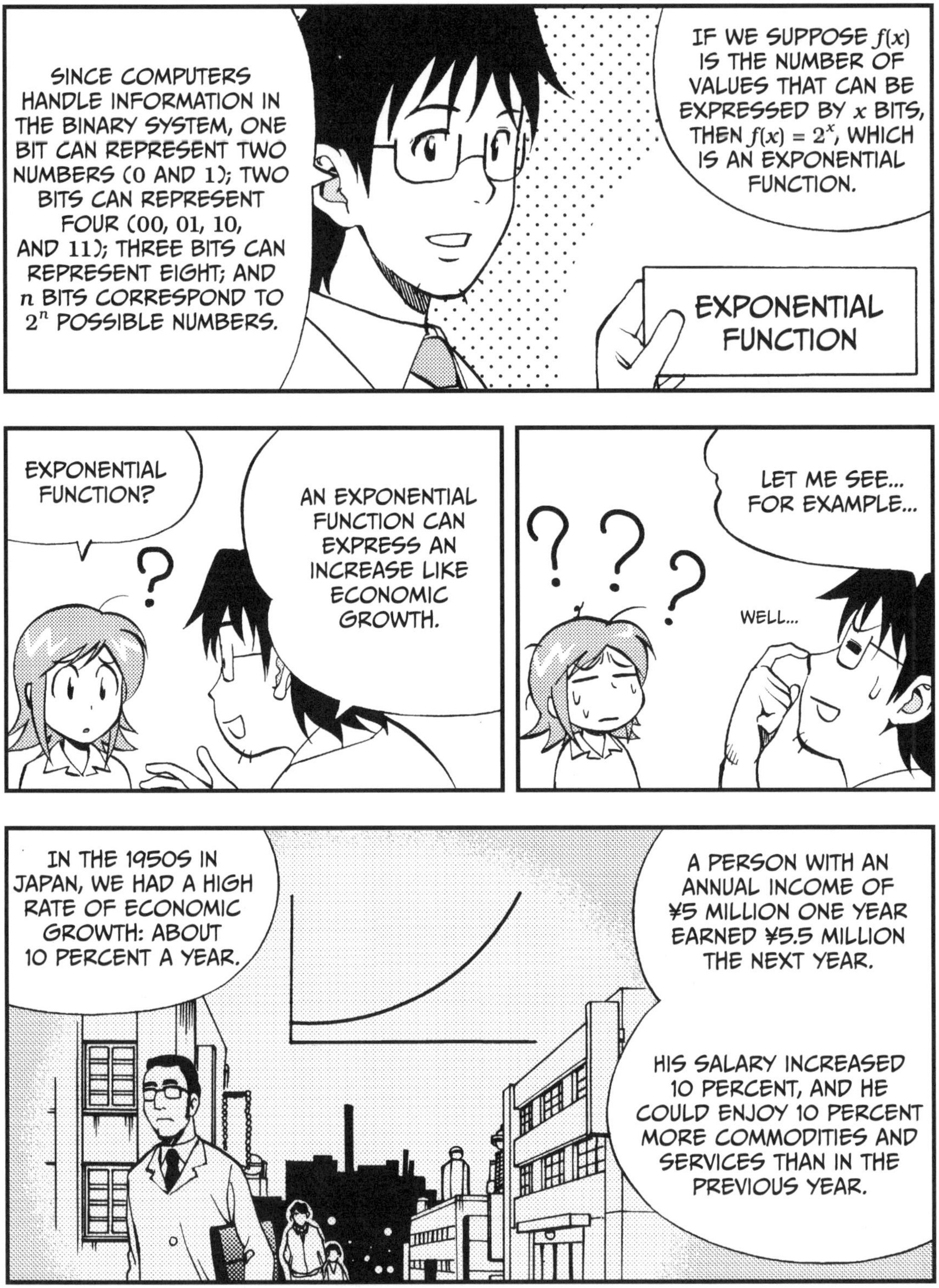
SINCE COMPUTERS HANDLE INFORMATION IN THE BINARY SYSTEM, ONE BIT CAN REPRESENT TWO NUMBERS (0 AND 1); TWO BITS CAN REPRESENT FOUR (00, 01, 10, AND 11); THREE BITS CAN REPRESENT EIGHT; AND n BITS CORRESPOND TO 2^n POSSIBLE NUMBERS.
IF WE SUPPOSE $f(x)$ IS THE NUMBER OF VALUES THAT CAN BE EXPRESSED BY x BITS, THEN $f(x) = 2^x$, WHICH IS AN EXPONENTIAL FUNCTION.
EXPONENTIAL FUNCTION
EXPONENTIAL FUNCTION?
AN EXPONENTIAL FUNCTION CAN EXPRESS AN INCREASE LIKE ECONOMIC GROWTH.
LET ME SEE... FOR EXAMPLE...
WELL...
IN THE 1950S IN JAPAN, WE HAD A HIGH RATE OF ECONOMIC GROWTH: ABOUT 10 PERCENT A YEAR.
A PERSON WITH AN ANNUAL INCOME OF ¥5 MILLION ONE YEAR EARNED ¥5.5 MILLION THE NEXT YEAR.
HIS SALARY INCREASED 10 PERCENT, AND HE COULD ENJOY 10 PERCENT MORE COMMODITIES AND SERVICES THAN IN THE PREVIOUS YEAR.

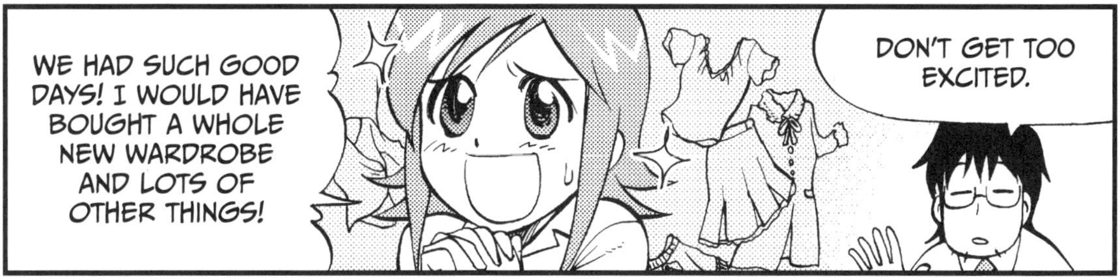
WE HAD SUCH GOOD DAYS! I WOULD HAVE BOUGHT A WHOLE NEW WARDROBE AND LOTS OF OTHER THINGS!
DON'T GET TOO EXCITED.

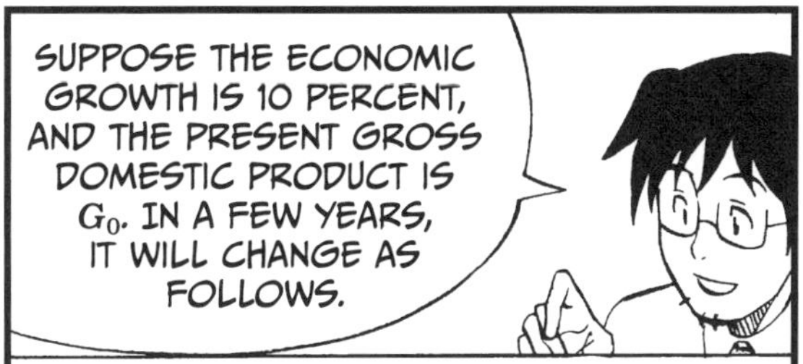
SUPPOSE THE ECONOMIC GROWTH IS 10 PERCENT, AND THE PRESENT GROSS DOMESTIC PRODUCT IS G_0. IN A FEW YEARS, IT WILL CHANGE AS FOLLOWS.

$G_1 = G_0 \times 1.1$
Gross domestic product after 1 year

$G_2 = G_1 \times 1.1 = G_0 \times 1.1^2$
Gross domestic product after 2 years

$G_3 = G_0 \times 1.1^3$
Gross domestic product after 3 years

$G_4 = G_0 \times 1.1^4$
Gross domestic product after 4 years

$G_5 = G_0 \times 1.1^5$
Gross domestic product after 5 years

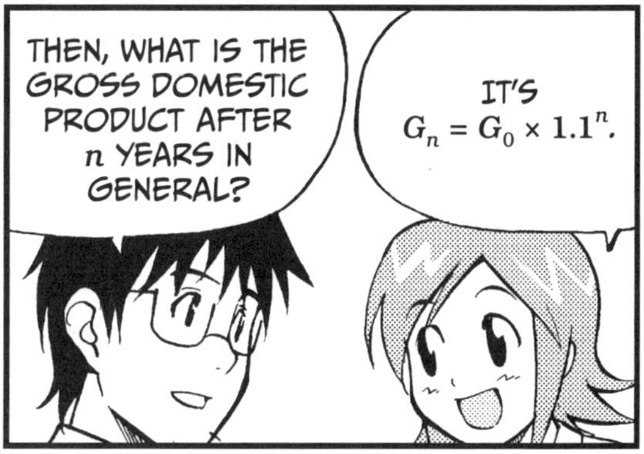
THEN, WHAT IS THE GROSS DOMESTIC PRODUCT AFTER n YEARS IN GENERAL?
IT'S $G_n = G_0 \times 1.1^n$.

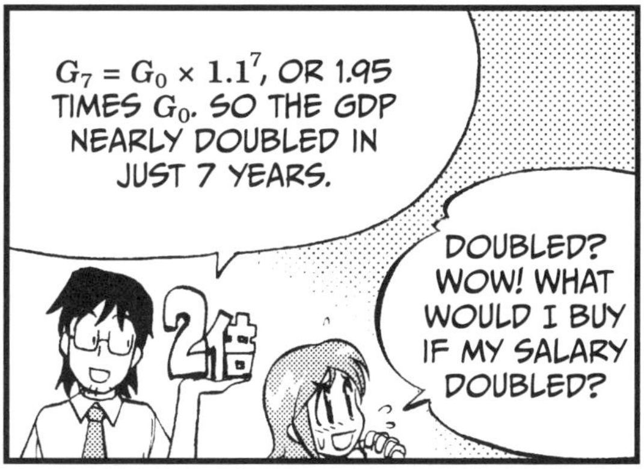
$G_7 = G_0 \times 1.1^7$, OR 1.95 TIMES G_0. SO THE GDP NEARLY DOUBLED IN JUST 7 YEARS.
DOUBLED? WOW! WHAT WOULD I BUY IF MY SALARY DOUBLED?

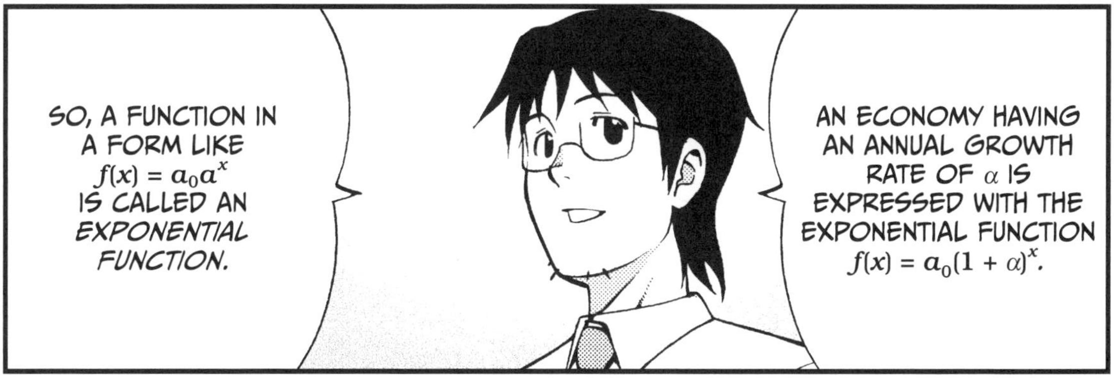
SO, A FUNCTION IN A FORM LIKE $f(x) = a_0 a^x$ IS CALLED AN *EXPONENTIAL FUNCTION*.
AN ECONOMY HAVING AN ANNUAL GROWTH RATE OF α IS EXPRESSED WITH THE EXPONENTIAL FUNCTION $f(x) = a_0(1 + \alpha)^x$.

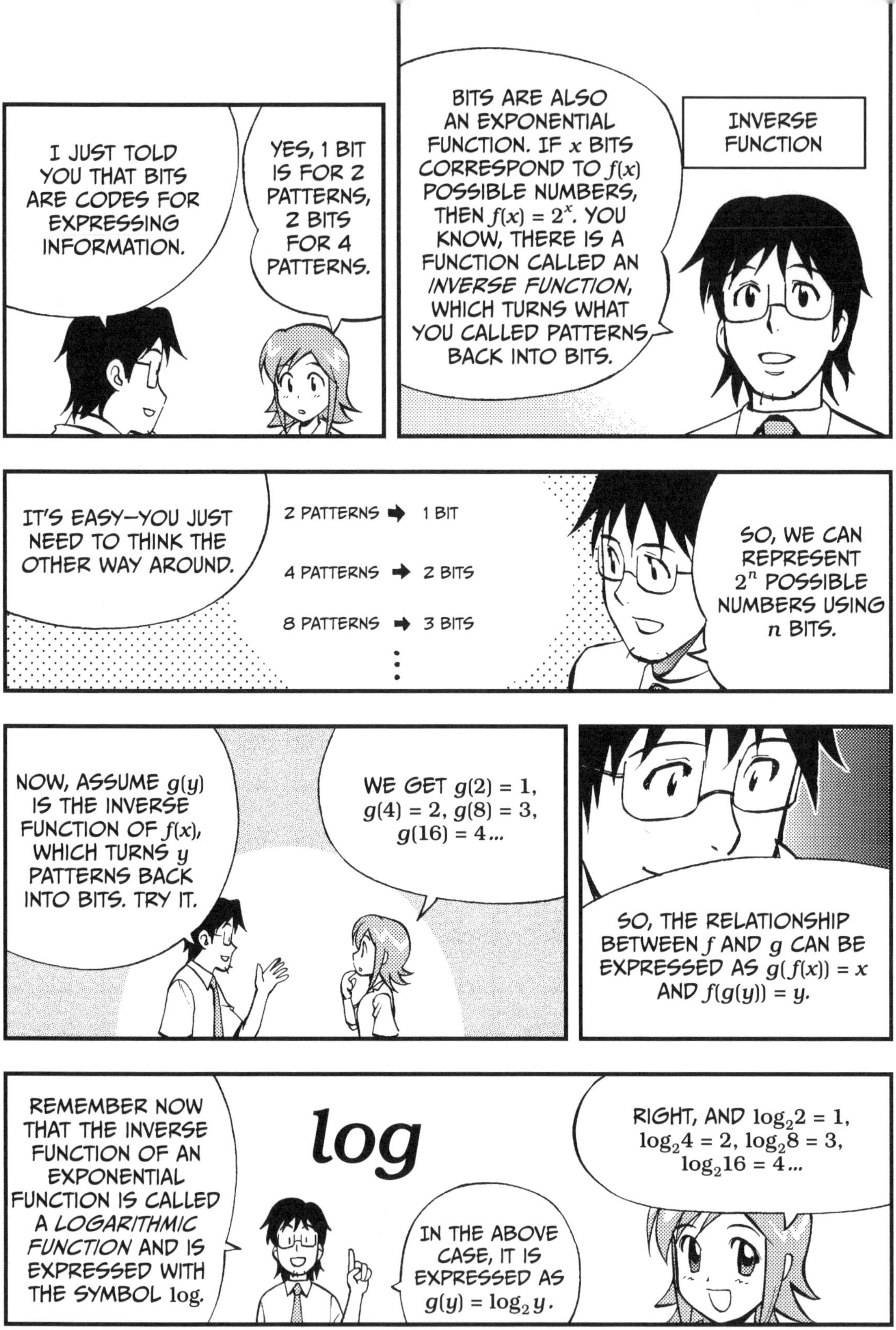

I JUST TOLD YOU THAT BITS ARE CODES FOR EXPRESSING INFORMATION.

YES, 1 BIT IS FOR 2 PATTERNS, 2 BITS FOR 4 PATTERNS.

BITS ARE ALSO AN EXPONENTIAL FUNCTION. IF x BITS CORRESPOND TO $f(x)$ POSSIBLE NUMBERS, THEN $f(x) = 2^x$. YOU KNOW, THERE IS A FUNCTION CALLED AN *INVERSE FUNCTION*, WHICH TURNS WHAT YOU CALLED PATTERNS BACK INTO BITS.

INVERSE FUNCTION

IT'S EASY—YOU JUST NEED TO THINK THE OTHER WAY AROUND.

2 PATTERNS ➡ 1 BIT

4 PATTERNS ➡ 2 BITS

8 PATTERNS ➡ 3 BITS

⋮

SO, WE CAN REPRESENT 2^n POSSIBLE NUMBERS USING n BITS.

NOW, ASSUME $g(y)$ IS THE INVERSE FUNCTION OF $f(x)$, WHICH TURNS y PATTERNS BACK INTO BITS. TRY IT.

WE GET $g(2) = 1$, $g(4) = 2$, $g(8) = 3$, $g(16) = 4 \ldots$

SO, THE RELATIONSHIP BETWEEN f AND g CAN BE EXPRESSED AS $g(f(x)) = x$ AND $f(g(y)) = y$.

REMEMBER NOW THAT THE INVERSE FUNCTION OF AN EXPONENTIAL FUNCTION IS CALLED A *LOGARITHMIC FUNCTION* AND IS EXPRESSED WITH THE SYMBOL log.

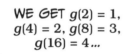

IN THE ABOVE CASE, IT IS EXPRESSED AS $g(y) = \log_2 y$.

RIGHT, AND $\log_2 2 = 1$, $\log_2 4 = 2$, $\log_2 8 = 3$, $\log_2 16 = 4 \ldots$

GENERALIZING EXPONENTIAL AND LOGARITHMIC FUNCTIONS

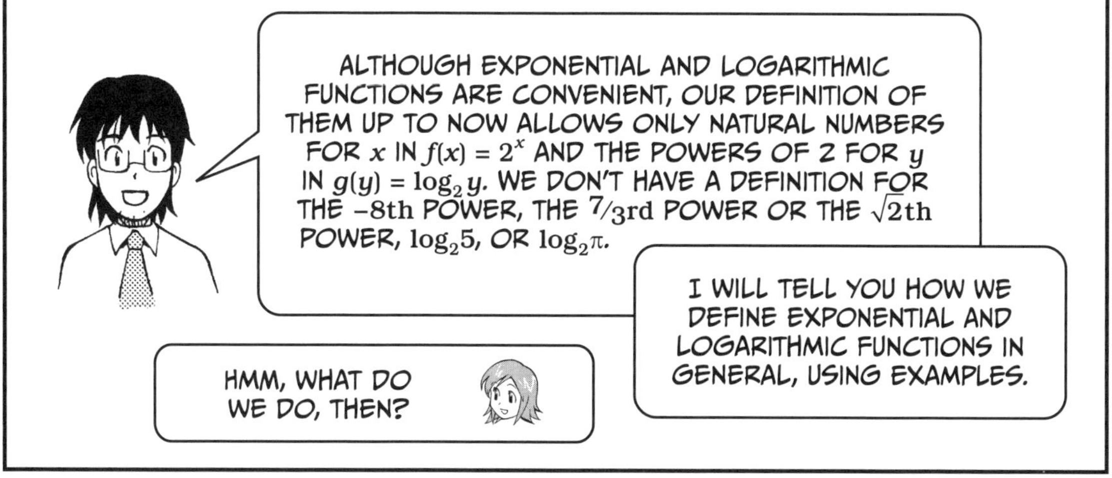
ALTHOUGH EXPONENTIAL AND LOGARITHMIC FUNCTIONS ARE CONVENIENT, OUR DEFINITION OF THEM UP TO NOW ALLOWS ONLY NATURAL NUMBERS FOR x IN $f(x) = 2^x$ AND THE POWERS OF 2 FOR y IN $g(y) = \log_2 y$. WE DON'T HAVE A DEFINITION FOR THE -8th POWER, THE $7/3$rd POWER OR THE $\sqrt{2}$th POWER, $\log_2 5$, OR $\log_2 \pi$.
HMM, WHAT DO WE DO, THEN?
I WILL TELL YOU HOW WE DEFINE EXPONENTIAL AND LOGARITHMIC FUNCTIONS IN GENERAL, USING EXAMPLES.

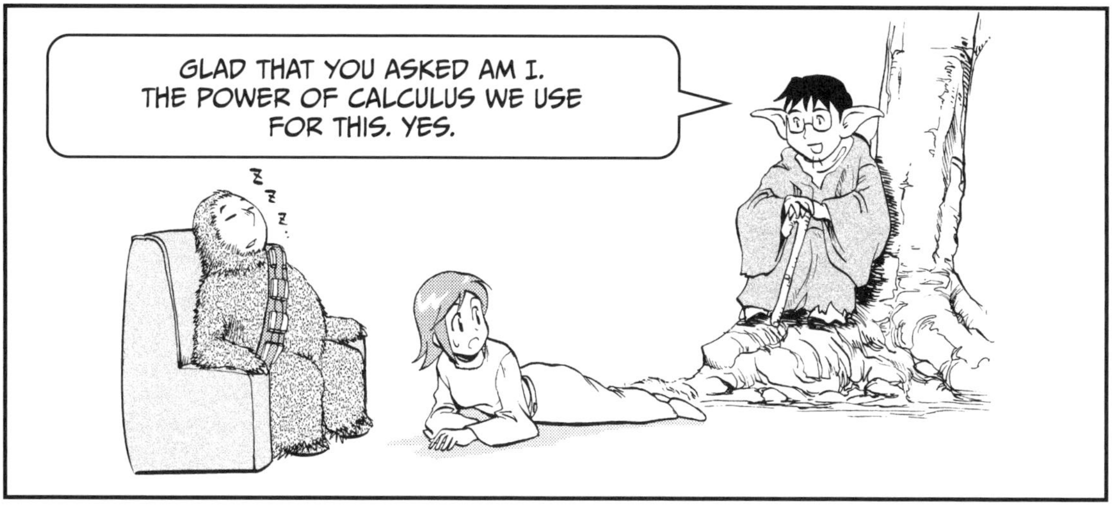
GLAD THAT YOU ASKED AM I. THE POWER OF CALCULUS WE USE FOR THIS. YES.
Z Z Z

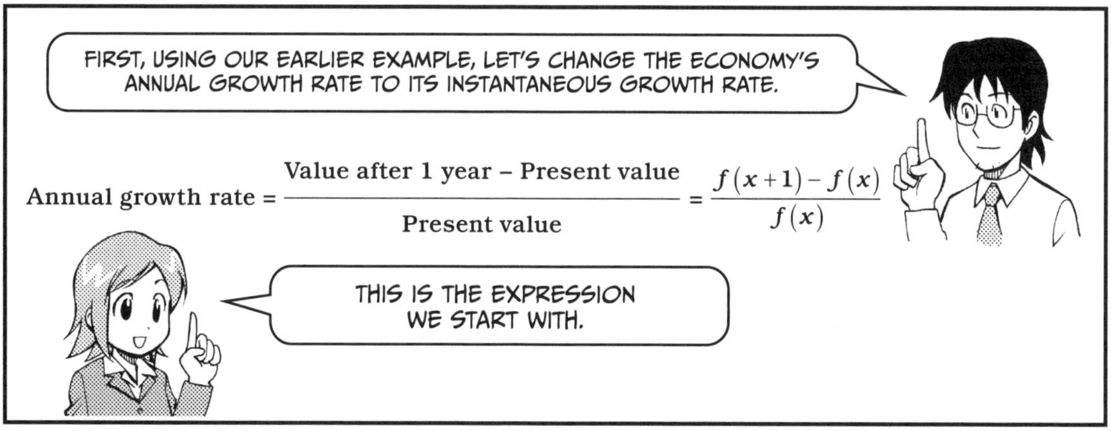
FIRST, USING OUR EARLIER EXAMPLE, LET'S CHANGE THE ECONOMY'S ANNUAL GROWTH RATE TO ITS INSTANTANEOUS GROWTH RATE.
$$\text{Annual growth rate} = \frac{\text{Value after 1 year} - \text{Present value}}{\text{Present value}} = \frac{f(x+1) - f(x)}{f(x)}$$
THIS IS THE EXPRESSION WE START WITH.

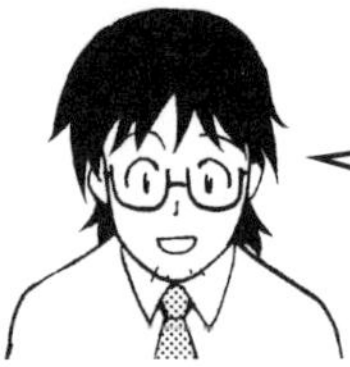
NOW WE DEVELOP THIS INTO THE INSTANTANEOUS GROWTH RATE, AS FOLLOWS.

Instantaneous growth rate

$$= \text{Idealization of } \left(\frac{\text{Value slightly later } - \text{ Present value}}{\text{Present value}} \div \text{Time elapsed} \right)$$

$$= \text{ Result obtained by letting } \varepsilon \to 0 \text{ in } \left(\frac{f(x+\varepsilon) - f(x)}{f(x)} \right) \frac{1}{\varepsilon}$$

$$= \lim_{\varepsilon \to 0} \frac{1}{f(x)} \left(\frac{f(x+\varepsilon) - f(x)}{\varepsilon} \right) = \frac{1}{f(x)} f'(x)$$

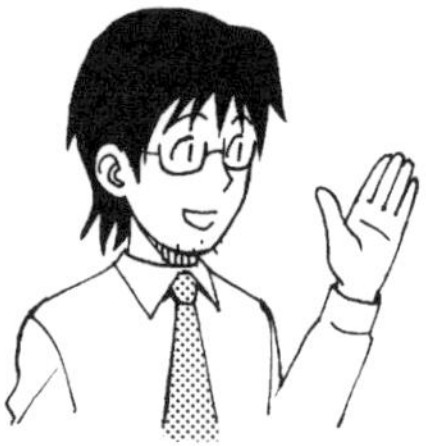
SO, WE DEFINE THE INSTANTANEOUS GROWTH RATE AS $\frac{f'(x)}{f(x)}$

Now, let's consider a function that satisfies the instantaneous growth rate when it is constant, or

$$\frac{f'(x)}{f(x)} = c \quad \text{where } c \text{ is a constant.}$$

Here we assume $c = 1$, and we will find $f(x)$ that satisfies

$$\frac{f'(x)}{f(x)} = 1$$

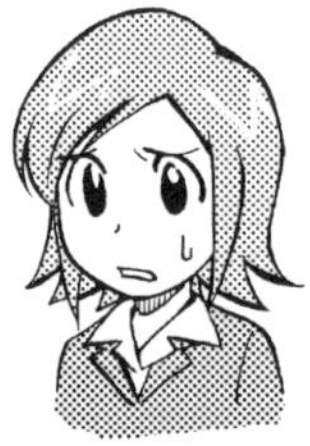
FIND $f(x)$? BUT HOW DO WE FIND IT?

1. We first guess this is an exponential function.

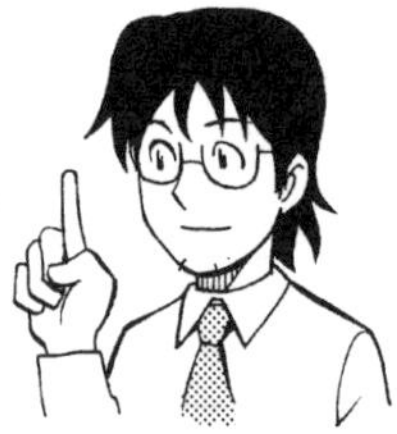
SINCE $f'(x) = f(x)$, ❶ $f'(0) = f(0)$
NOW, RECALL THAT WHEN h WAS CLOSE ENOUGH TO ZERO, WE HAD $f(h) \approx f'(0)(h-0) + f(0)$

From ❶, we have $f(h) \approx f(0)h + f(0)$ and get

❷ $\quad f(h) \approx f(0)(h+1)$

If x is close enough to h, we have

$$f(x) \approx f'(h)(x-h) + f(h)$$

Replacing x with $2h$ and using $f'(h) = f(h)$,

$$f(2h) \approx f'(h)(2h-h) + f(h)$$

$$f(2h) \approx f(h)(h) + f(h)$$

$$f(2h) \approx f(h)(h+1)$$

We'll then substitute $f(h) = f(0)(h+1)$ into our equation.

$$f(2h) = f(0)(h+1)(h+1)$$

$$f(2h) = f(0)(h+1)^2$$

In the same way, we substitute $3h$, $4h$, $5h$, ..., for x and allow $mh = 1$.

$$f(1) = f(mh) \approx f(0)(h+1)^m$$

Similarly,

$$f(2) = f(2mh) \approx f(0)(h+1)^{2m} = f(0)\left\{(1+h)^m\right\}^2$$

$$f(3) = f(3mh) \approx f(0)(h+1)^{3m} = f(0)\left\{(1+h)^m\right\}^3$$

Thus, we get

$$f(n) \approx f(0)a^n \quad \text{where we used } a = (1+h)^m$$

which is suggestive of an exponential function.[*]

---

[*] Since $mh = 1$, $h = \dfrac{1}{m}$. Then, $f(1) \approx f(0)\left(1+\dfrac{1}{m}\right)^m$. If we let $m \to \infty$ here, $\left(1+\dfrac{1}{m}\right)^m \to e$, or *Euler's number*, a number about equal to 2.718. Thus, $f(1) = f(0) \times e$, which is consistent with the discussion on page 141.

2. Next we will find out that $f(x)$ surely exists and what it is like.

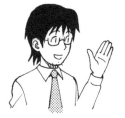

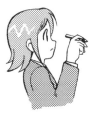

EXPRESS THE INVERSE FUNCTION OF $y = f(x)$ AS $x = g(y)$.

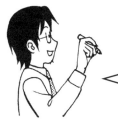

FROM $f'(x) = f(x)$ INDICATED ON PAGE 136, THE DERIVATIVE OF $f(x)$ IS ITSELF. BUT THIS DOES NOT HELP US. WHAT IS THE DERIVATIVE OF $g(y)$ THEN?

❸ $$g'(y) = \frac{1}{f'(x)}$$

Since we get this generally,*

❹ $$g'(y) = \frac{1}{f'(x)} = \frac{1}{f(x)} = \frac{1}{y}$$

we get this result, which shows that the derivative of the inverse function $g(y)$ is explicitly given by $\frac{1}{y}$.

Now, we can use the Fundamental Theorem of Calculus. It gives

❺ $$\int_1^{\alpha} \frac{1}{y}\,dy = g(\alpha) - g(1)$$

Since we now know $g'(y) = \frac{1}{y}$, function $g(\alpha)$ is found to be a function obtained by integrating $\frac{1}{y}$ from 1 to α.

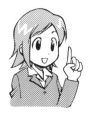

If we assume $g(1) = 0$ here . . .

WE GET $g(\alpha) = \int_1^{\alpha} \frac{1}{y}\,dy$

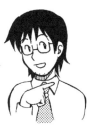

GOOD! NOW, LET'S DRAW THE GRAPH OF $z = \frac{1}{y}$!

* As shown on page 75, if the inverse function of $y = f(x)$ is $x = g(y)$, $f'(x)\,g'(y) = 1$.

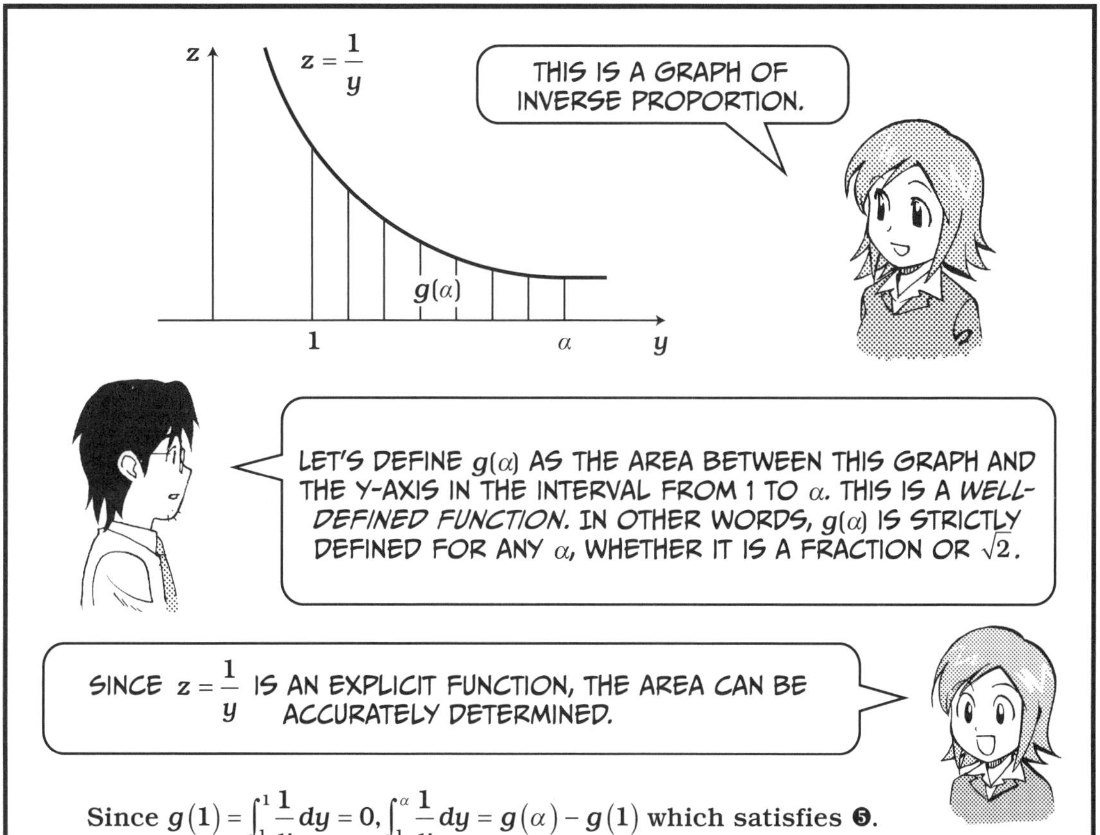
z
$z = \frac{1}{y}$
$g(\alpha)$
1
α
y
THIS IS A GRAPH OF INVERSE PROPORTION.
LET'S DEFINE $g(\alpha)$ AS THE AREA BETWEEN THIS GRAPH AND THE Y-AXIS IN THE INTERVAL FROM 1 TO α. THIS IS A *WELL-DEFINED FUNCTION*. IN OTHER WORDS, $g(\alpha)$ IS STRICTLY DEFINED FOR ANY α, WHETHER IT IS A FRACTION OR $\sqrt{2}$.
SINCE $z = \frac{1}{y}$ IS AN EXPLICIT FUNCTION, THE AREA CAN BE ACCURATELY DETERMINED.

Since $g(1) = \int_1^1 \frac{1}{y}\,dy = 0, \int_1^\alpha \frac{1}{y}\,dy = g(\alpha) - g(1)$ which satisfies ❺.

Thus, we have found out the inverse function $g(y)$, the area under the curve, which also gives the original function $f(x)$.

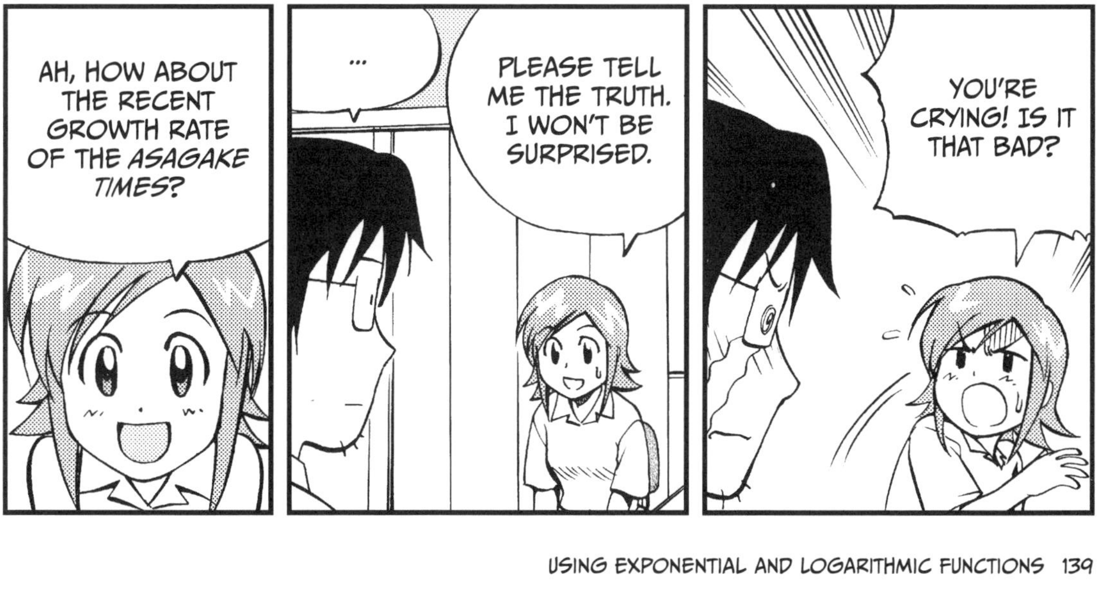
AH, HOW ABOUT THE RECENT GROWTH RATE OF THE *ASAGAKE TIMES*?
...
PLEASE TELL ME THE TRUTH. I WON'T BE SURPRISED.
YOU'RE CRYING! IS IT THAT BAD?

SUMMARY OF EXPONENTIAL AND LOGARITHMIC FUNCTIONS

❶ $\dfrac{f'(x)}{f(x)}$ is thought to be the growth rate.

❷ $y = f(x)$ which satisfies $\dfrac{f'(x)}{f(x)} = 1$ is the function that has a constant growth rate of 1.

This is an exponential function and satisfies

$$f'(x) = f(x)$$

❸ If the inverse function of $y = f(x)$ is given by $x = g(y)$, we have

$$g'(y) = \frac{1}{y} \quad \star$$

❹ If we define $g(\alpha)$, we can find the area of $h(y) = \dfrac{1}{y}$,

$$g(\alpha) = \int_1^\alpha \frac{1}{y}\,dy$$

The inverse function of $f(x)$ is the function that satisfies $\star$ and $g(1) = 0$.

❺

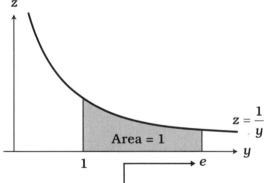

e is an irrational number that is about 2.7178.

We define e (the base of the natural logarithm) as y that satisfies $g(y) = 1$. That is, it is the α for which the area between the 1 / y curve and the y-axis in the interval from 1 to α equals 1.

Since $f(x)$ is an exponential function, we can write, using constant a_0,

$$f(x) = a_0 a^x$$

Since $f(g(1)) = f(0) = a_0 a^0 = a_0$ and $f(g(1)) = 1$, we get

$$f(g(1)) = 1 = a$$

And so we know

$$f(x) = a^x$$

Similarly, since

$$f(g(e)) = f(1) = a^1 \quad \text{and}$$

$$f(g(e)) = e$$

$$e = a^1$$

Thus, we have $f(x) = e^x$.

The inverse function $g(y)$ of this is $\log_e y$, which can be simply written as $\ln y$ ($\ln$ stands for the natural logarithm).

Now let's rewrite ❷ through ❹ in terms of e^x and $\ln y$.

❻ $f'(x) = f(x) \Leftrightarrow \left(e^x\right)' = e^x$

❼ $g'(y) = \dfrac{1}{y} \Leftrightarrow \left(\ln y\right)' = \dfrac{1}{y}$

❽ $g(\alpha) = \int_1^\alpha \dfrac{1}{y}\, dy \Leftrightarrow \ln y = \int_1^y \dfrac{1}{y}\, dy$

❾   To define 2^x, a function of bits, for any real number x, we look at

$$f(x) = e^{(\ln 2)x} \quad (x \text{ is any real number})$$

The reason is as follows. Because e^x and $\ln y$ are inverse functions to each other,

$$e^{\ln 2} = 2$$

Therefore, for any natural number x, we have

$$f(x) = \left(e^{\ln 2}\right)^x = 2^x$$

MORE APPLICATIONS OF THE FUNDAMENTAL THEOREM

Other functions can be expressed in the form of $f(x) = x^\alpha$. Some of them are

$$\frac{1}{x} = x^{-1}, \frac{1}{x^2} = x^{-2}, \frac{1}{x^3} = x^{-3}, \ldots$$

For such functions in general, the formula we found earlier holds true.

FORMULA 4-2: THE POWER RULE OF DIFFERENTIATION

$$f(x) = x^\alpha \qquad f'(x) = \alpha x^{\alpha - 1}$$

EXAMPLE:

For $f(x) = \dfrac{1}{x^3}$, $f'(x) = \left(x^{-3}\right)' = -3x^{-4} = -\dfrac{3}{x^4}$

For $f(x) = \sqrt[4]{x}$, $f'(x) = \left(x^{\frac{1}{4}}\right)' = \dfrac{1}{4}x^{-\frac{3}{4}} = \dfrac{1}{4\sqrt[4]{x^3}}$

PROOF:

Let's express $f(x)$ in terms of e. Noting $e^{\ln x} = x$, we have

$$f(x) = x^\alpha = \left(e^{\ln x}\right)^\alpha = e^{\alpha \ln x}$$

Thus,

$$\ln f(x) = \alpha \ln x$$

Differentiating both sides, remembering that the derivative of $\ln w = \dfrac{1}{w}$, and applying the chain rule,

$$\frac{1}{f(x)} \times f'(x) = \alpha \times \frac{1}{x}$$

Therefore,

$$f'(x) = \alpha \times \frac{1}{x} \times f(x) = \alpha \times \frac{1}{x} \times x^\alpha = \alpha x^{\alpha - 1}$$

INTEGRATION BY PARTS

If $h(x) = f(x) g(x)$, we get from the product rule of differentiation,

$$h'(x) = f'(x)g(x) + f(x)g'(x)$$

Thus, since the function (the antiderivative) that gives $f'(x) g(x) + f(x) g'(x)$ after differentiation is $f(x) g(x)$, we obtain from the Fundamental Theorem of Calculus,

$$\int_a^b \{ f'(x)g(x) + f(x)g'(x) \} dx = f(b)g(b) - f(a)g(a)$$

Using the sum rule of integration, we obtain the following formula.

FORMULA 4-3: INTEGRATION BY PARTS

$$\int_a^b f'(x)g(x)\,dx + \int_a^b f(x)g'(x)\,dx = f(b)g(b) - f(a)g(a)$$

As an example, let's calculate:

$$\int_0^\pi x \sin x \, dx$$

We guess the integral's answer will be a similar form to $x \cos x$, so we say $f(x) = x$ and $g(x) = \cos x$. So we try,

$$\int_0^\pi x' \cos x \, dx + \int_0^\pi x (\cos x)' \, dx = f(x)g(x) \big|_0^\pi$$

We can evaluate that

$$= f(\pi)g(\pi) - f(0)g(0)$$

Substituting in our original functions of $f(x)$ and $g(x)$, we find that

$$= \pi \cos \pi - 0 \cos 0 = \pi(-1) - 0 = -\pi$$

We can use this result in our first equation.

$$\int_0^\pi x' \cos x \, dx + \int_0^\pi x (\cos x)' \, dx = -\pi$$

We then get:

$$\int_0^\pi \cos x \, dx + \int_0^\pi x(-\sin x) \, dx = -\pi$$

Rearranging it further by pulling out the negatives, we find:

$$\int_0^\pi \cos x \, dx - \int_0^\pi x \sin x \, dx = -\pi$$

And you can see here that we have the original integral, but now we have it in terms that we can actually solve! We solve for our original function:

$$\int_0^\pi x \sin x \, dx = \int_0^\pi \cos x \, dx + \pi$$

Remember that $\int \cos x \, dx = \sin x$, and you can see that

$$\int_0^\pi x \sin x \, dx = \sin x \Big|_0^\pi + \pi$$

$$= \sin \pi - \sin 0 + \pi$$

$$= 0 - 0 + \pi = \pi$$

There you have it.

EXERCISES

1. tan x is a function defined as sin x / cos x. Obtain the derivative of tan x.

2. Calculate

$$\int_0^{\frac{\pi}{4}} \frac{1}{\cos^2 x} \, dx$$

3. Obtain such x that makes $f(x) = xe^x$ minimum.

4. Calculate

$$\int_1^e 2x \ln x \, dx$$

A clue: Suppose $f(x) = x^2$ and $g(x) = \ln x$, and use integration by parts.

5

LET'S LEARN ABOUT TAYLOR EXPANSIONS!

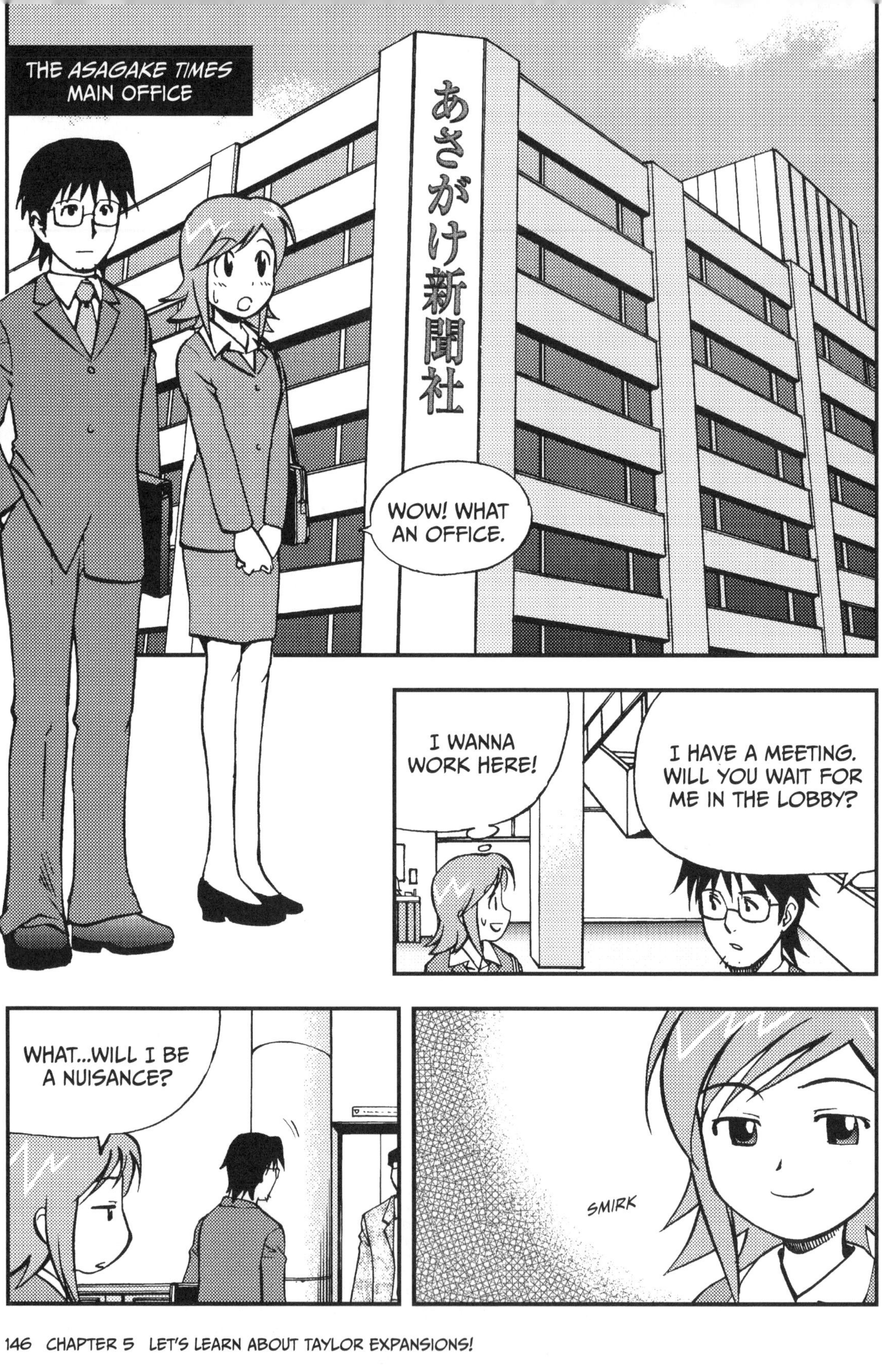
THE ASAGAKE TIMES MAIN OFFICE
あさがけ新聞社
WOW! WHAT AN OFFICE.
I WANNA WORK HERE!
I HAVE A MEETING. WILL YOU WAIT FOR ME IN THE LOBBY?
WHAT...WILL I BE A NUISANCE?
SMIRK

IMITATING WITH POLYNOMIALS

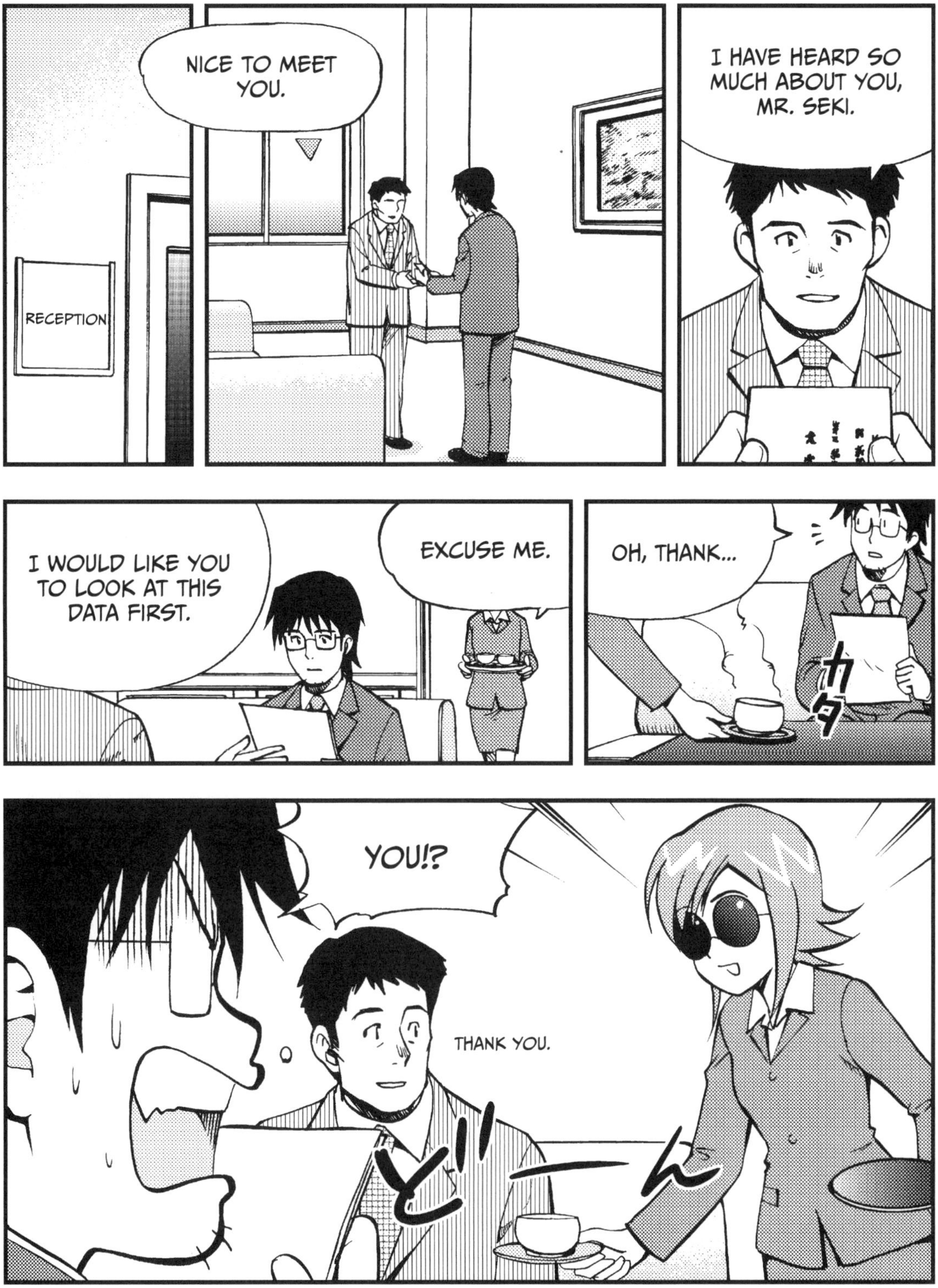

RECEPTION
NICE TO MEET YOU.
I HAVE HEARD SO MUCH ABOUT YOU, MR. SEKI.
I WOULD LIKE YOU TO LOOK AT THIS DATA FIRST.
EXCUSE ME.
OH, THANK...
カタ
YOU!?
THANK YOU.

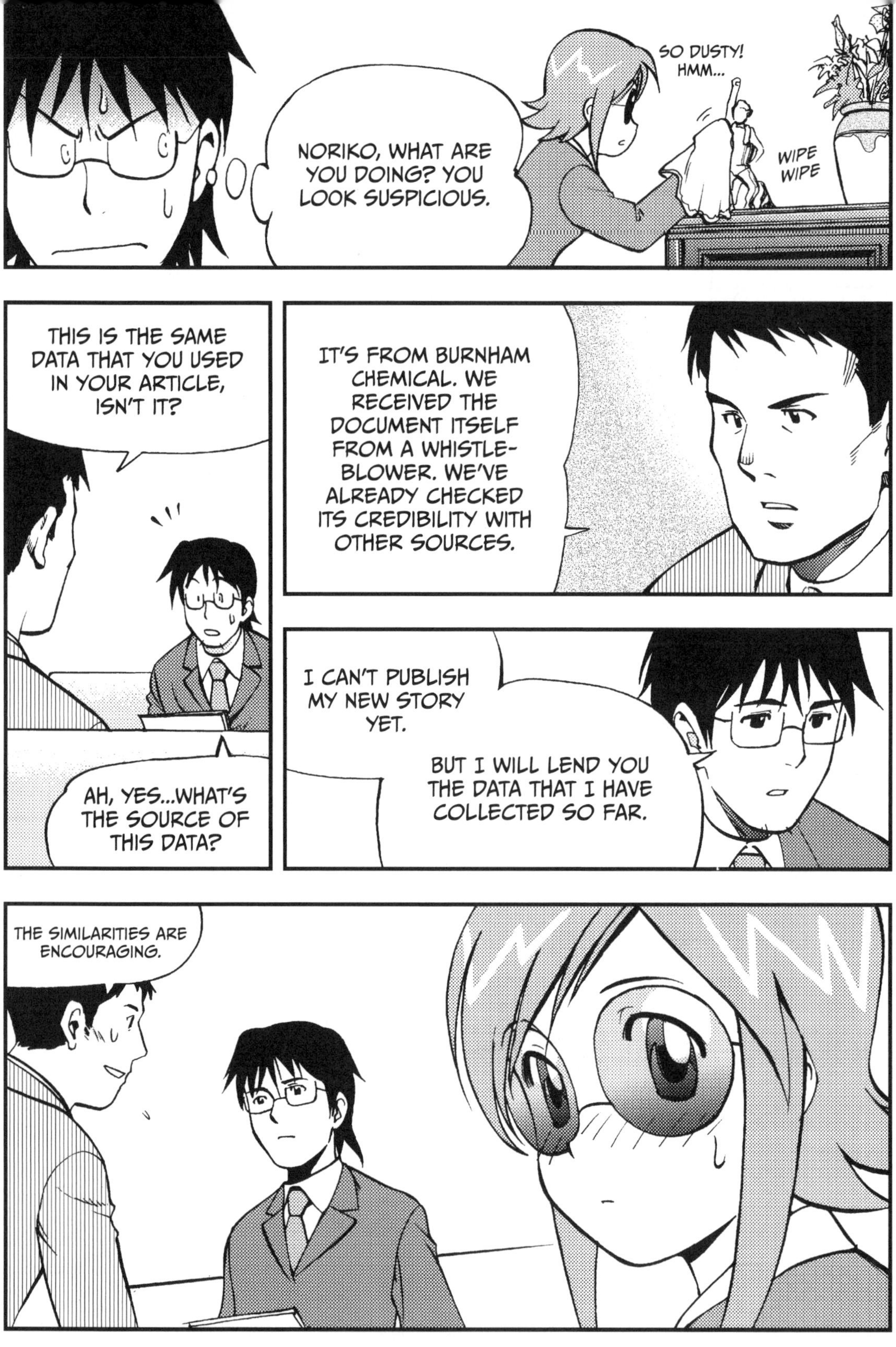
NORIKO, WHAT ARE YOU DOING? YOU LOOK SUSPICIOUS.
SO DUSTY! HMM...
WIPE WIPE
THIS IS THE SAME DATA THAT YOU USED IN YOUR ARTICLE, ISN'T IT?
AH, YES...WHAT'S THE SOURCE OF THIS DATA?
IT'S FROM BURNHAM CHEMICAL. WE RECEIVED THE DOCUMENT ITSELF FROM A WHISTLE-BLOWER. WE'VE ALREADY CHECKED ITS CREDIBILITY WITH OTHER SOURCES.
I CAN'T PUBLISH MY NEW STORY YET.
BUT I WILL LEND YOU THE DATA THAT I HAVE COLLECTED SO FAR.
THE SIMILARITIES ARE ENCOURAGING.

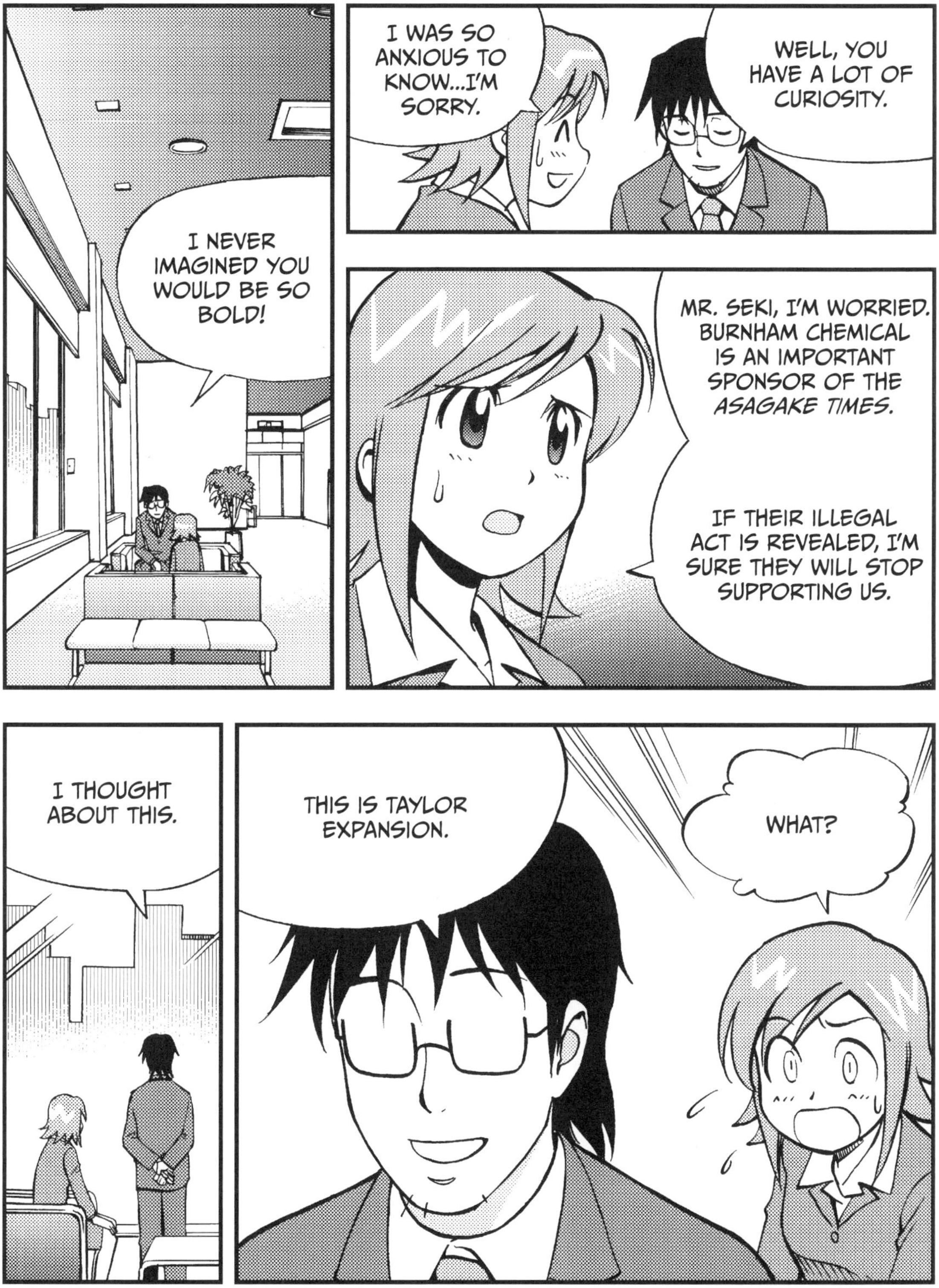
I NEVER IMAGINED YOU WOULD BE SO BOLD!
I WAS SO ANXIOUS TO KNOW...I'M SORRY.
WELL, YOU HAVE A LOT OF CURIOSITY.
MR. SEKI, I'M WORRIED. BURNHAM CHEMICAL IS AN IMPORTANT SPONSOR OF THE *ASAGAKE TIMES*.
IF THEIR ILLEGAL ACT IS REVEALED, I'M SURE THEY WILL STOP SUPPORTING US.
I THOUGHT ABOUT THIS.
THIS IS TAYLOR EXPANSION.
WHAT?

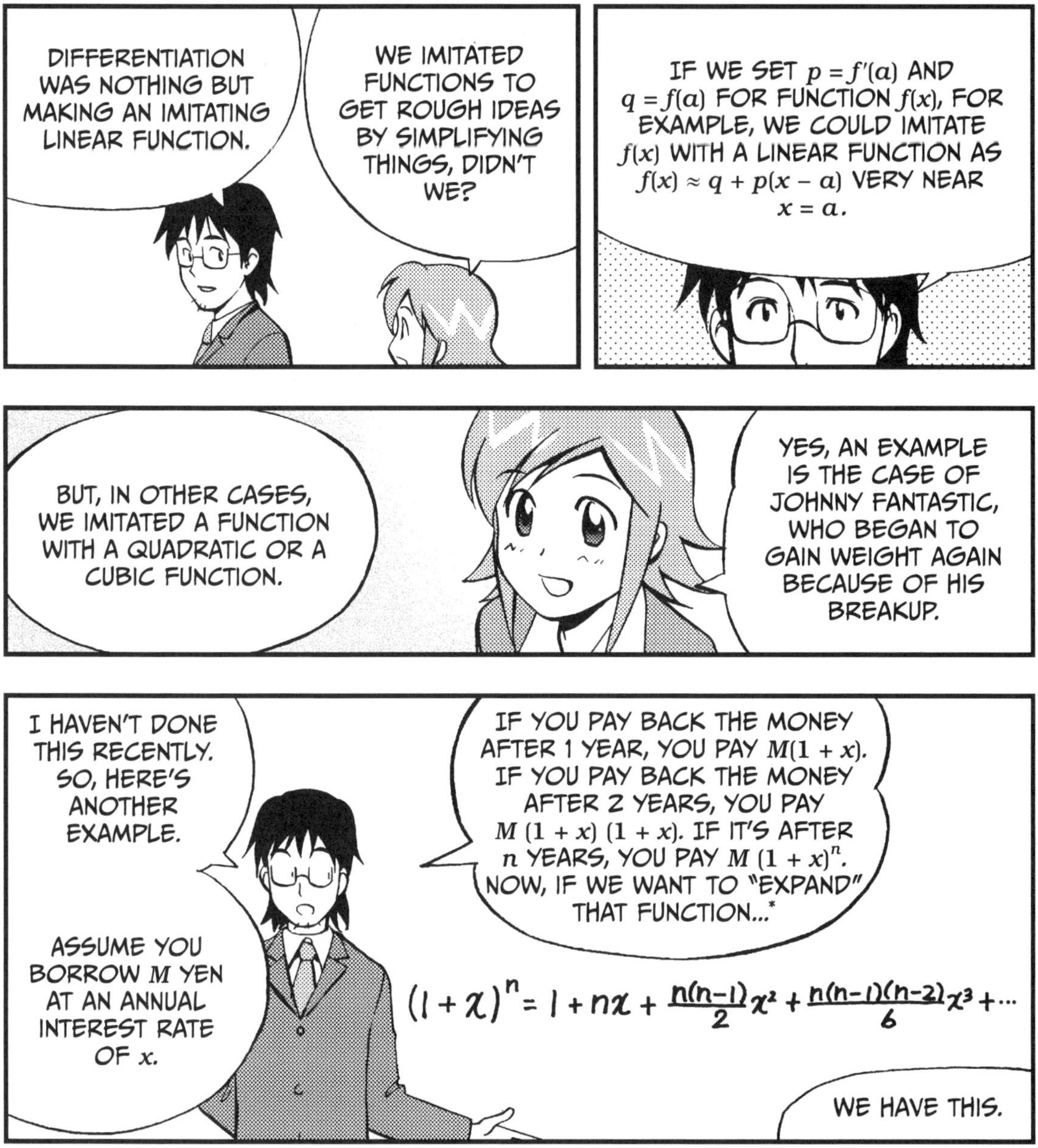
DIFFERENTIATION WAS NOTHING BUT MAKING AN IMITATING LINEAR FUNCTION.
WE IMITATED FUNCTIONS TO GET ROUGH IDEAS BY SIMPLIFYING THINGS, DIDN'T WE?
IF WE SET $p = f'(a)$ AND $q = f(a)$ FOR FUNCTION $f(x)$, FOR EXAMPLE, WE COULD IMITATE $f(x)$ WITH A LINEAR FUNCTION AS $f(x) \approx q + p(x - a)$ VERY NEAR $x = a$.
BUT, IN OTHER CASES, WE IMITATED A FUNCTION WITH A QUADRATIC OR A CUBIC FUNCTION.
YES, AN EXAMPLE IS THE CASE OF JOHNNY FANTASTIC, WHO BEGAN TO GAIN WEIGHT AGAIN BECAUSE OF HIS BREAKUP.
I HAVEN'T DONE THIS RECENTLY. SO, HERE'S ANOTHER EXAMPLE.
ASSUME YOU BORROW M YEN AT AN ANNUAL INTEREST RATE OF x.
IF YOU PAY BACK THE MONEY AFTER 1 YEAR, YOU PAY $M(1 + x)$. IF YOU PAY BACK THE MONEY AFTER 2 YEARS, YOU PAY $M(1 + x)(1 + x)$. IF IT'S AFTER n YEARS, YOU PAY $M(1 + x)^n$. NOW, IF WE WANT TO "EXPAND" THAT FUNCTION...*
$(1+x)^n = 1 + nx + \frac{n(n-1)}{2}x^2 + \frac{n(n-1)(n-2)}{6}x^3 + \dots$
WE HAVE THIS.

$$(1+x)^n = 1 + {}_nC_1 x + {}_nC_2 x^2 + {}_nC_3 x^3 + \dots + {}_nC_n x^n$$

* THIS IS THE FORMULA OF BINOMIAL EXPANSION, WHERE ${}_nC_r = \dfrac{n!}{r!(n-r)!}$ AND ${}_nC_1 = n$

$${}_nC_2 = \frac{n(n-1)}{2}, {}_nC_3 = \frac{n(n-1)(n-2)}{6}, \dots, {}_nC_r = \frac{n(n-1)\dots\{n-(r-1)\}}{r!}$$

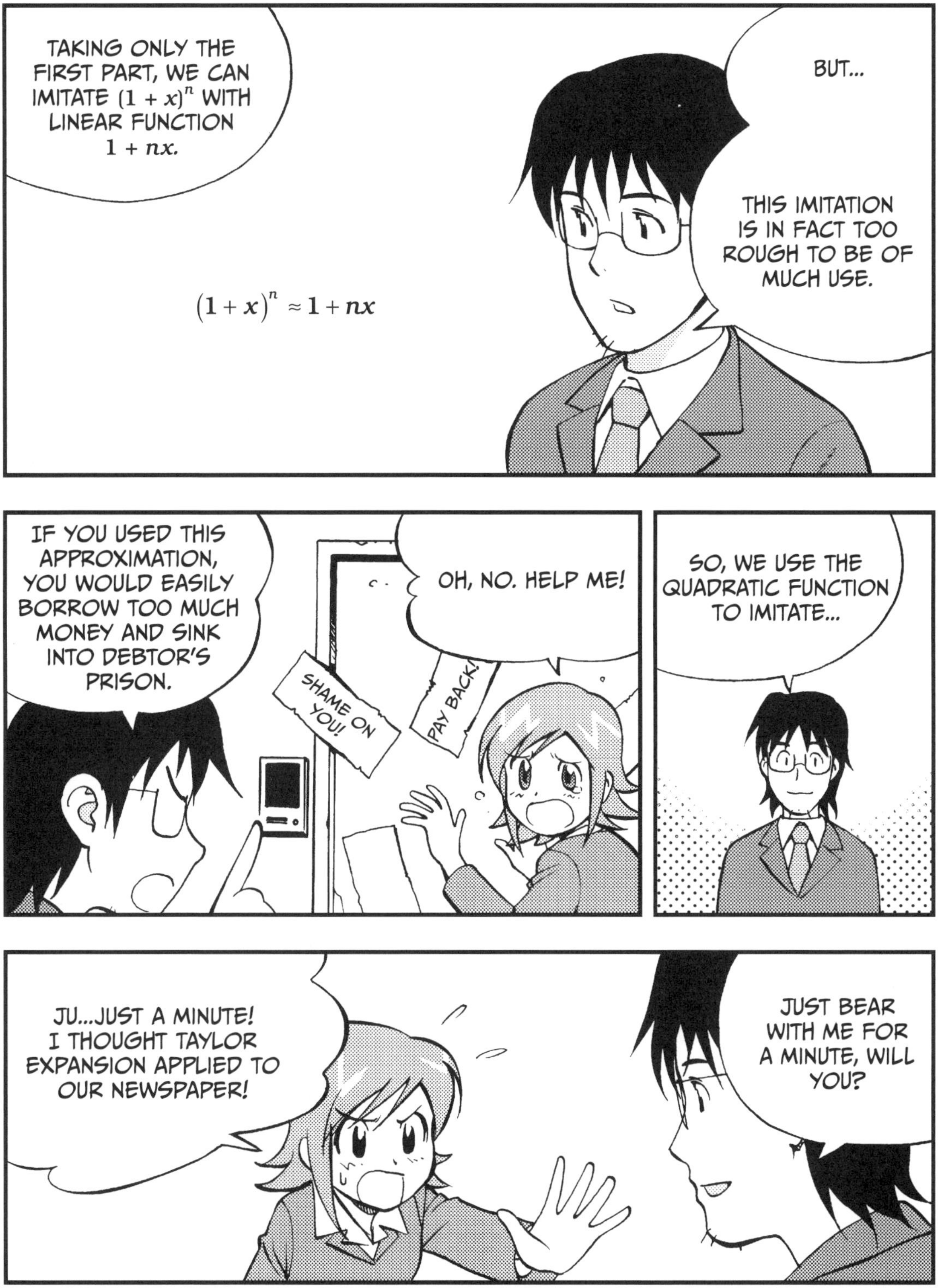
TAKING ONLY THE FIRST PART, WE CAN IMITATE $(1 + x)^n$ WITH LINEAR FUNCTION $1 + nx$.
$(1+x)^n \approx 1+nx$
BUT...
THIS IMITATION IS IN FACT TOO ROUGH TO BE OF MUCH USE.
IF YOU USED THIS APPROXIMATION, YOU WOULD EASILY BORROW TOO MUCH MONEY AND SINK INTO DEBTOR'S PRISON.
OH, NO. HELP ME!
SHAME ON YOU!
PAY BACK!
SO, WE USE THE QUADRATIC FUNCTION TO IMITATE...
JU...JUST A MINUTE! I THOUGHT TAYLOR EXPANSION APPLIED TO OUR NEWSPAPER!
JUST BEAR WITH ME FOR A MINUTE, WILL YOU?

FORMULA 5-1: THE FORMULA OF QUADRATIC APPROXIMATION

$$(1+x)^n \approx 1 + nx + \frac{n(n-1)}{2}x^2$$

> IF WE MODIFY THIS EXPRESSION A LITTLE, WE GET A VERY INTERESTING LAW.

For any pair of n and x that satisfy $nx = 0.7$, we get

$$(1+x)^n \approx 1 + nx + \frac{n(n-1)}{2}x^2 \approx 1 + nx + \frac{1}{2}(nx)^2 - \underbrace{\frac{1}{2}nx^2}$$

$$\approx 1 + 0.7 + \frac{1}{2} \times 0.7^2 = 1.945 \approx 2 \qquad \boxed{\text{Nearly zero, so we neglect it.}}$$

In short, if $nx = 0.7$, $(1 + x)^n$ is almost 2. This can be written as a law as follows.

LAW OF DEBT HELL

When years to repay loan × interest rate = 0.7, the amount you will repay is about twice as much as you borrowed.

> ABOUT TWICE IF BORROWED FOR 35 YEARS AT 2 PERCENT
>
> ABOUT TWICE IF BORROWED FOR 7 YEARS AT 10 PERCENT
>
> ABOUT TWICE IF BORROWED FOR 2 YEARS AT 35 PERCENT

> OH, NO!! THIS IS TERRIBLE!!

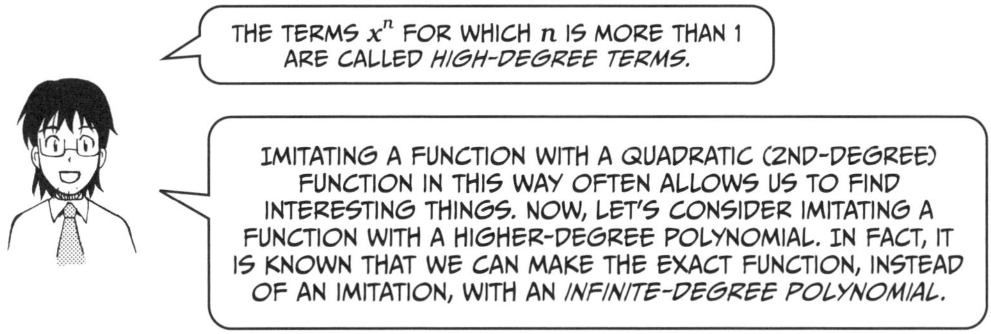
THE TERMS x^n FOR WHICH n IS MORE THAN 1 ARE CALLED *HIGH-DEGREE TERMS.*
IMITATING A FUNCTION WITH A QUADRATIC (2ND-DEGREE) FUNCTION IN THIS WAY OFTEN ALLOWS US TO FIND INTERESTING THINGS. NOW, LET'S CONSIDER IMITATING A FUNCTION WITH A HIGHER-DEGREE POLYNOMIAL. IN FACT, IT IS KNOWN THAT WE CAN MAKE THE EXACT FUNCTION, INSTEAD OF AN IMITATION, WITH AN *INFINITE-DEGREE POLYNOMIAL.*

For example, if we set $f(x) = \dfrac{1}{1-x}$, we get

❶ $$f(x) = \frac{1}{1-x} = 1 + x + x^2 + x^3 + x^4 + \ldots \quad \text{(continues infinitely)}$$

Note this is = instead of ≈.

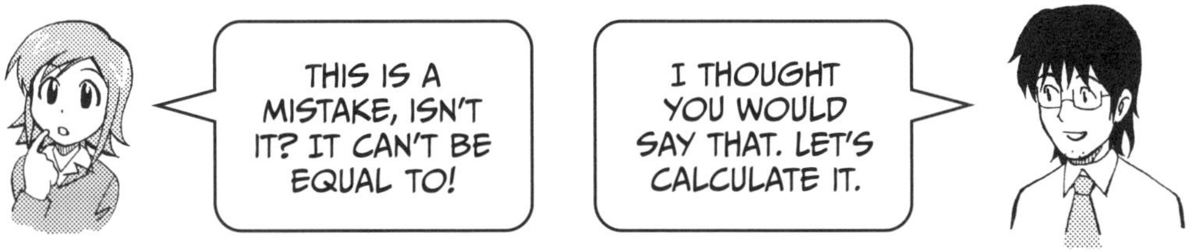
THIS IS A MISTAKE, ISN'T IT? IT CAN'T BE EQUAL TO!
I THOUGHT YOU WOULD SAY THAT. LET'S CALCULATE IT.

Suppose $x = 0.1$. We get

$$f(0.1) = \frac{1}{1-0.1} = \frac{1}{0.9} = \frac{10}{9}$$

$$\begin{aligned}\text{Right side } &= 1 + 0.1 + 0.1^2 + 0.1^3 + 0.1^4 + \ldots \\ &= 1 + 0.1 + 0.01 + 0.001 + 0.0001 + \ldots \\ &= 1.111111\ldots\end{aligned}$$

If we actually calculate 10/9 by long division, we will obtain the same result.

```
    1.111...
 9 ) 10
      9
     --
      10
       9
      --
       10
        9
       --
        10
         9
        --
```

When a general function $f(x)$ (provided it is differentiable infinitely many times) can be expressed as

$$f(x) = a_0 + a_1 x + a_2 x^2 + a_3 x^3 + \ldots + a_n x^n + \ldots$$

the right side is called the *Taylor expansion* of $f(x)$ (about $x = 0$).

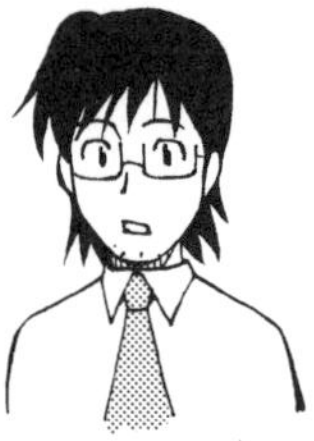
THIS MEANS THAT $f(x)$ PERFECTLY COINCIDES WITH AN INFINITE-DEGREE POLYNOMIAL IN A DEFINITE INTERVAL INCLUDING $x = 0$. IT SHOULD BE NOTED, HOWEVER, THAT THE RIGHT SIDE MAY BECOME MEANINGLESS BECAUSE IT MAY NOT HAVE A SINGLE DEFINED VALUE OUTSIDE THE INTERVAL.
FOR EXAMPLE, SUBSTITUTING $x = 2$ IN BOTH SIDES OF EXPRESSION ❶,

$$\text{Left side} = \frac{1}{1-2} = -1$$

$$\text{Right side} = 1 + 2 + 4 + 8 + 16 + \ldots$$

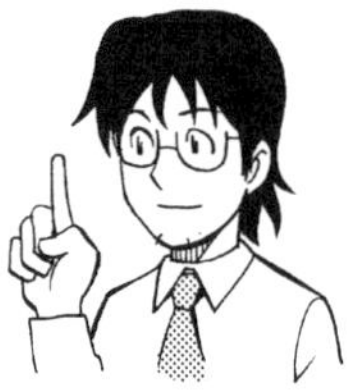
SEE? THE TWO SIDES ARE NOT EQUAL.

It turns out that expression ❶ is correct for all x satisfying $-1 < x < 1$, which is the allowed interval of a Taylor expansion. In technical terms, the interval $-1 < x < 1$ is called the *circle of convergence*.

HOW TO OBTAIN A TAYLOR EXPANSION

When we have

> ❷ $\quad f(x) = a_0 + a_1 x + a_2 x^2 + a_3 x^3 + \ldots + a_n x^n + \ldots$

let's find the coefficient a_n.

Substituting $x = 0$ in the above equation and noting $f(0) = a_0$, we find that the 0th-degree coefficient a_0 is $f(0)$.

We then differentiate ❷.

> ❸ $\quad f'(x) = a_1 + 2a_2 x + 3a_3 x^2 + \ldots + n a_n x^{n-1} + \ldots$

Substituting $x = 0$ in ❸ and noting $f'(0) = a_1$, we find that the 1st-degree coefficient a_1 is $f'(0)$.

We differentiate ❸ to get

> ❹ $\quad f''(x) = 2a_2 + 6a_3 x + \ldots + n(n-1)a_n x^{n-2} + \ldots$

Substituting $x = 0$ in ❹, we find that the 2nd-degree coefficient a_2 is $\frac{1}{2} f''(0)$.

Differentiating ❹, we get

$$f'''(x) = 6a_3 + \ldots + n(n-1)(n-2)a_n x^{n-3} + \ldots$$

From this, we find that the 3rd-degree coefficient a_3 is $\frac{1}{6} f'''(0)$.

Repeating this differentiation operation n times, we get

$$f^{(n)}(x) = n(n-1)\ldots \times 2 \times 1 a_n + \ldots$$

where $f^{(n)}(x)$ is the expression obtained after differentiating $f(x)$ n times.

From this result, we find

> nth-degree coefficient $\quad a_n = \dfrac{1}{n!} f^{(n)}(0)$

$n!$ is read "n factorial" and means $n \times (n-1) \times (n-2) \times \ldots \times 2 \times 1$.

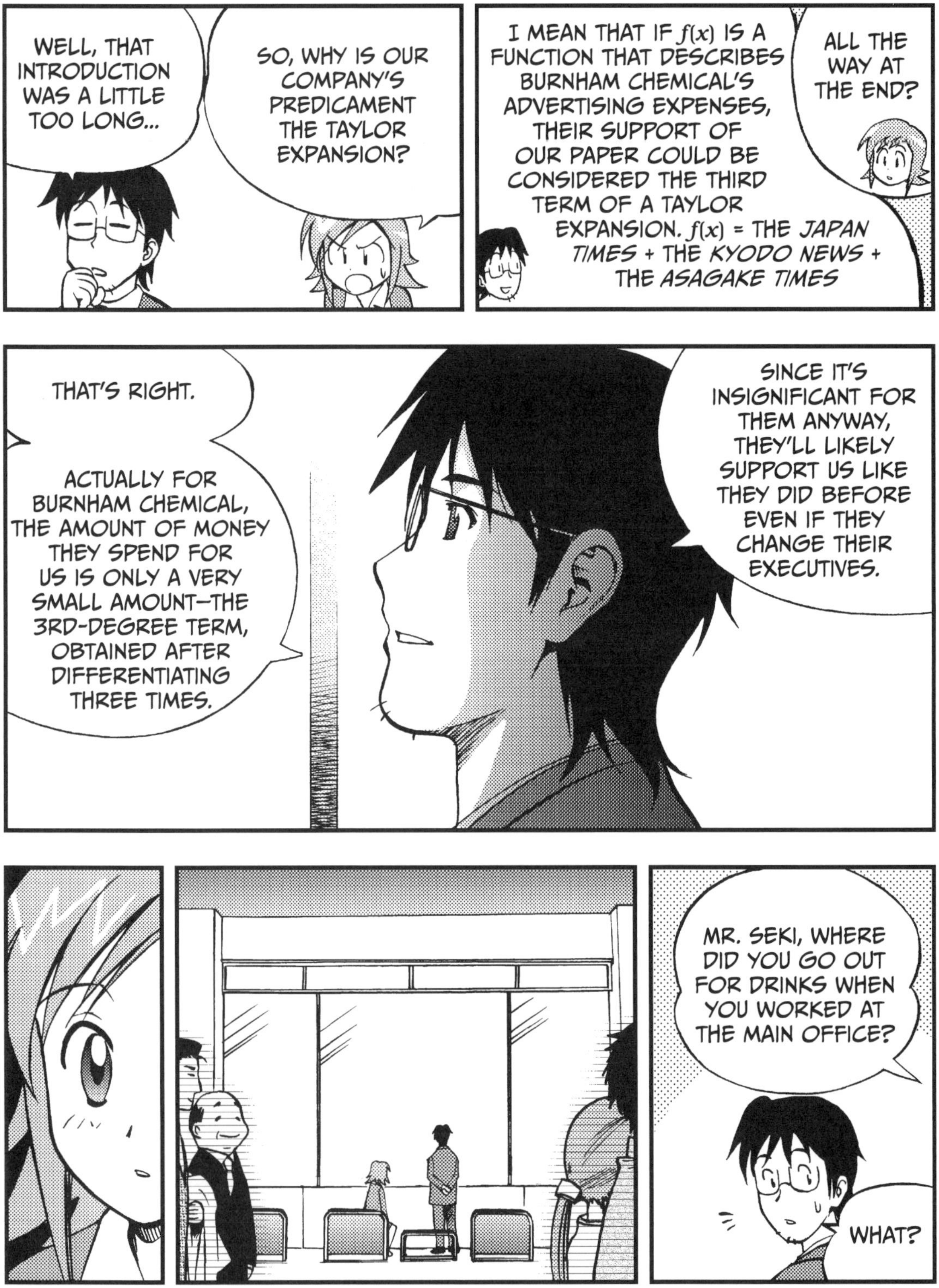
WELL, THAT INTRODUCTION WAS A LITTLE TOO LONG...
SO, WHY IS OUR COMPANY'S PREDICAMENT THE TAYLOR EXPANSION?
I MEAN THAT IF $f(x)$ IS A FUNCTION THAT DESCRIBES BURNHAM CHEMICAL'S ADVERTISING EXPENSES, THEIR SUPPORT OF OUR PAPER COULD BE CONSIDERED THE THIRD TERM OF A TAYLOR EXPANSION. $f(x)$ = THE *JAPAN TIMES* + THE *KYODO NEWS* + THE *ASAGAKE TIMES*
ALL THE WAY AT THE END?
THAT'S RIGHT.
ACTUALLY FOR BURNHAM CHEMICAL, THE AMOUNT OF MONEY THEY SPEND FOR US IS ONLY A VERY SMALL AMOUNT—THE 3RD-DEGREE TERM, OBTAINED AFTER DIFFERENTIATING THREE TIMES.
SINCE IT'S INSIGNIFICANT FOR THEM ANYWAY, THEY'LL LIKELY SUPPORT US LIKE THEY DID BEFORE EVEN IF THEY CHANGE THEIR EXECUTIVES.
MR. SEKI, WHERE DID YOU GO OUT FOR DRINKS WHEN YOU WORKED AT THE MAIN OFFICE?
WHAT?

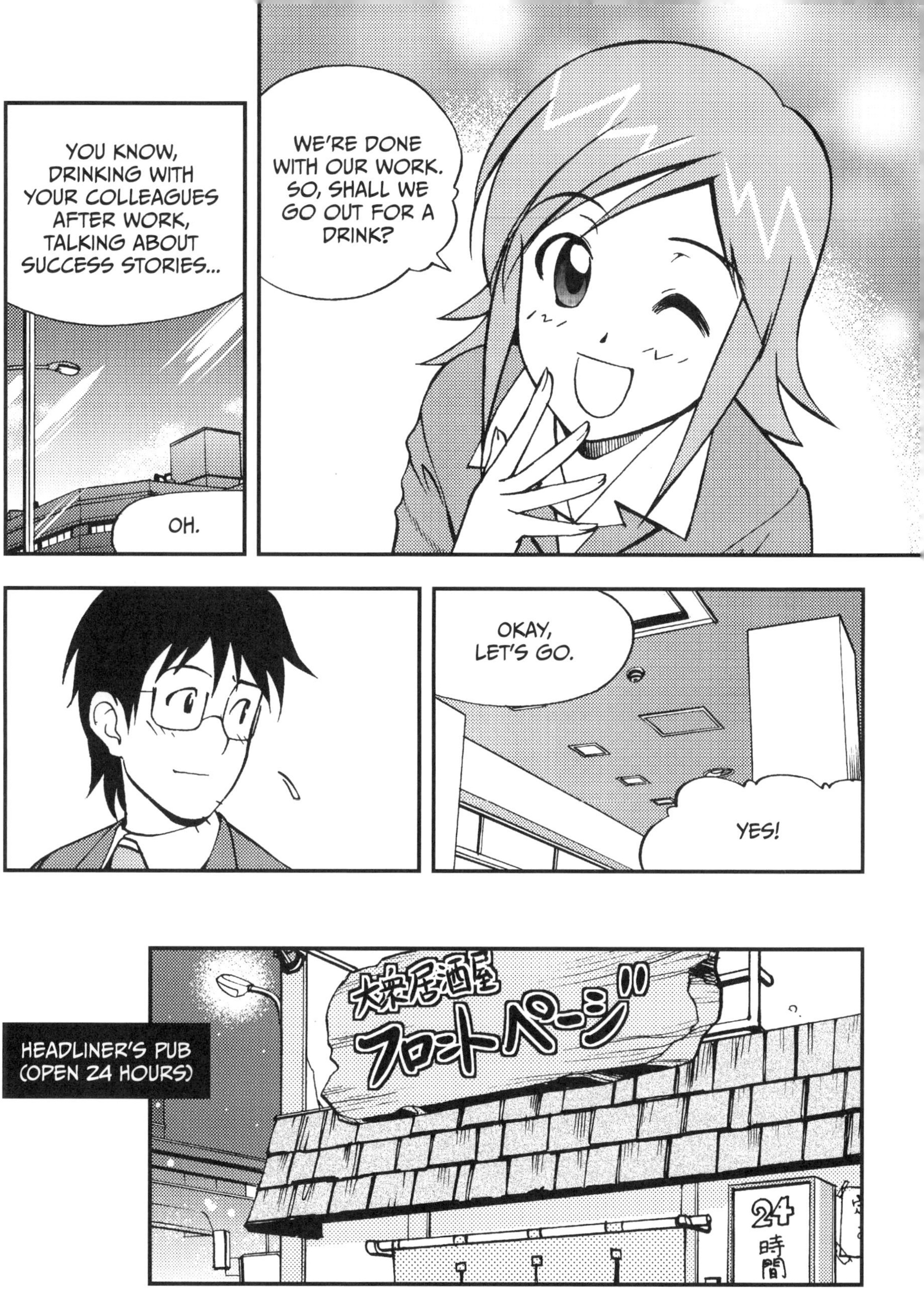
YOU KNOW, DRINKING WITH YOUR COLLEAGUES AFTER WORK, TALKING ABOUT SUCCESS STORIES...
OH.
WE'RE DONE WITH OUR WORK. SO, SHALL WE GO OUT FOR A DRINK?
OKAY, LET'S GO.
YES!
HEADLINER'S PUB (OPEN 24 HOURS)
大衆居酒屋
フロントページ
24
時間

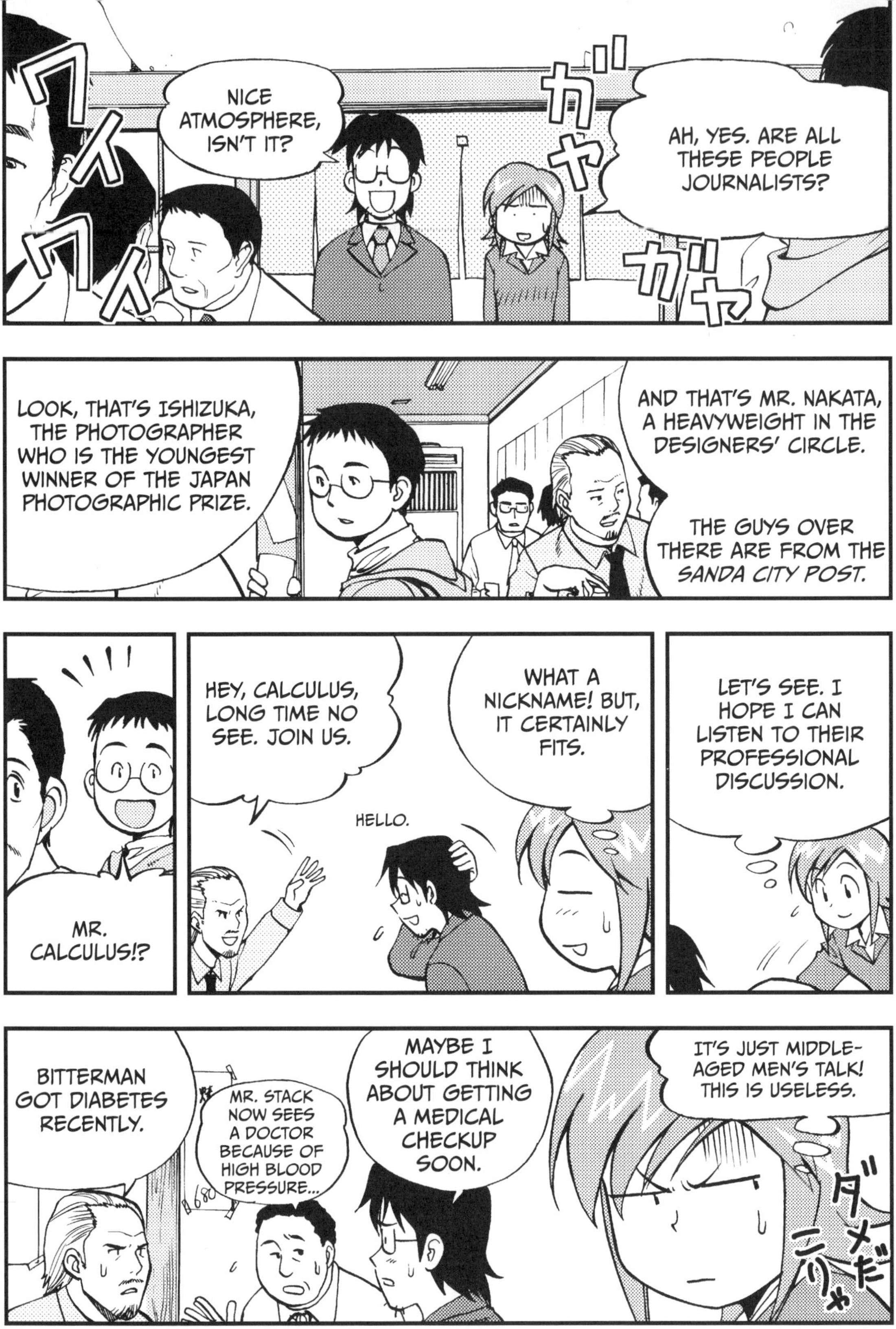
ワイワイ
NICE ATMOSPHERE, ISN'T IT?
ガヤガヤ
AH, YES. ARE ALL THESE PEOPLE JOURNALISTS?
LOOK, THAT'S ISHIZUKA, THE PHOTOGRAPHER WHO IS THE YOUNGEST WINNER OF THE JAPAN PHOTOGRAPHIC PRIZE.
AND THAT'S MR. NAKATA, A HEAVYWEIGHT IN THE DESIGNERS' CIRCLE.
THE GUYS OVER THERE ARE FROM THE *SANDA CITY POST*.
MR. CALCULUS!?
HEY, CALCULUS, LONG TIME NO SEE. JOIN US.
HELLO.
WHAT A NICKNAME! BUT, IT CERTAINLY FITS.
LET'S SEE. I HOPE I CAN LISTEN TO THEIR PROFESSIONAL DISCUSSION.
BITTERMAN GOT DIABETES RECENTLY.
MR. STACK NOW SEES A DOCTOR BECAUSE OF HIGH BLOOD PRESSURE...
MAYBE I SHOULD THINK ABOUT GETTING A MEDICAL CHECKUP SOON.
IT'S JUST MIDDLE-AGED MEN'S TALK! THIS IS USELESS.
ダメだこりゃ

FORMULA 5-2: THE FORMULA OF TAYLOR EXPANSION

If $f(x)$ has a Taylor expansion about $x = 0$, it is given by

$$f(x) = f(0) + \frac{1}{1!}f'(0)x + \frac{1}{2!}f''(0)x^2 + \frac{1}{3!}f'''(0)x^3 + \ldots + \frac{1}{n!}f^{(n)}(0)x^n + \ldots$$

For the above,

$f(0)$	0th-degree constant term	$a_0 = f(0)$
$f'(0)x$	1st-degree term	$a_1 = f'(0)$
$\frac{1}{2!}f''(0)x^2$	2nd-degree term	$a_2 = \frac{1}{2}f''(0)$
$\frac{1}{3!}f'''(0)x^3$	3rd-degree term	$a_3 = \frac{1}{6}f'''(0)$

For the moment, we forget about the conditions for having Taylor expansion and the circle of convergence.

Using this formula, we check ❶ on page 153.

$$f(x) = \frac{1}{1-x}, f'(x) = \frac{1}{(1-x)^2}, f''(x) = \frac{2}{(1-x)^3}, f'''(x) = \frac{6}{(1-x)^4}, \ldots$$

$$f(0) = 1, f'(0) = 1, f''(0) = 2, f'''(0) = 6, \ldots, f^{(n)}(0) = n!$$

Thus, we have

$$\begin{aligned} f(x) &= f(0) + \frac{1}{1!}f'(0)x + \frac{1}{2!}f''(0)x^2 + \frac{1}{3!}f'''(0)x^3 + \ldots + \frac{1}{n!}f^{(n)}(0)x^n + \ldots \\ &= 1 + x + \frac{1}{2!}\times 2x^2 + \frac{1}{3!}\times 6x^3 + \ldots + \frac{1}{n!}n!x^n + \ldots \\ &= 1 + x + x^2 + x^3 + \ldots x^n + \ldots \end{aligned}$$

THEY COINCIDE!

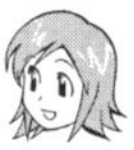

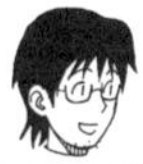

THE FORMULA ABOVE IS FOR AN INFINITE-DEGREE POLYNOMIAL THAT COINCIDES WITH THE ORIGINAL NEAR $x = 0$. THE FORMULA FOR A POLYNOMIAL THAT COINCIDES NEAR $x = a$ IS GENERALLY GIVEN AS FOLLOWS. TRY THE EXERCISE ON PAGE 178 TO CHECK THIS!

$$\begin{aligned} f(x) = f(a) &+ \frac{1}{1!}f'(a)(x-a) + \frac{1}{2!}f''(a)(x-a)^2 \\ &+ \frac{1}{3!}f'''(a)(x-a)^3 + \ldots + \frac{1}{n!}f^{(n)}(a)(x-a)^n + \ldots \end{aligned}$$

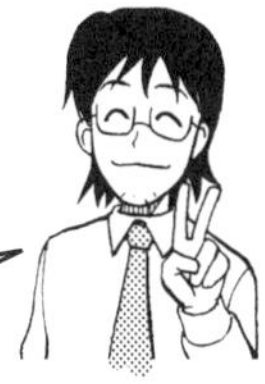

TAYLOR EXPANSION IS A SUPERIOR IMITATING FUNCTION.

TAYLOR EXPANSION OF VARIOUS FUNCTIONS

[1] TAYLOR EXPANSION OF A SQUARE ROOT

We set $f(x) = \sqrt{1+x} = (1+x)^{\frac{1}{2}}$.

Thus, from $f'(x) = \frac{1}{2}(1+x)^{-\frac{1}{2}}$

$$f''(x) = -\frac{1}{2} \times \frac{1}{2}(1+x)^{-\frac{3}{2}}$$

$$f'''(x) = \frac{1}{2} \times \frac{1}{2} \times \frac{3}{2}(1+x)^{-\frac{5}{2}}, \ldots$$

$$f'(0) = \frac{1}{2}, f''(0) = -\frac{1}{4}, f'''(0) = \frac{3}{8}, \ldots$$

$$f(x) = \sqrt{1+x}$$

$$= 1 + \frac{1}{2}x + \frac{1}{2!} \times \left(-\frac{1}{4}\right)x^2 + \frac{1}{3!} \times \frac{3}{8}x^3 + \ldots$$

$$\sqrt{1+x} = 1 + \frac{1}{2}x - \frac{1}{8}x^2 + \frac{1}{16}x^3 \ldots$$

[2] TAYLOR EXPANSION OF EXPONENTIAL FUNCTION e^x

If we set $f(x) = e^x$,

$$f'(x) = e^x, f''(x) = e^x, f'''(x) = e^x, \ldots$$

So, from

$$e^x = 1 + \frac{1}{1!}x + \frac{1}{2!}x^2 + \frac{1}{3!}x^3 + \frac{1}{4!}x^4 + \ldots + \frac{1}{n!}x^n + \ldots$$

Substituting $x = 1$, we get

$$e = 1 + \frac{1}{1!} + \frac{1}{2!} + \frac{1}{3!} + \frac{1}{4!} + \ldots + \frac{1}{n!} + \ldots$$

IN CHAPTER 4, WE LEARNED THAT e IS ABOUT 2.7. HERE, WE HAVE OBTAINED THE EXPRESSION TO CALCULATE IT EXACTLY.

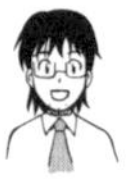

[3] TAYLOR EXPANSION OF LOGARITHMIC FUNCTION $\ln(1+x)$

We set $f(x) = \ln(x+1)$

$$f'(x) = \frac{1}{1+x} = (1+x)^{-1}$$

$$f''(x) = -(1+x)^{-2}, f^{(3)}(x) = 2(1+x)^{-3},$$

$$f^{(4)}(x) = -6(1+x)^{-4}, \ldots$$

$$f(0) = 0, f'(0) = 1, f''(0) = -1, f^{(3)}(0) = 2!,$$

$$f^{(4)}(0) = -3!, \ldots$$

Thus, we have

$$\ln(1+x) =$$

$$0 + x - \frac{1}{2}x^2 + \frac{1}{3!} \times 2!x^3 - \frac{1}{4}3!x^4 + \ldots$$

$$\ln(1+x) =$$

$$x - \frac{1}{2}x^2 + \frac{1}{3}x^3 - \frac{1}{4}x^4 + \ldots + (-1)^{n+1}\frac{1}{n}x^n + \ldots$$

[4] TAYLOR EXPANSION OF TRIGONOMETRIC FUNCTIONS

We set $f(x) = \cos x$.

$$f'(x) = -\sin x, f''(x) = -\cos x, f^{(3)}(x) = \sin x, f^{(4)}(x) = \cos x, \ldots$$

From

$$f(0) = 1, f'(0) = 0, f''(0) = -1,$$

$$f^{(3)}(0) = 0, f^{(4)}(0) = 1, \ldots$$

Thus,

$$\cos x = 1 + 0x - \frac{1}{2!} \times 1 \times x^2 + \frac{1}{3!} \times 0 \times x^3 + \frac{1}{4!} \times 1 \times x^4 + \ldots$$

$$\cos x = 1 - \frac{1}{2!}x^2 + \frac{1}{4!}x^4 + \ldots + (-1)^n \frac{1}{(2n)!}x^{2n} + \ldots$$

Similarly,

$$\sin x = x - \frac{1}{3!}x^3 + \frac{1}{5!}x^5 + \ldots + (-1)^{n-1}\frac{1}{(2n-1)!}x^{2n-1} + \ldots$$

WHAT DOES TAYLOR EXPANSION TELL US?

TAYLOR EXPANSION REPLACES COMPLICATED FUNCTIONS WITH POLYNOMIALS. CAN YOU DRAW THE GRAPH OF, FOR EXAMPLE, $\ln(1 + x)$?

AFTER ALL, IT IS NECESSARY TO APPROXIMATE OR IMITATE FUNCTIONS TO UNCOVER THEIR COMPLICATED WORLD, ISN'T IT?

LET'S USE $\ln(1 + x) = x - \dfrac{1}{2}x^2 + \dfrac{1}{3}x^3 - \dfrac{1}{4}x^4 + \ldots$, AN EXAMPLE GIVEN ABOVE, TO SEE WHAT WE CAN GAIN FROM A TAYLOR EXPANSION.

$$\ln(1 + x) = 0 + x - \frac{1}{2}x^2 + \frac{1}{3}x^3 - \frac{1}{4}x^4 + \ldots$$

1 Linear approx. — 3 Cubic approx.

0th degree approx. — 2 Quadratic approx.

FIRST, 0th-DEGREE APPROXIMATION. $\ln(1 + x) \approx 0$ NEAR $x = 0$. WHAT DOES THIS MEAN?

AH, WELL...IT MEANS THAT THE VALUE OF $f(x)$ IS 0 AT $x = 0$ AND IT PASSES THROUGH POINT $(0, 0)$.

THAT'S RIGHT. NEXT IS LINEAR APPROXIMATION. YOU SEE THAT $y = f(x)$ ROUGHLY RESEMBLES $y = x$ NEAR $x = 0$? SO, THIS MEANS THAT THE FUNCTION IS INCREASING AT $x = 0$. (NOTE: THE EQUATION OF A TANGENT LINE = LINEAR APPROXIMATION.)

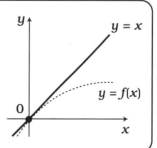

WE'LL NOW TAKE ONE MORE STEP TO QUADRATIC APPROXIMATION. LET'S CONSIDER THE GRAPH OF
$\ln(1+x) \approx x - \frac{1}{2}x^2$
AROUND $x = 0$. NORIKO, WHAT DOES THIS MEAN?
y
$y = f(x)$
-1
0
x
$y = x - \frac{1}{2}x^2$
THIS MEANS THAT $y = f(x)$ ROUGHLY RESEMBLES $y = x - \frac{1}{2}x^2$ NEAR $x = 0$ AND ITS GRAPH IS CONCAVE DOWN AT $x = 0$. (QUADRATIC APPROXIMATION ALLOWS US TO FIND HOW IT IS CURVED AT $x = a$.)
LET'S USE CUBIC APPROXIMATION AS THE LAST PUSH!! NEAR $x = 0$,
$\ln(1+x) \approx x - \frac{1}{2}x^2 + \frac{1}{3}x^3$
(CUBIC APPROXIMATION FURTHER CORRECTS THE ERROR IN QUADRATIC APPROXIMATION.)
$y = \frac{1}{3}x^3 - \frac{1}{2}x^2 + x$
y
$y = f(x)$
0
x
NOW, MR. SEKI, ON TO THE NEXT BAR!
!

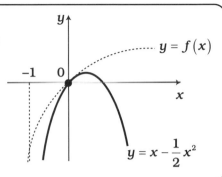

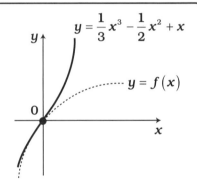

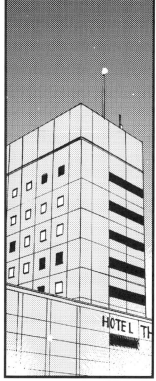

THIS IS BETTER! WE CAN TALK MORE QUIETLY AT THIS HOTEL BAR.

WE SHOULD HAVE JUST COME HERE IN THE FIRST PLACE.

YOU COULD HAVE TALKED MORE WITH THE GUYS AT THAT PUB.

WELL, THEY ALL SEEMED BRILLIANT. I FELT...I WOULD SAY, SOMEWHAT INFERIOR.

SO WHAT'S YOUR DEAL, MR. SEKI?

ALL OF THOSE PEOPLE ARE FAMOUS—THEY'VE ALL WON JOURNALISTIC PRIZES.

BUT, I COULD IMMEDIATELY TELL THAT THEY RESPECTED YOU.

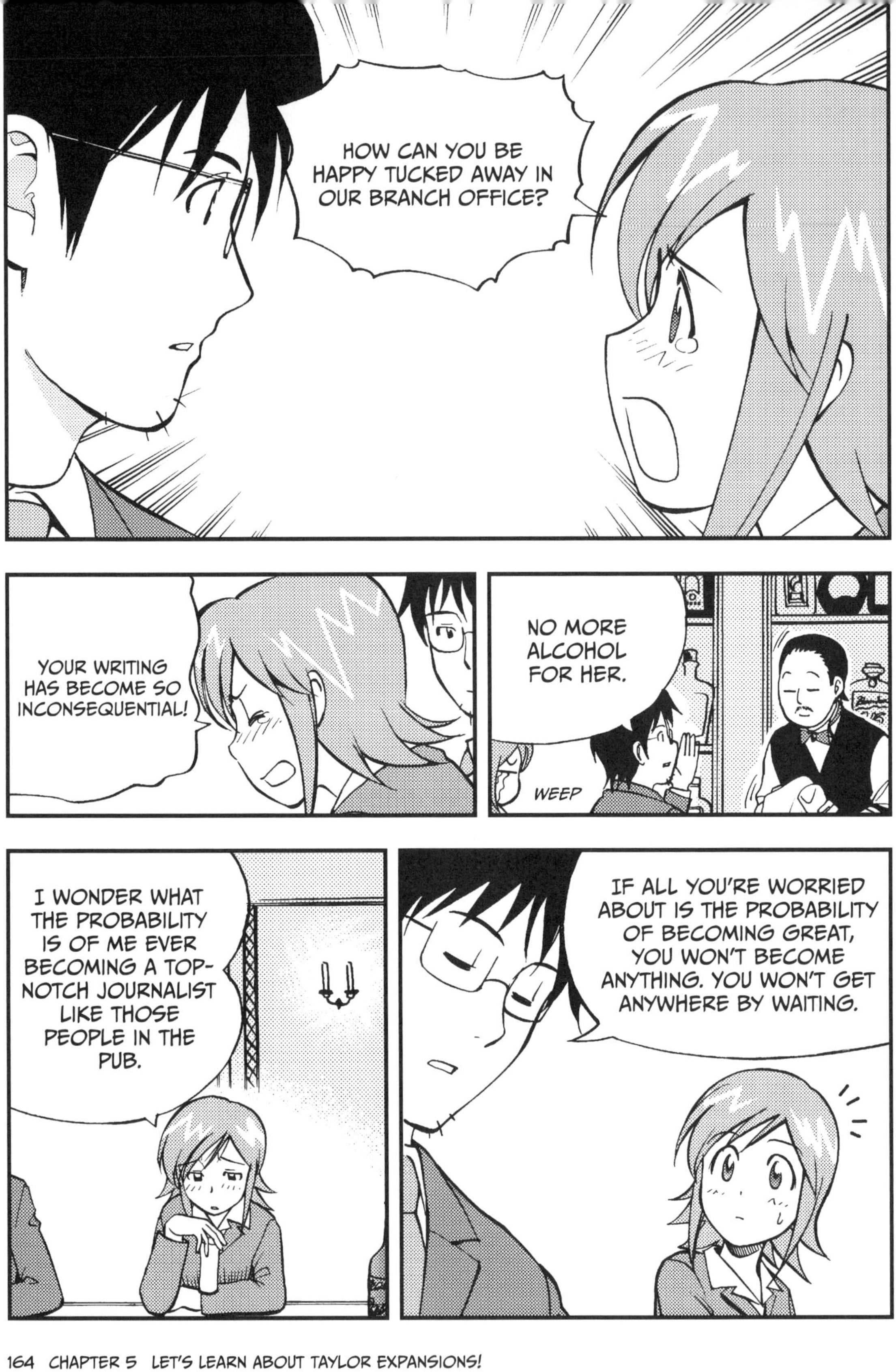
HOW CAN YOU BE HAPPY TUCKED AWAY IN OUR BRANCH OFFICE?
YOUR WRITING HAS BECOME SO INCONSEQUENTIAL!
NO MORE ALCOHOL FOR HER.
WEEP
I WONDER WHAT THE PROBABILITY IS OF ME EVER BECOMING A TOP-NOTCH JOURNALIST LIKE THOSE PEOPLE IN THE PUB.
IF ALL YOU'RE WORRIED ABOUT IS THE PROBABILITY OF BECOMING GREAT, YOU WON'T BECOME ANYTHING. YOU WON'T GET ANYWHERE BY WAITING.

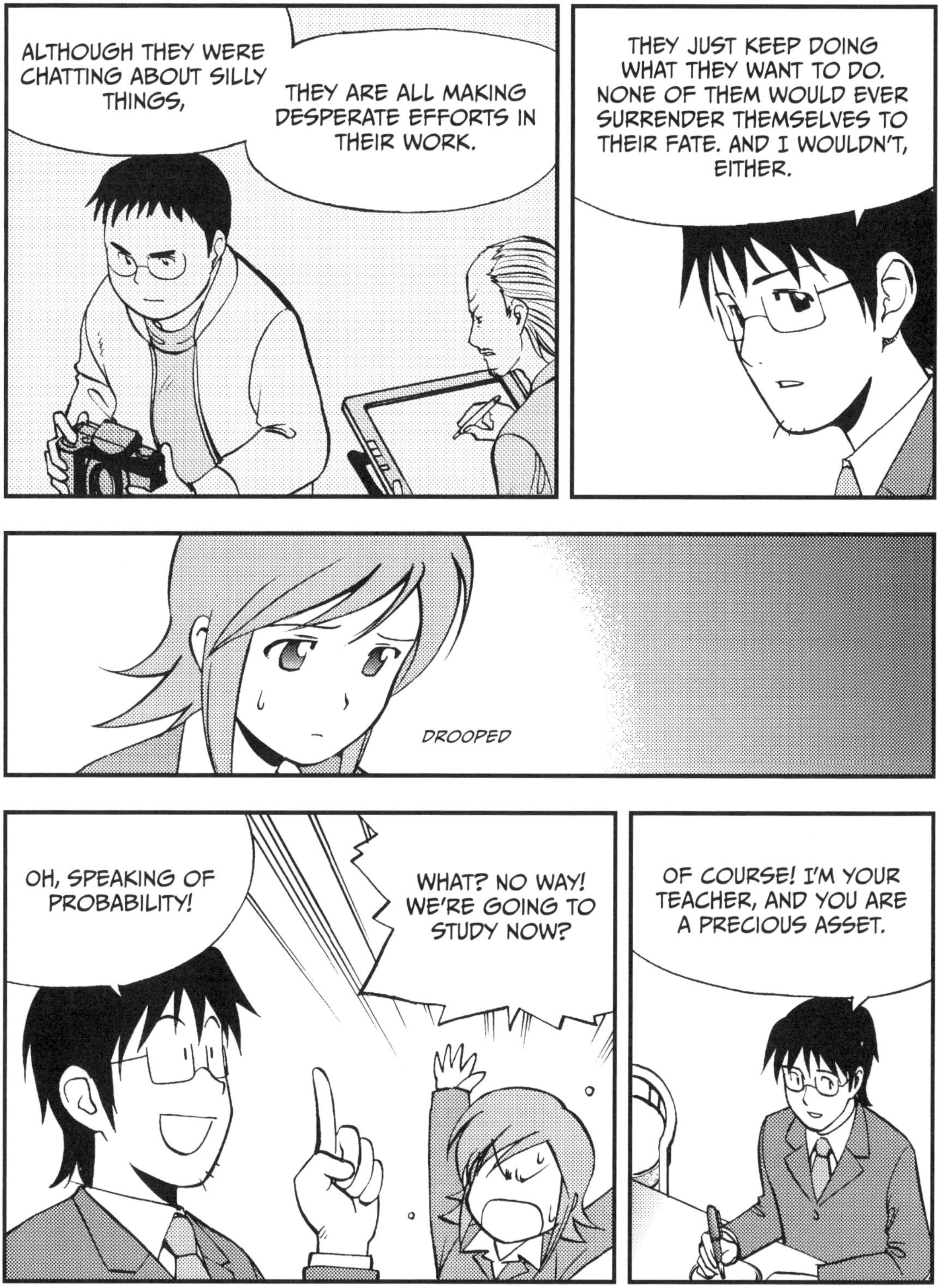
ALTHOUGH THEY WERE CHATTING ABOUT SILLY THINGS,
THEY ARE ALL MAKING DESPERATE EFFORTS IN THEIR WORK.
THEY JUST KEEP DOING WHAT THEY WANT TO DO. NONE OF THEM WOULD EVER SURRENDER THEMSELVES TO THEIR FATE. AND I WOULDN'T, EITHER.
DROOPED
OH, SPEAKING OF PROBABILITY!
WHAT? NO WAY! WE'RE GOING TO STUDY NOW?
OF COURSE! I'M YOUR TEACHER, AND YOU ARE A PRECIOUS ASSET.

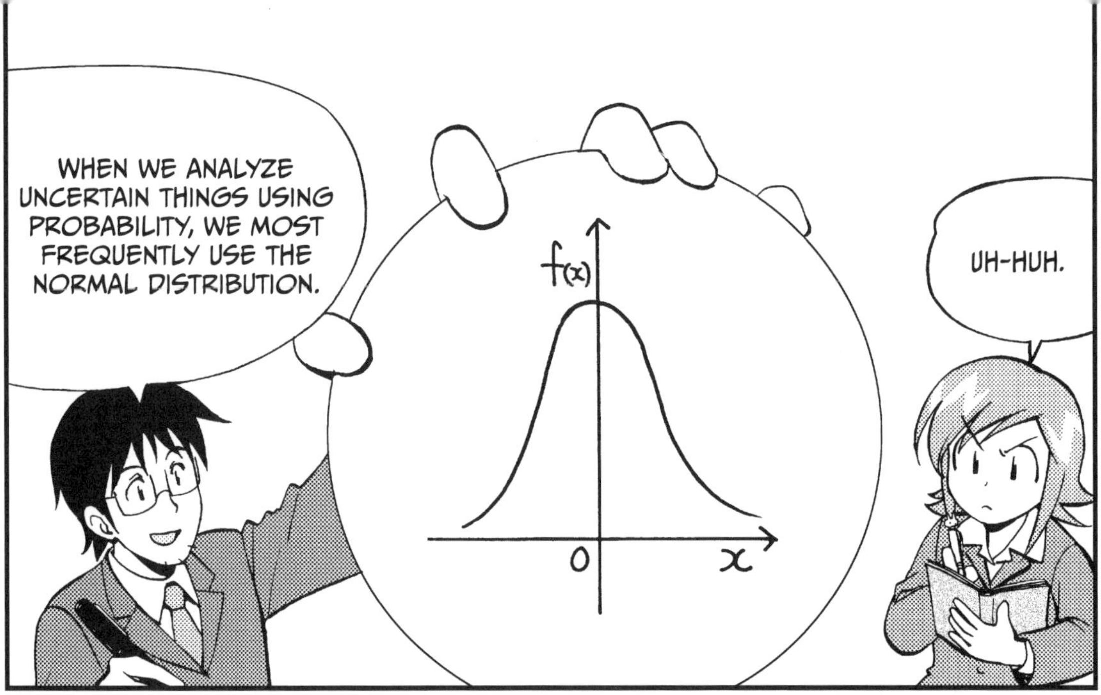
WHEN WE ANALYZE UNCERTAIN THINGS USING PROBABILITY, WE MOST FREQUENTLY USE THE NORMAL DISTRIBUTION.
f(x)
0
x
UH-HUH.

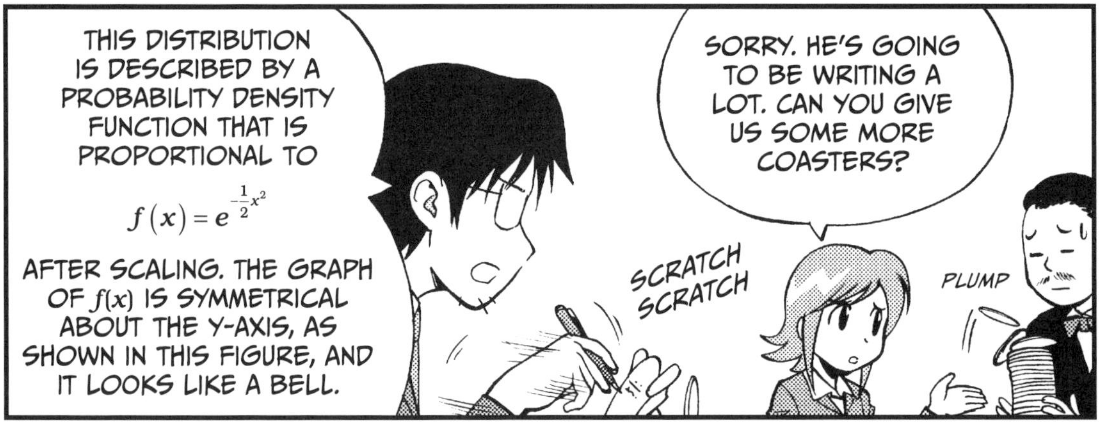
THIS DISTRIBUTION IS DESCRIBED BY A PROBABILITY DENSITY FUNCTION THAT IS PROPORTIONAL TO
$f(x) = e^{-\frac{1}{2}x^2}$
AFTER SCALING. THE GRAPH OF $f(x)$ IS SYMMETRICAL ABOUT THE Y-AXIS, AS SHOWN IN THIS FIGURE, AND IT LOOKS LIKE A BELL.
SCRATCH SCRATCH
SORRY. HE'S GOING TO BE WRITING A LOT. CAN YOU GIVE US SOME MORE COASTERS?
PLUMP

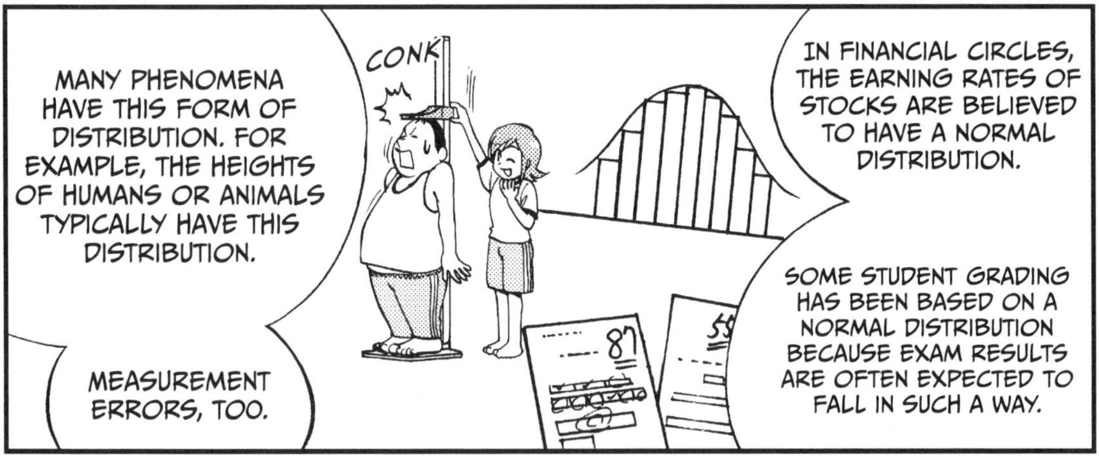
MANY PHENOMENA HAVE THIS FORM OF DISTRIBUTION. FOR EXAMPLE, THE HEIGHTS OF HUMANS OR ANIMALS TYPICALLY HAVE THIS DISTRIBUTION.
MEASUREMENT ERRORS, TOO.
CONK
IN FINANCIAL CIRCLES, THE EARNING RATES OF STOCKS ARE BELIEVED TO HAVE A NORMAL DISTRIBUTION.
SOME STUDENT GRADING HAS BEEN BASED ON A NORMAL DISTRIBUTION BECAUSE EXAM RESULTS ARE OFTEN EXPECTED TO FALL IN SUCH A WAY.

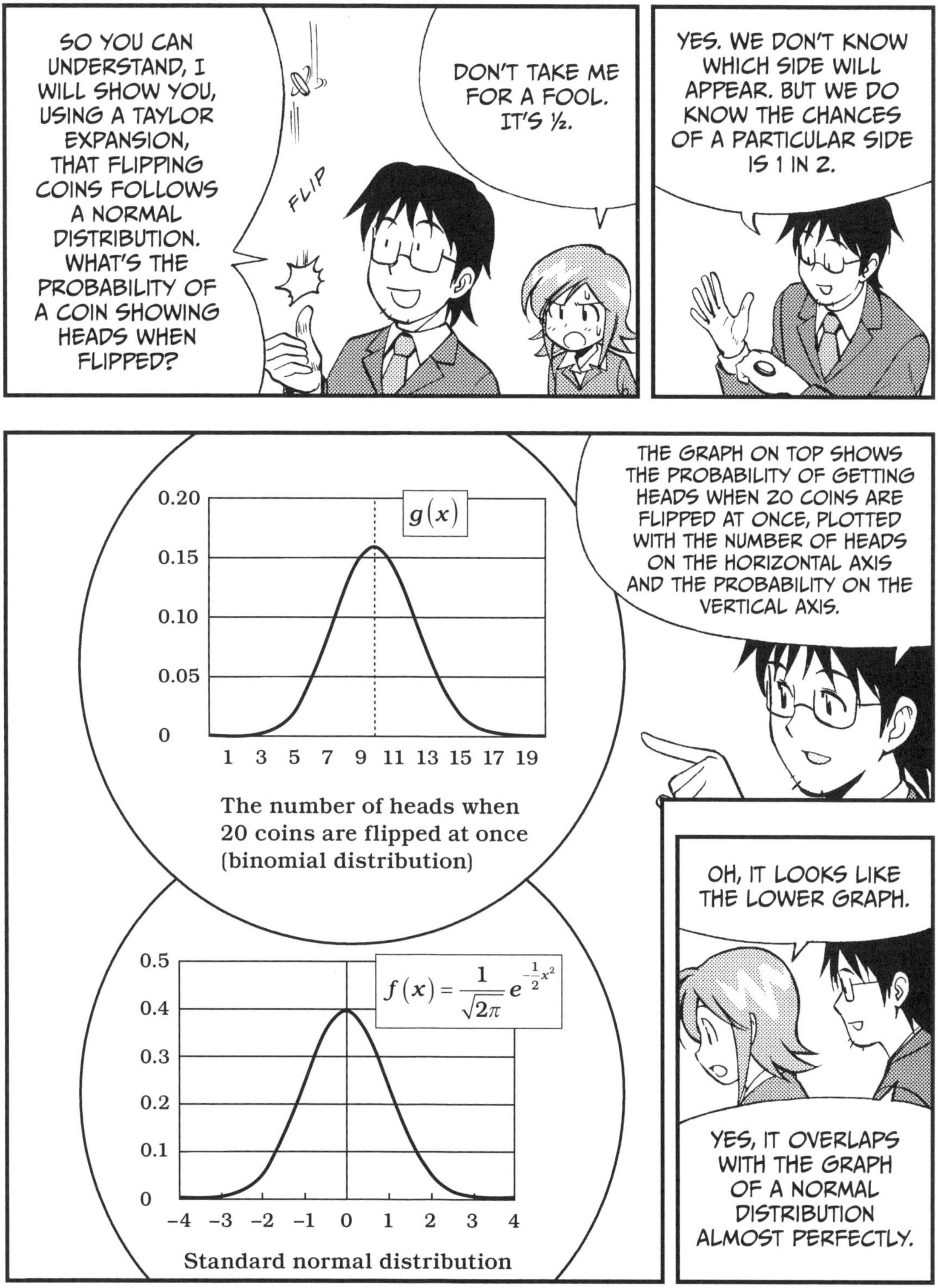
SO YOU CAN UNDERSTAND, I WILL SHOW YOU, USING A TAYLOR EXPANSION, THAT FLIPPING COINS FOLLOWS A NORMAL DISTRIBUTION. WHAT'S THE PROBABILITY OF A COIN SHOWING HEADS WHEN FLIPPED?
FLIP
DON'T TAKE ME FOR A FOOL. IT'S ½.
YES. WE DON'T KNOW WHICH SIDE WILL APPEAR. BUT WE DO KNOW THE CHANCES OF A PARTICULAR SIDE IS 1 IN 2.
$g(x)$
0.20
0.15
0.10
0.05
0
1 3 5 7 9 11 13 15 17 19
The number of heads when 20 coins are flipped at once (binomial distribution)
$f(x) = \frac{1}{\sqrt{2\pi}} e^{-\frac{1}{2}x^2}$
0.5
0.4
0.3
0.2
0.1
0
−4 −3 −2 −1 0 1 2 3 4
Standard normal distribution
THE GRAPH ON TOP SHOWS THE PROBABILITY OF GETTING HEADS WHEN 20 COINS ARE FLIPPED AT ONCE, PLOTTED WITH THE NUMBER OF HEADS ON THE HORIZONTAL AXIS AND THE PROBABILITY ON THE VERTICAL AXIS.
OH, IT LOOKS LIKE THE LOWER GRAPH.
YES, IT OVERLAPS WITH THE GRAPH OF A NORMAL DISTRIBUTION ALMOST PERFECTLY.

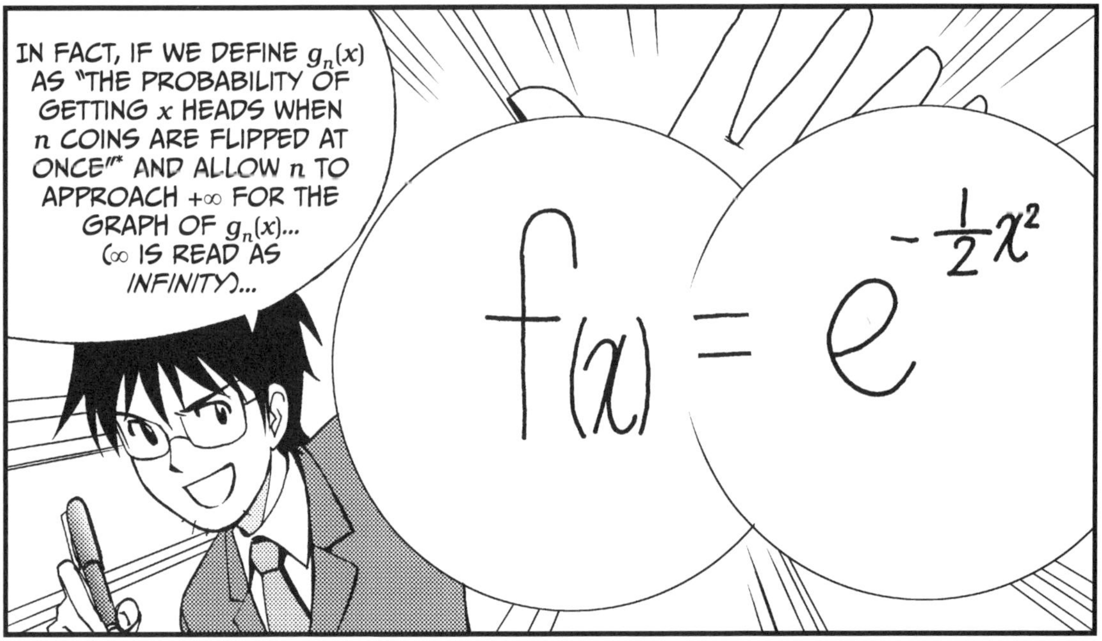
IN FACT, IF WE DEFINE $g_n(x)$ AS "THE PROBABILITY OF GETTING x HEADS WHEN n COINS ARE FLIPPED AT ONCE"* AND ALLOW n TO APPROACH $+\infty$ FOR THE GRAPH OF $g_n(x)$... (∞ IS READ AS *INFINITY*)...
$f(x) = e^{-\frac{1}{2}x^2}$

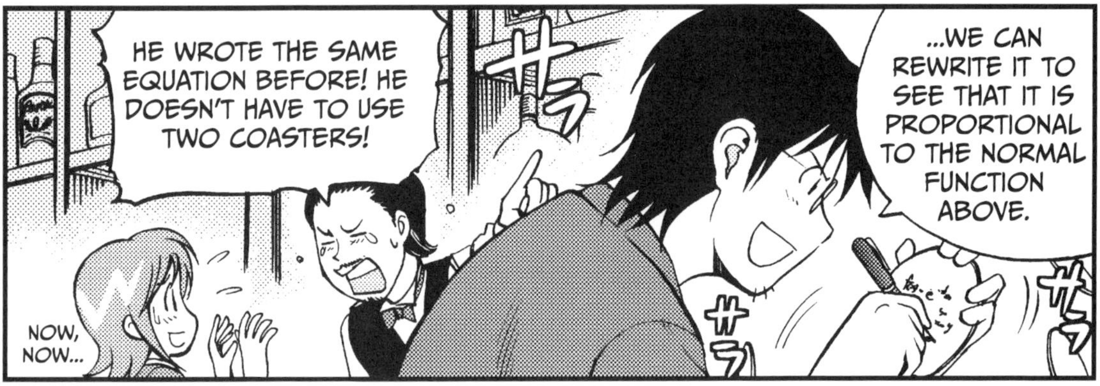
HE WROTE THE SAME EQUATION BEFORE! HE DOESN'T HAVE TO USE TWO COASTERS!
...WE CAN REWRITE IT TO SEE THAT IT IS PROPORTIONAL TO THE NORMAL FUNCTION ABOVE.
NOW, NOW...

* The distribution of such probabilities as that of getting x heads when n coins are flipped is generally called the *binomial distribution*.

For example, let's find the probability of getting 3 heads when 5 coins are flipped. The probability of getting HHTHT (H: heads, T: tails) is

$$\frac{1}{2}\times\frac{1}{2}\times\frac{1}{2}\times\frac{1}{2}\times\frac{1}{2}=\left(\frac{1}{2}\right)^5$$

Since there are ${}_5C_3$ ways of getting 3 heads and 2 tails, it is ${}_5C_3\left(\frac{1}{2}\right)^5$. The general expression is ${}_nC_x\left(\frac{1}{2}\right)^n$. We will show that if n is very large, the binomial distribution is the normal distribution.

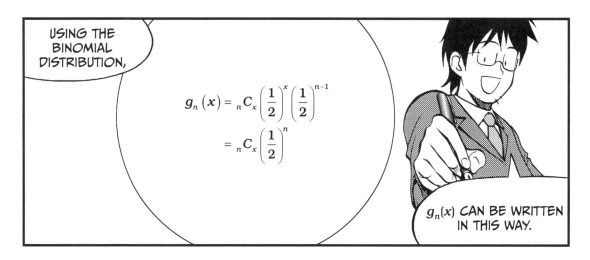

USING THE BINOMIAL DISTRIBUTION,

$$g_n(x) = {}_nC_x \left(\frac{1}{2}\right)^x \left(\frac{1}{2}\right)^{n-1}$$
$$= {}_nC_x \left(\frac{1}{2}\right)^n$$

$g_n(x)$ CAN BE WRITTEN IN THIS WAY.

SINCE THE GRAPH OF $f(x)$ IS SYMMETRICAL ABOUT $x = 0$ AND $g_n(x)$ ABOUT $x = \frac{1}{2}$ ...

SCRIBBLE

SCRIBBLE

AH, SO MANY COASTERS...

WE CONSIDER $g_n(\frac{n}{2})$ INSTEAD OF $g_n(x)$.

FIRST...

$$g_n\left(\frac{n}{2}\right) = {}_nC_{\frac{n}{2}} \left(\frac{1}{2}\right)^n$$

DIVIDING $g_n(x)$ BY THIS...

$$h_n(x) = \frac{g_n(x)}{g_n\left(\frac{n}{2}\right)} = \frac{{}_nC_x}{{}_nC_{\frac{n}{2}}}$$

WE GET h_n, THE SCALED FUNCTION

SINCE

$$_nC_x = \frac{n!}{x!(n-x)!}$$

SO THEN...

$$_nC_{\frac{n}{2}} = \frac{n!}{\left(\frac{n}{2}\right)!\left(\frac{n}{2}\right)!}$$

DIVIDE:

$$h_n(x) = \left(\frac{n!}{x!(n-x)!}\right) \times \left(\frac{\left(\frac{n}{2}\right)!\left(\frac{n}{2}\right)!}{n!}\right) = \frac{\left(\frac{n}{2}\right)!\left(\frac{n}{2}\right)!}{x!(n-x)!}$$

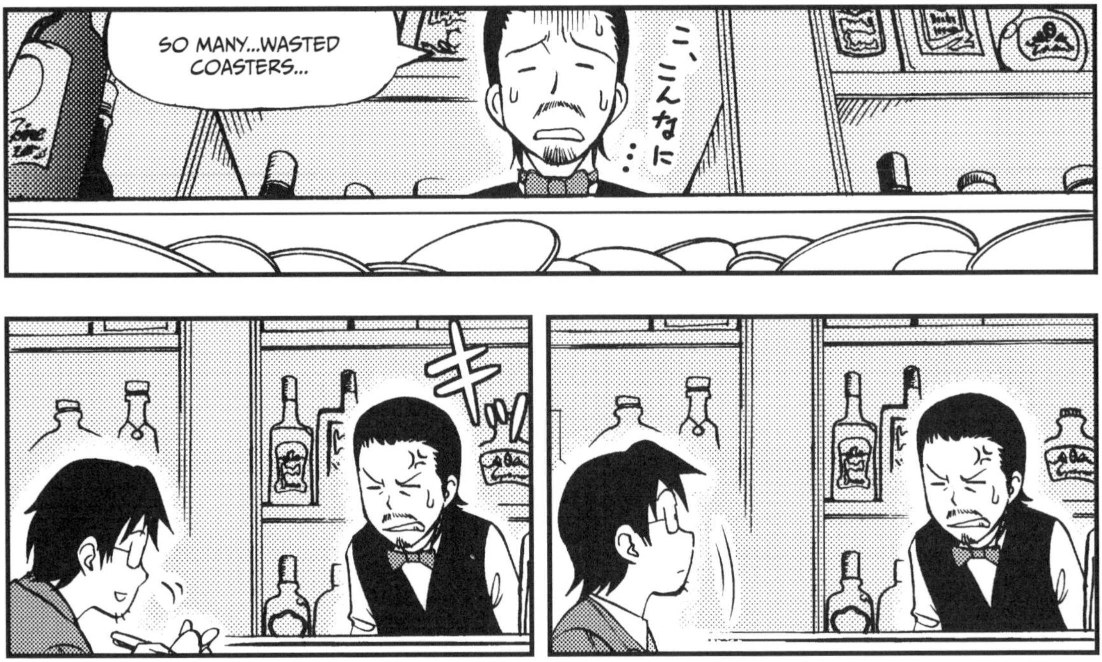
SO MANY...WASTED COASTERS...
こ、こんなに…

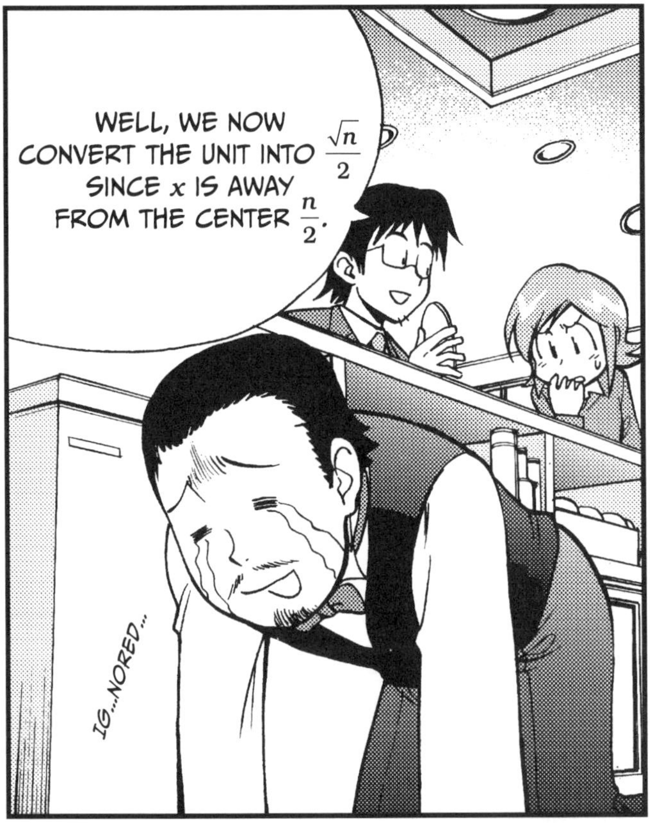
WELL, WE NOW CONVERT THE UNIT INTO $\frac{\sqrt{n}}{2}$ SINCE x IS AWAY FROM THE CENTER $\frac{n}{2}$.
IG...NORED...

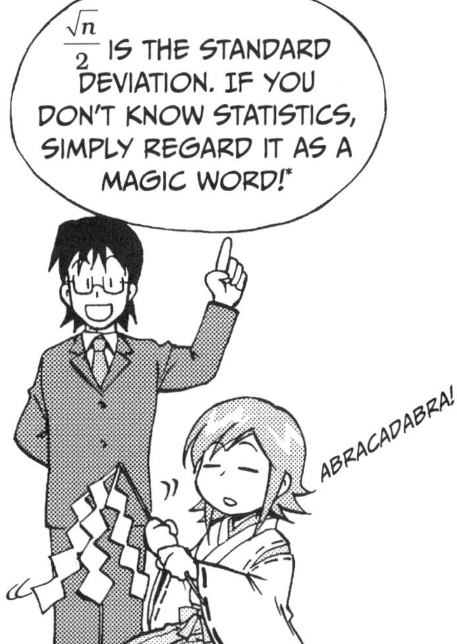
$\frac{\sqrt{n}}{2}$ IS THE STANDARD DEVIATION. IF YOU DON'T KNOW STATISTICS, SIMPLY REGARD IT AS A MAGIC WORD!*
ABRACADABRA!

* STANDARD DEVIATION IS AN INDEX WE USE TO DESCRIBE THE SCATTERING OF DATA.

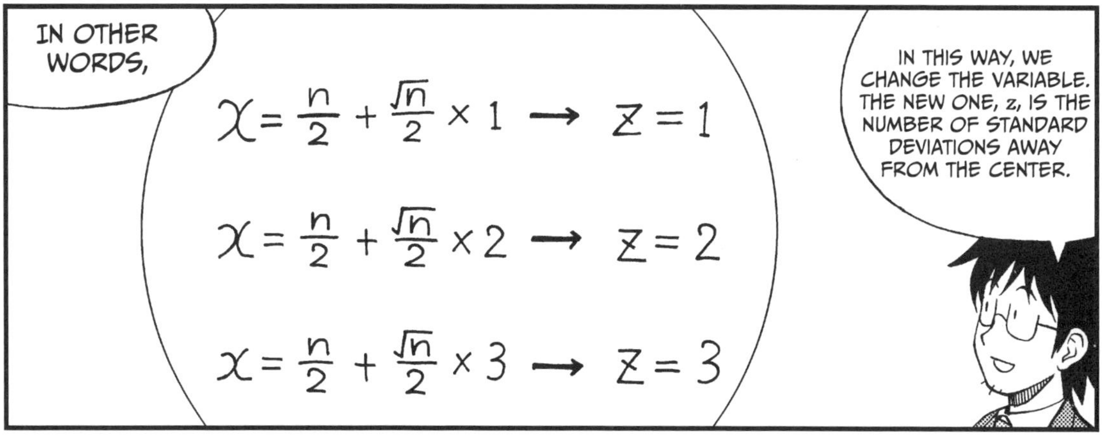
IN OTHER WORDS,
$x = \frac{n}{2} + \frac{\sqrt{n}}{2} \times 1 \rightarrow z = 1$
$x = \frac{n}{2} + \frac{\sqrt{n}}{2} \times 2 \rightarrow z = 2$
$x = \frac{n}{2} + \frac{\sqrt{n}}{2} \times 3 \rightarrow z = 3$
IN THIS WAY, WE CHANGE THE VARIABLE. THE NEW ONE, z, IS THE NUMBER OF STANDARD DEVIATIONS AWAY FROM THE CENTER.

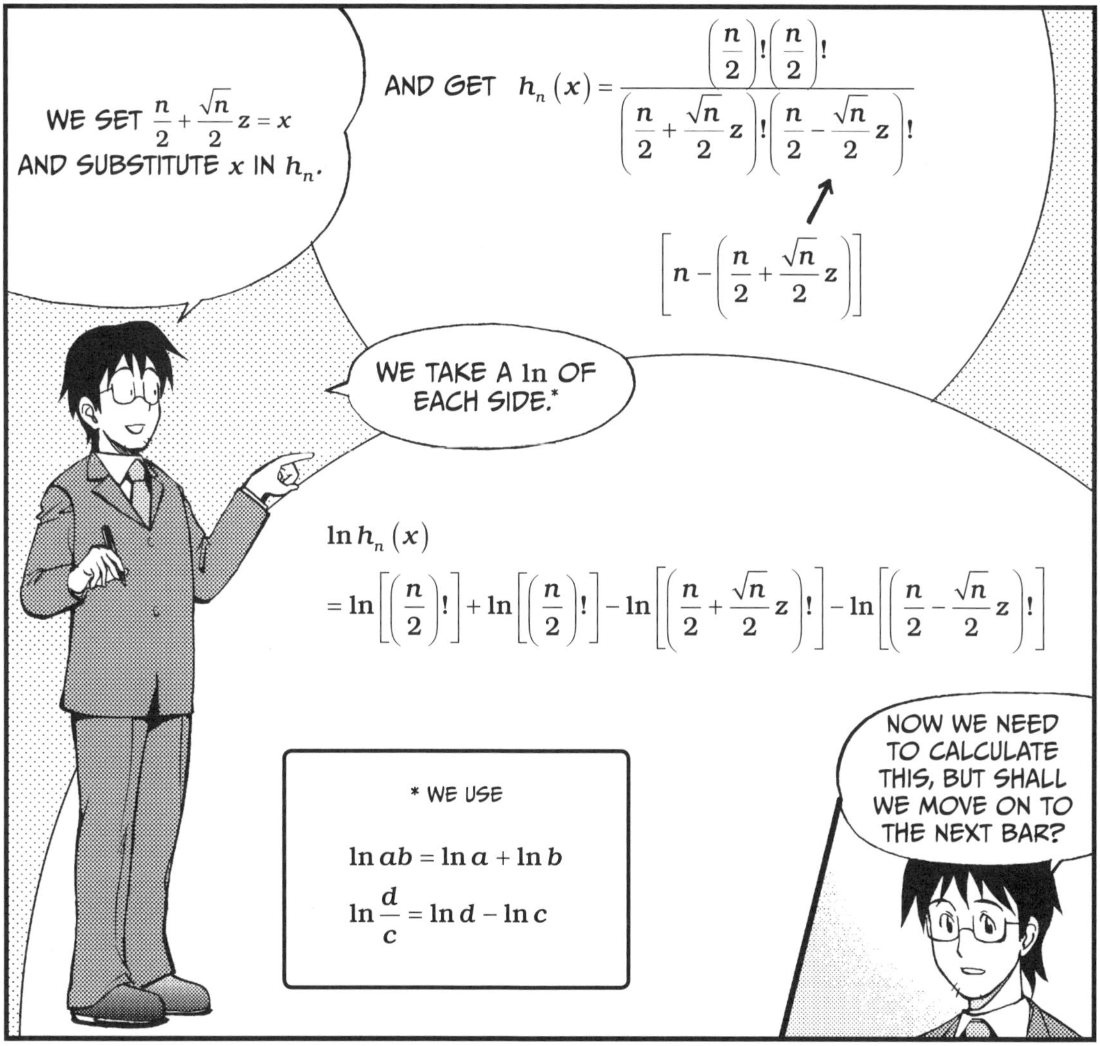
WE SET $\frac{n}{2} + \frac{\sqrt{n}}{2} z = x$ AND SUBSTITUTE x IN h_n.
AND GET $h_n(x) = \dfrac{\left(\frac{n}{2}\right)!\left(\frac{n}{2}\right)!}{\left(\frac{n}{2} + \frac{\sqrt{n}}{2} z\right)!\left(\frac{n}{2} - \frac{\sqrt{n}}{2} z\right)!}$
$\left[n - \left(\frac{n}{2} + \frac{\sqrt{n}}{2} z\right)\right]$
WE TAKE A ln OF EACH SIDE.*
$\ln h_n(x)$
$= \ln\left[\left(\frac{n}{2}\right)!\right] + \ln\left[\left(\frac{n}{2}\right)!\right] - \ln\left[\left(\frac{n}{2} + \frac{\sqrt{n}}{2} z\right)!\right] - \ln\left[\left(\frac{n}{2} - \frac{\sqrt{n}}{2} z\right)!\right]$
* WE USE
$\ln ab = \ln a + \ln b$
$\ln \frac{d}{c} = \ln d - \ln c$
NOW WE NEED TO CALCULATE THIS, BUT SHALL WE MOVE ON TO THE NEXT BAR?

THANK YOU. I THINK WE'RE DONE.

PHEW

THERE ARE ONLY A FEW LEFT!!

READY TO GO?

I GUESS I SHOULD BE HAPPY I STILL HAVE SOME...

POSITIVE THINKING

Approximating $\ln(m!)$

$$\ln m! = \ln 1 + \ln 2 + \ln 3 + \ldots + \ln m$$

If we pack rectangles in the graph of $\ln x$, as shown here, we get

$$\ln 2 + \ldots + \ln m \approx \int_1^m \ln x\, dx$$

$$\left(x \ln x - x\right)' = \ln x + x \times \frac{1}{x} - 1 = \ln x$$

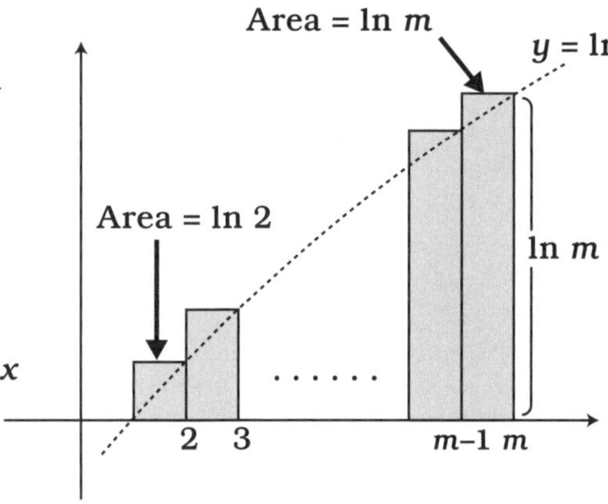

Area = $\ln m$

$y = \ln x$

Area = $\ln 2$

$\ln m$

2 3 $\ldots\ldots$ $m-1$ m

Thus,

$$\int_1^m \ln x\, dx = \left(m \ln m - m\right) - \left(1 \ln 1 - 1\right)$$

$$= m \ln m - m + 1$$

Since we will use this where m is very large, $m \ln m$ is the important term. $-m + 1$ is much smaller, so we'll ignore it. Therefore, we can consider roughly that $\ln m! = m \ln m$.

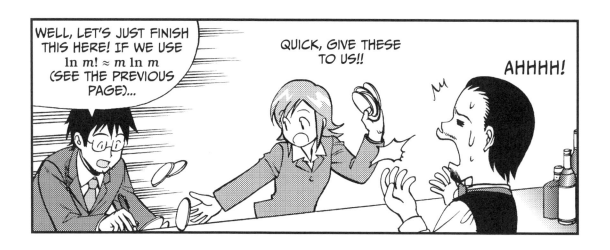

WELL, LET'S JUST FINISH THIS HERE! IF WE USE $\ln m! \approx m \ln m$ (SEE THE PREVIOUS PAGE)...
QUICK, GIVE THESE TO US!!
AHHHH!

$$\ln h_n(x) \approx \frac{n}{2}\ln\frac{n}{2} + \frac{n}{2}\ln\frac{n}{2} - \left(\frac{n}{2} + \frac{\sqrt{n}}{2}z\right)\ln\left(\frac{n}{2} + \frac{\sqrt{n}}{2}z\right) - \left(\frac{n}{2} - \frac{\sqrt{n}}{2}z\right)\ln\left(\frac{n}{2} - \frac{\sqrt{n}}{2}z\right)$$

AFTER A LOT OF ALGEBRA, WE GET

$$\ln h_n(x) \approx -\left[\left(\frac{n}{2} + \frac{\sqrt{n}}{2}z\right)\ln\left(1 + \frac{\sqrt{n}}{n}z\right) + \left(\frac{n}{2} - \frac{\sqrt{n}}{2}z\right)\ln\left(1 - \frac{\sqrt{n}}{n}z\right)\right]$$

WE USED, E.G., $\ln\left(\frac{n}{2} + \frac{\sqrt{n}}{2}z\right) = \ln\left\{\frac{n}{2}\left(1 + \frac{\sqrt{n}}{n}z\right)\right\} = \ln\frac{n}{2} + \ln\left(1 + \frac{\sqrt{n}}{n}z\right)$

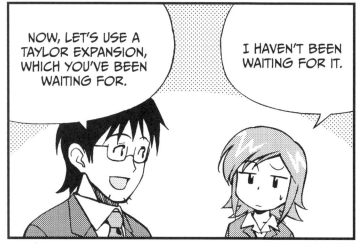

NOW, LET'S USE A TAYLOR EXPANSION, WHICH YOU'VE BEEN WAITING FOR.
I HAVEN'T BEEN WAITING FOR IT.

...
SQUEAK SQUEAK
JUST TAKE THEM

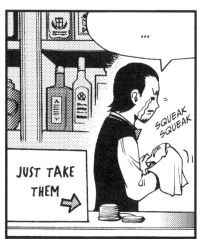

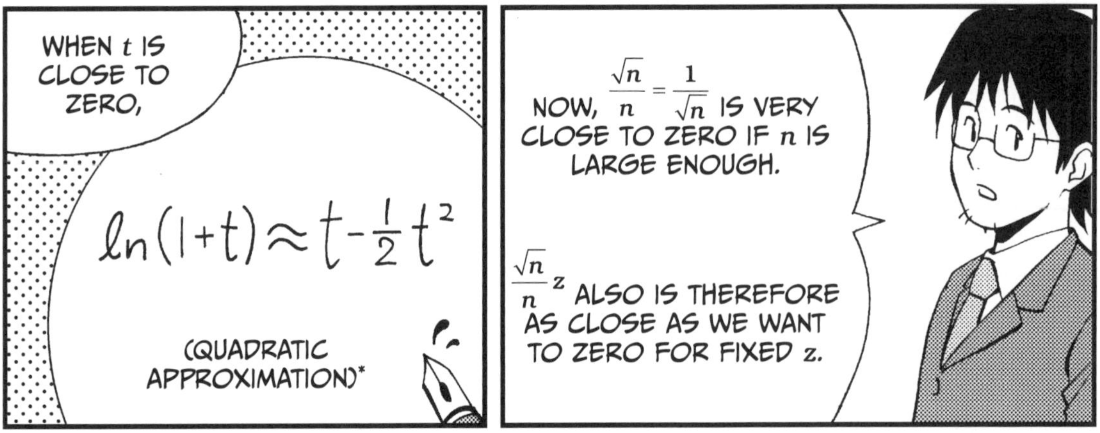
WHEN t IS CLOSE TO ZERO,
$\ln(1+t) \approx t - \frac{1}{2}t^2$
(QUADRATIC APPROXIMATION)*
NOW, $\frac{\sqrt{n}}{n} = \frac{1}{\sqrt{n}}$ IS VERY CLOSE TO ZERO IF n IS LARGE ENOUGH.
$\frac{\sqrt{n}}{n}z$ ALSO IS THEREFORE AS CLOSE AS WE WANT TO ZERO FOR FIXED z.

* SEE PAGE 161.

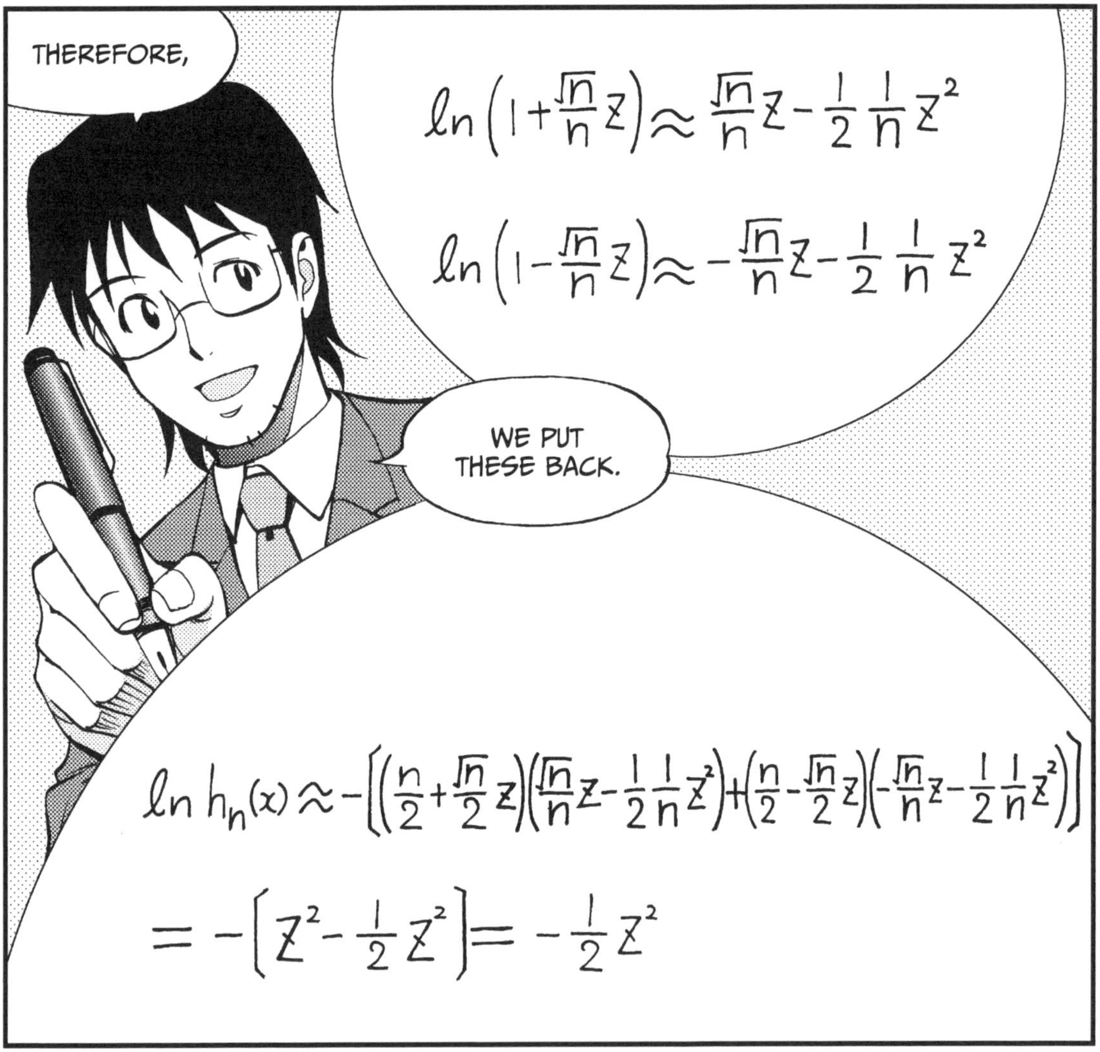
THEREFORE,
$\ln\left(1+\frac{\sqrt{n}}{n}z\right) \approx \frac{\sqrt{n}}{n}z - \frac{1}{2}\frac{1}{n}z^2$
$\ln\left(1-\frac{\sqrt{n}}{n}z\right) \approx -\frac{\sqrt{n}}{n}z - \frac{1}{2}\frac{1}{n}z^2$
WE PUT THESE BACK.
$\ln h_n(x) \approx -\left[\left(\frac{n}{2}+\frac{\sqrt{n}}{2}z\right)\left(\frac{\sqrt{n}}{n}z - \frac{1}{2}\frac{1}{n}z^2\right) + \left(\frac{n}{2}-\frac{\sqrt{n}}{2}z\right)\left(-\frac{\sqrt{n}}{n}z - \frac{1}{2}\frac{1}{n}z^2\right)\right]$
$= -\left[z^2 - \frac{1}{2}z^2\right] = -\frac{1}{2}z^2$

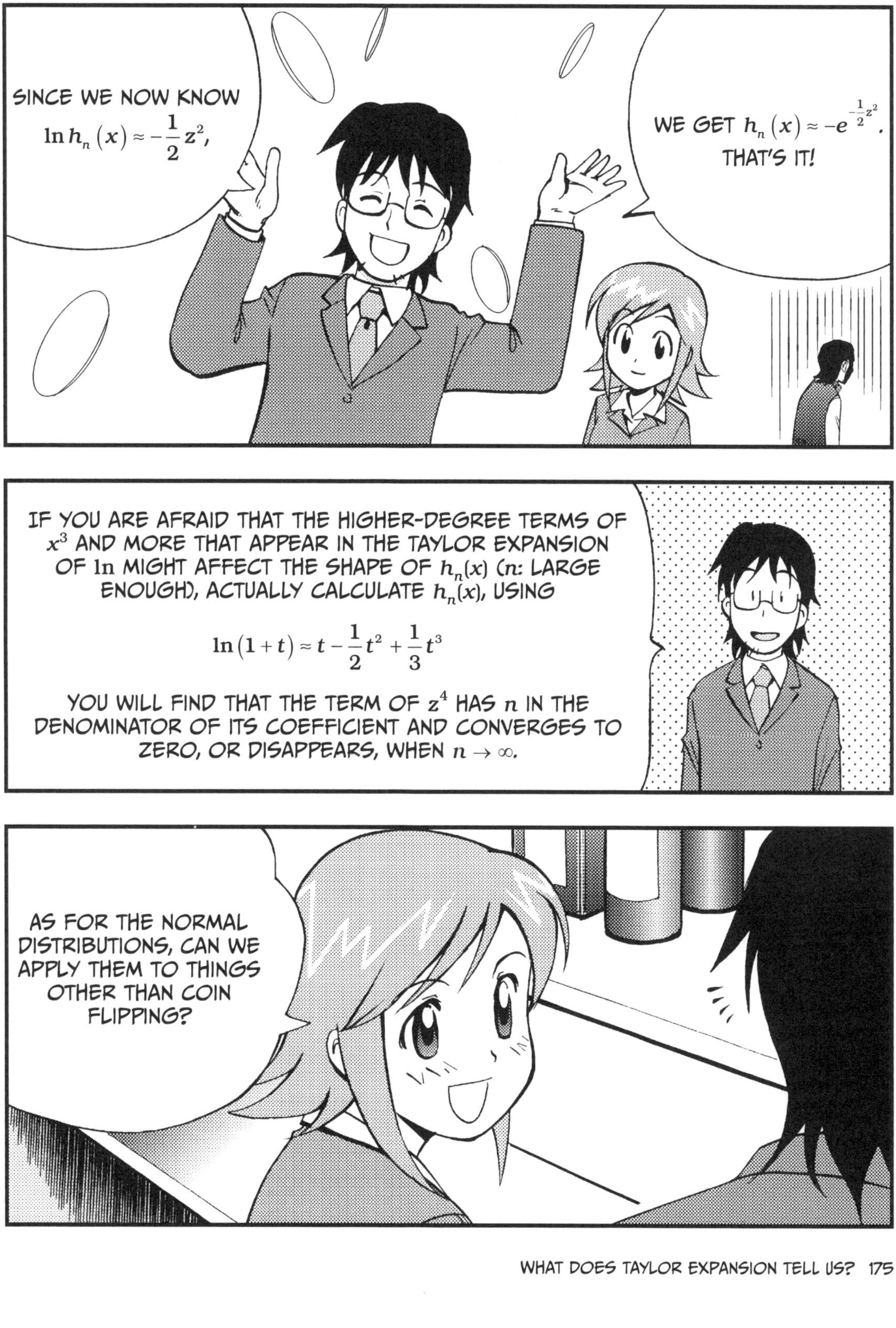
SINCE WE NOW KNOW $\ln h_n(x) \approx -\frac{1}{2}z^2$,
WE GET $h_n(x) \approx -e^{-\frac{1}{2}z^2}$. THAT'S IT!
IF YOU ARE AFRAID THAT THE HIGHER-DEGREE TERMS OF x^3 AND MORE THAT APPEAR IN THE TAYLOR EXPANSION OF ln MIGHT AFFECT THE SHAPE OF $h_n(x)$ (n: LARGE ENOUGH), ACTUALLY CALCULATE $h_n(x)$, USING
$\ln(1+t) \approx t - \frac{1}{2}t^2 + \frac{1}{3}t^3$
YOU WILL FIND THAT THE TERM OF z^4 HAS n IN THE DENOMINATOR OF ITS COEFFICIENT AND CONVERGES TO ZERO, OR DISAPPEARS, WHEN $n \to \infty$.
AS FOR THE NORMAL DISTRIBUTIONS, CAN WE APPLY THEM TO THINGS OTHER THAN COIN FLIPPING?

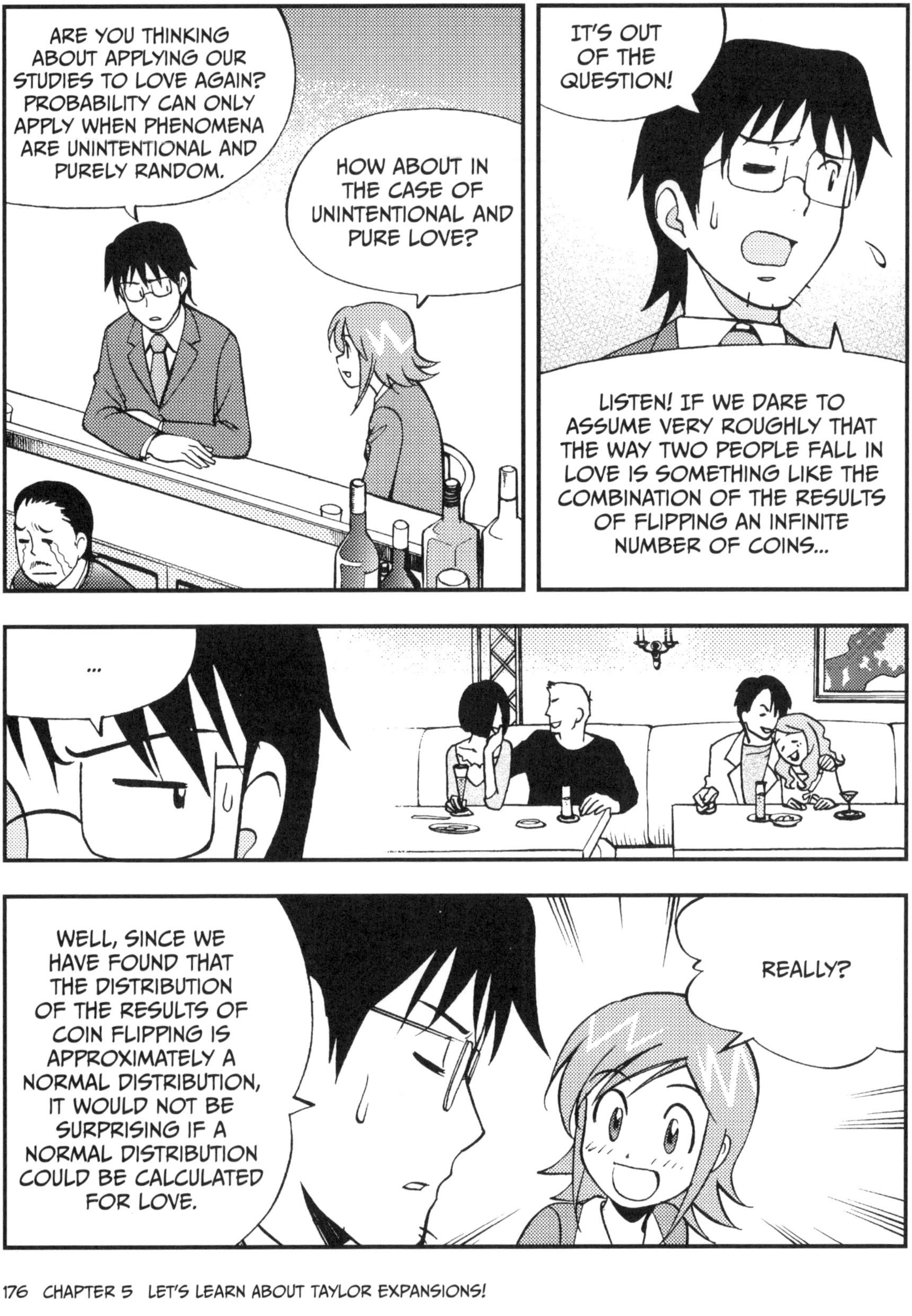
ARE YOU THINKING ABOUT APPLYING OUR STUDIES TO LOVE AGAIN? PROBABILITY CAN ONLY APPLY WHEN PHENOMENA ARE UNINTENTIONAL AND PURELY RANDOM.
HOW ABOUT IN THE CASE OF UNINTENTIONAL AND PURE LOVE?
IT'S OUT OF THE QUESTION!
LISTEN! IF WE DARE TO ASSUME VERY ROUGHLY THAT THE WAY TWO PEOPLE FALL IN LOVE IS SOMETHING LIKE THE COMBINATION OF THE RESULTS OF FLIPPING AN INFINITE NUMBER OF COINS...
...
WELL, SINCE WE HAVE FOUND THAT THE DISTRIBUTION OF THE RESULTS OF COIN FLIPPING IS APPROXIMATELY A NORMAL DISTRIBUTION, IT WOULD NOT BE SURPRISING IF A NORMAL DISTRIBUTION COULD BE CALCULATED FOR LOVE.
REALLY?

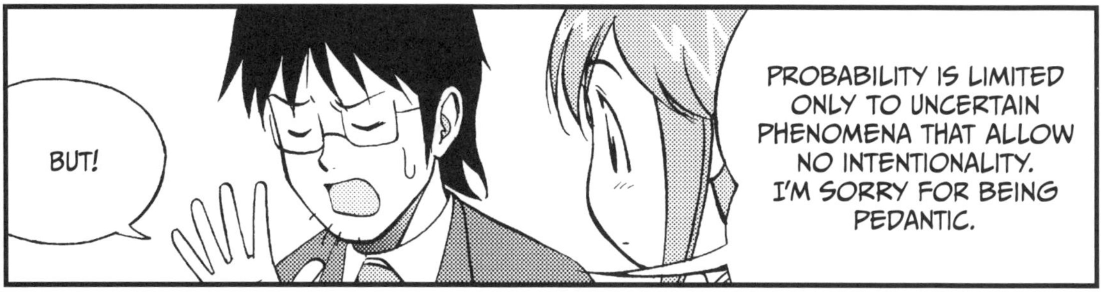
BUT!
PROBABILITY IS LIMITED ONLY TO UNCERTAIN PHENOMENA THAT ALLOW NO INTENTIONALITY. I'M SORRY FOR BEING PEDANTIC.

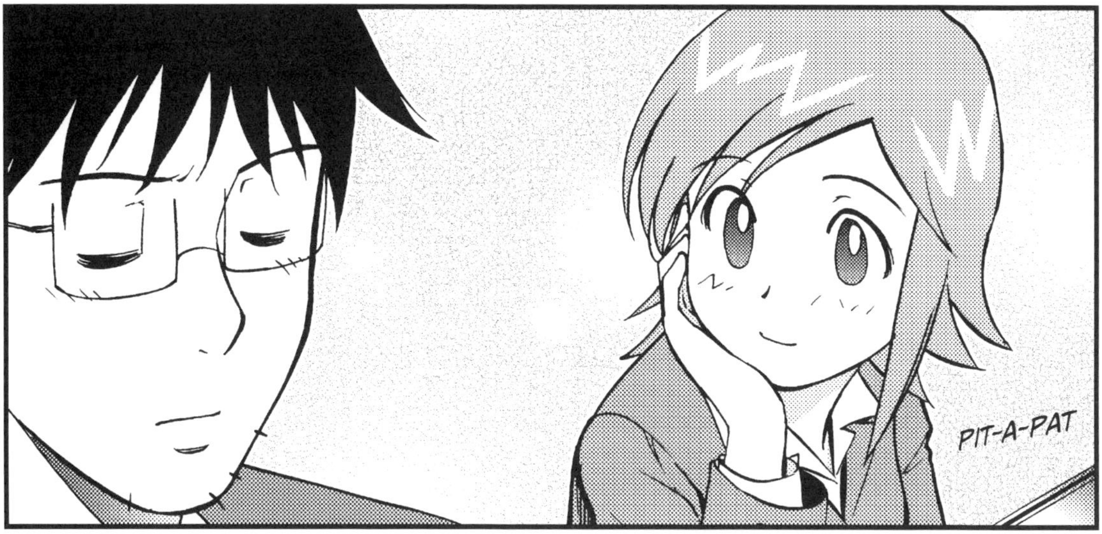
PIT-A-PAT

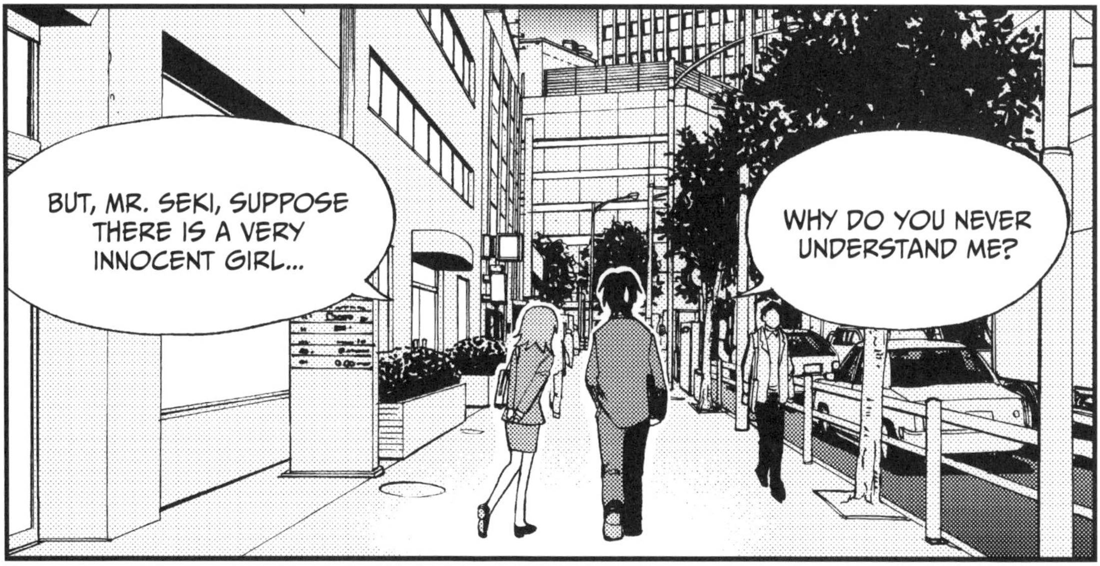
BUT, MR. SEKI, SUPPOSE THERE IS A VERY INNOCENT GIRL...
WHY DO YOU NEVER UNDERSTAND ME?

EXERCISES

1. Obtain the Taylor expansion of $f(x) = e^{-x}$ at $x = 0$.

2. Obtain the quadratic approximation of $f(x) = \dfrac{1}{\cos x}$ at $x = 0$.

3. Derive for yourself the formula for the Taylor expansion of $f(x)$ centered at $x = 1$, which is given on page 159. In other words, work out what c_n must be in the equation:

$$f(x) = c_0 + c_1(x - a) + c_2(x - a)^2 + \ldots + c_n(x - a)^n$$

6

LET'S LEARN ABOUT PARTIAL DIFFERENTIATION!

WHAT ARE MULTIVARIABLE FUNCTIONS?

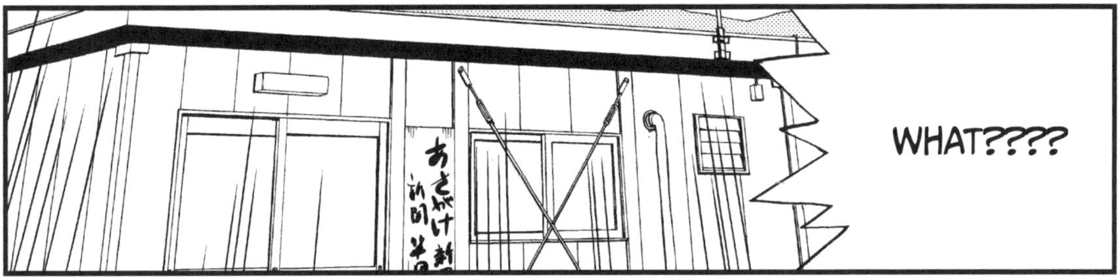
WHAT????

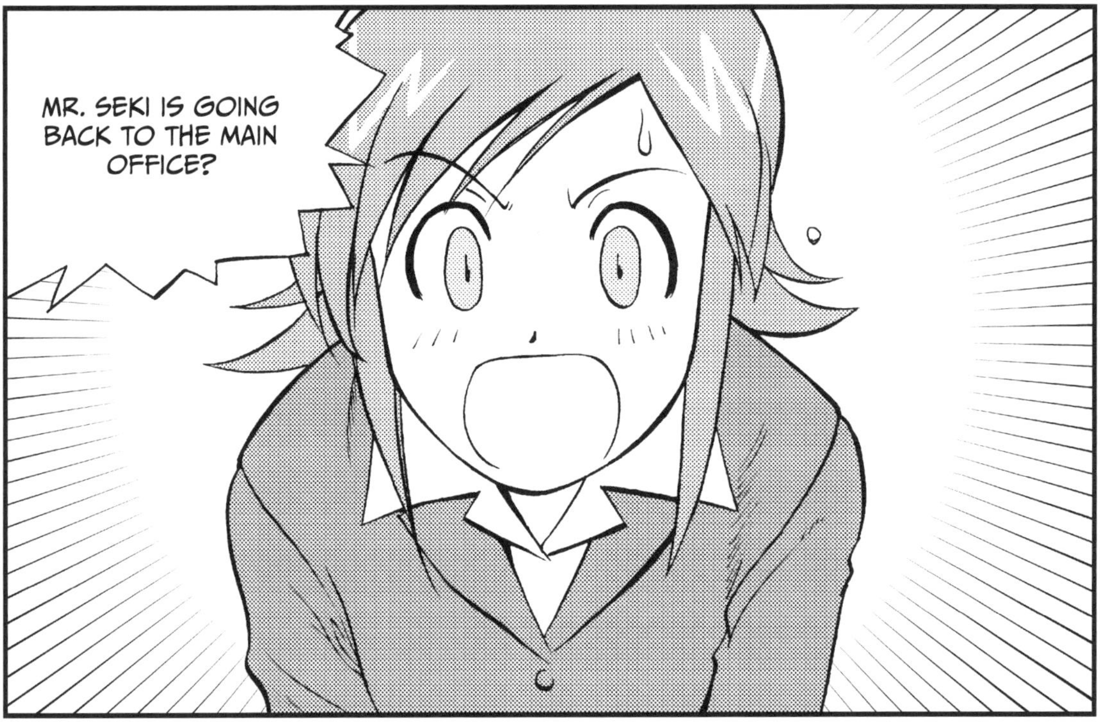
MR. SEKI IS GOING BACK TO THE MAIN OFFICE?

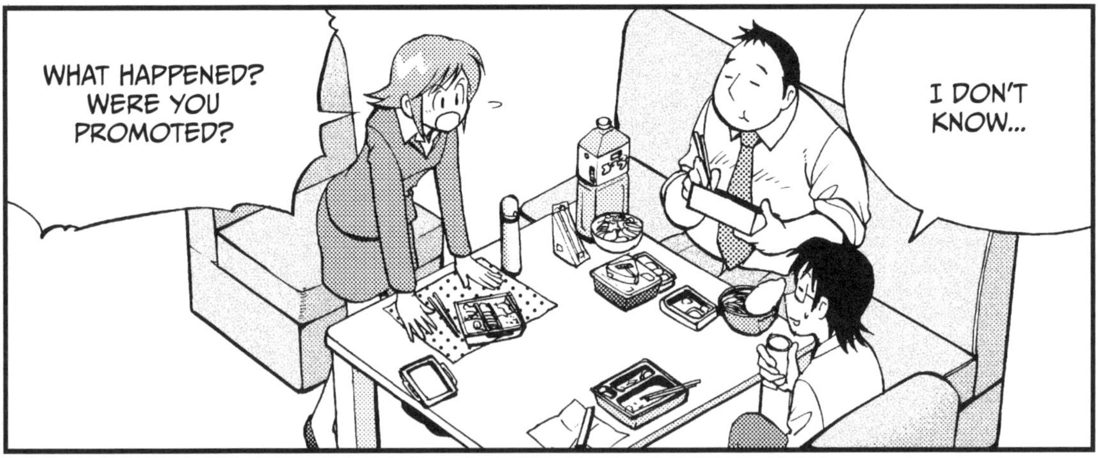
WHAT HAPPENED? WERE YOU PROMOTED?
I DON'T KNOW...

BUT YOU TOLD ME, "AN EFFECT OCCURS BECAUSE IT HAS A CAUSE."

YOU'VE BEEN TEACHING ME EVERY DAY! I EVEN HAD NIGHTMARES ABOUT IT!

CAUSE AND EFFECT...I REMEMBER THAT. WE TALKED ABOUT THAT IN ONE OF OUR FIRST LESSONS.

IT'S TRUE THAT WE HAVE BEEN EXPLORING SIMPLE FUNCTIONS THAT HAVE A CAUSE AND AN EFFECT.

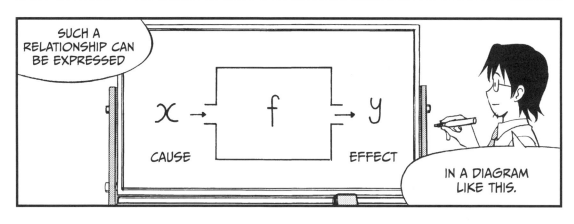

SUCH A RELATIONSHIP CAN BE EXPRESSED

$x \rightarrow$ f $\rightarrow y$

CAUSE

EFFECT

IN A DIAGRAM LIKE THIS.

BUT THIS TRANSFER HAS REMINDED ME THAT THE WORLD IS NOT SO SIMPLE, AFTER ALL.

YES.

SLURP

I GUESS MY TRANSFER TO THE MAIN OFFICE HAS BEEN BROUGHT ABOUT AS A COMBINED RESULT OF SEVERAL CAUSES.

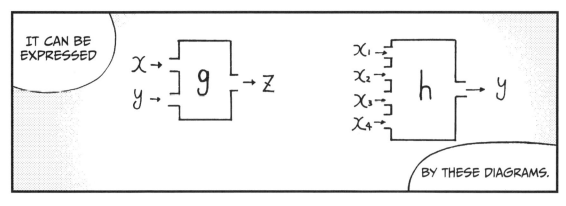

IT CAN BE EXPRESSED

$x \rightarrow$ | g | $\rightarrow z$
$y \rightarrow$

$x_1 \rightarrow$
$x_2 \rightarrow$ | h | $\rightarrow y$
$x_3 \rightarrow$
$x_4 \rightarrow$

BY THESE DIAGRAMS.

IN THE CASE OF MR. SEKI, x IS EXCELLENT WRITING, y IS HARD-HITTING REPORTING, AND z IS TRANSFER TO THE MAIN OFFICE. IS THAT RIGHT?

WELL, I DON'T KNOW THE REASONS FOR MY TRANSFER YET.

IN THE CASE OF NORIKO, x_1 IS LAST MONTH'S BLUNDER, x_2 IS THIS MONTH'S BLUNDER, AND x_3 AND x_4 ARE POOR GROOMING AND HYGIENE, WHICH MAKES y HER DEMOTION TO WRITING OBITUARIES.

SHUT UP, YOU DUMB OX!

SQUEEZE

ALL RIGHT, THAT'S ENOUGH. NORIKO, WE DON'T HAVE MUCH TIME LEFT.

LET'S LEARN THE BASICS QUICKLY.

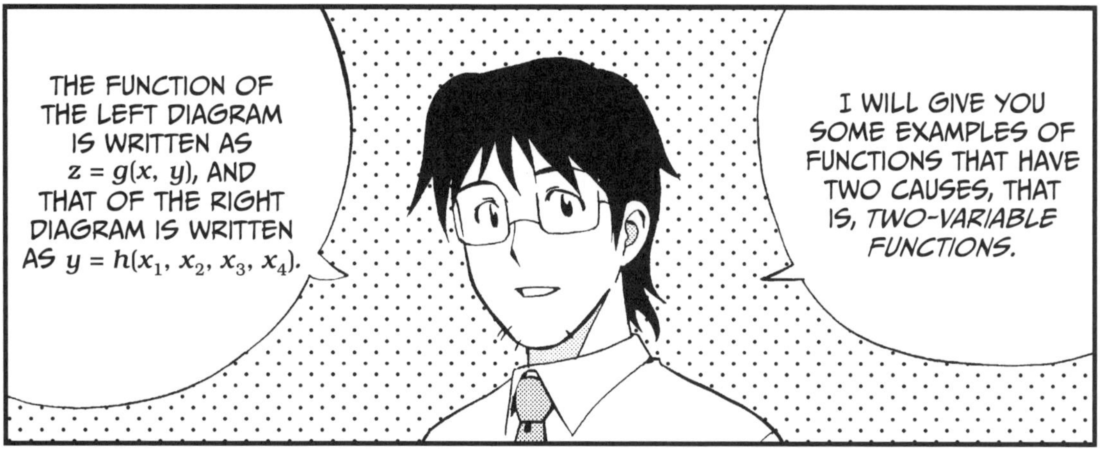
THE FUNCTION OF THE LEFT DIAGRAM IS WRITTEN AS $z = g(x, y)$, AND THAT OF THE RIGHT DIAGRAM IS WRITTEN AS $y = h(x_1, x_2, x_3, x_4)$.
I WILL GIVE YOU SOME EXAMPLES OF FUNCTIONS THAT HAVE TWO CAUSES, THAT IS, *TWO-VARIABLE FUNCTIONS*.

EXAMPLE 1

Assume that an object is at height $h(v, t)$ in meters after t seconds when it is thrown vertically upward from the ground with velocity v. Then, $h(v, t)$ is given by

$$h(v,t) = vt - 4.9t^2$$

EXAMPLE 2

The concentration $f(x, y)$ of sugar syrup obtained by dissolving y grams of sugar in x grams of water is given by

$$f(x,y) = \frac{y}{x+y} \times 100$$

EXAMPLE 3

When the amount of equipment and machinery (called *capital*) in a nation is represented with K and the amount of labor by L, we assume that the total production of commodities (GDP: Gross Domestic Product) is given by $Y(L, K)$.

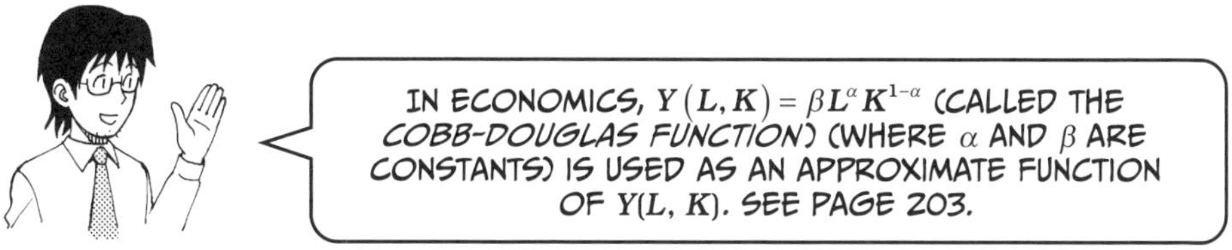
IN ECONOMICS, $Y(L,K) = \beta L^{\alpha} K^{1-\alpha}$ (CALLED THE *COBB-DOUGLAS FUNCTION*) (WHERE α AND β ARE CONSTANTS) IS USED AS AN APPROXIMATE FUNCTION OF $Y(L, K)$. SEE PAGE 203.

EXAMPLE 4

In physics, when the pressure of an ideal gas is given by P and its volume by V, its temperature T is known to be a function of P and V as $T(P, V)$. And it is given by

$$T(P,V) = \gamma PV$$

THE BASICS OF VARIABLE LINEAR FUNCTIONS

WHAT DO YOU THINK WE DO TO EXAMINE THE PROPERTIES OF THESE COMPLICATED TWO-VARIABLE FUNCTIONS?

DO WE USE IMITATING LINEAR FUNCTIONS?

HMM

WELL, YES. BUT SINCE WE NOW HAVE TWO-VARIABLE FUNCTIONS, WE HAVE TO USE TWO-VARIABLE LINEAR FUNCTIONS.

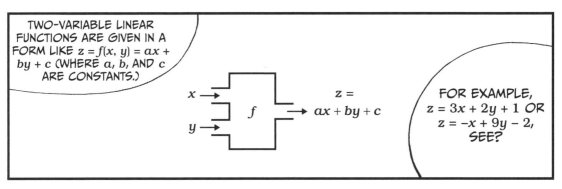

TWO-VARIABLE LINEAR FUNCTIONS ARE GIVEN IN A FORM LIKE $z = f(x, y) = ax + by + c$ (WHERE a, b, AND c ARE CONSTANTS.)

$x \rightarrow$

$y \rightarrow$

f

$\rightarrow$ $z = ax + by + c$

FOR EXAMPLE, $z = 3x + 2y + 1$ OR $z = -x + 9y - 2$, SEE?

NOW, LET'S SEE WHAT THEIR GRAPHS LOOK LIKE. SINCE THEY HAVE TWO INPUTS (x AND y) AND AN OUTPUT (z), IT IS NATURAL TO USE 3-DIMENSIONAL COORDINATES.

HMM

WELL, JUST THINK OF AN IMAGE IN WHICH THE X-Y PLANE IS THE FLOOR AND THE Z-AXIS IS A PILLAR.

A PILLAR...

IS SOMETHING WRONG?

OH! NO, NOTHING. LET'S CONTINUE.

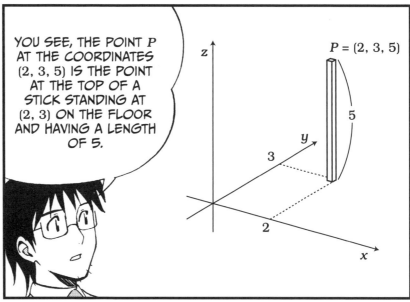

YOU SEE, THE POINT P AT THE COORDINATES $(2, 3, 5)$ IS THE POINT AT THE TOP OF A STICK STANDING AT $(2, 3)$ ON THE FLOOR AND HAVING A LENGTH OF 5.

$P = (2, 3, 5)$

NOW, WHAT DO YOU THINK THE GRAPH OF THE TWO-VARIABLE LINEAR FUNCTION $z = f(x, y) = ax + by + c$ LOOKS LIKE?

LET'S DRAW THE GRAPH OF $z = f(x, y) = 3x + 2y + 1$ AS AN EXAMPLE.

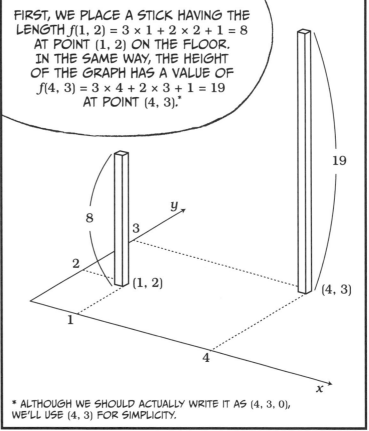

FIRST, WE PLACE A STICK HAVING THE LENGTH $f(1, 2) = 3 \times 1 + 2 \times 2 + 1 = 8$ AT POINT $(1, 2)$ ON THE FLOOR. IN THE SAME WAY, THE HEIGHT OF THE GRAPH HAS A VALUE OF $f(4, 3) = 3 \times 4 + 2 \times 3 + 1 = 19$ AT POINT $(4, 3)$.*

* ALTHOUGH WE SHOULD ACTUALLY WRITE IT AS $(4, 3, 0)$, WE'LL USE $(4, 3)$ FOR SIMPLICITY.

IN THE SAME WAY, WE PUT UP 16 STICKS AT 16 POINTS (x, y) SATISFYING $1 \leq x \leq 4$ AND $1 \leq y \leq 4$, WHICH ARE SHOWN IN THIS FIGURE.

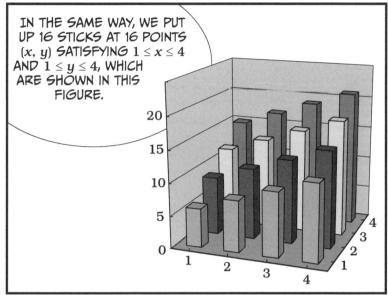

LOOKING AT THIS FIGURE, YOU CAN VAGUELY SEE THAT THE GRAPH FORMS A PLANE, CAN'T YOU?

YES, I SEE IT!

NOW, LET'S LOOK AT THE PILLARS ON THE NEAREST SIDE.

THEIR HEIGHTS ARE, BEGINNING FROM THE LEFT, $f(1, 1) = 6, f(2, 1) = 9, f(3, 1) = 12$, AND $f(4, 1) = 15$.

THESE POINTS FORM A LINE WHOSE SLOPE IS 3, WHICH IS INTUITIVE BECAUSE IF y IS A CONSTANT ($y = 1$) IN $z = f(x, y) = 3x + 2y + 1$, WE GET $z = 3x + 2 \times 1 + 1 = 3x + 3$.

NEXT, LET'S LOOK AT THE HEIGHTS OF THE STICKS RIGHT BEHIND THE FIRST ONES. THEIR HEIGHTS ARE $f(1, 2) = 8$, $f(2, 2) = 11, f(3, 2) = 14$, AND $f(4, 2) = 17$, EACH OF WHICH IS HIGHER THAN THE STICK IN FRONT OF IT BY 2.

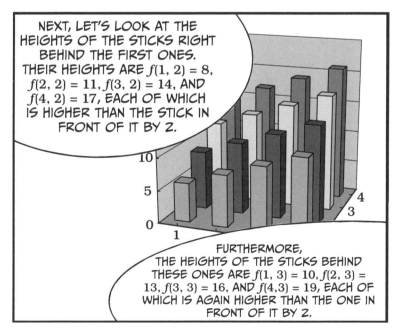

FURTHERMORE, THE HEIGHTS OF THE STICKS BEHIND THESE ONES ARE $f(1, 3) = 10, f(2, 3) = 13, f(3, 3) = 16$, AND $f(4,3) = 19$, EACH OF WHICH IS AGAIN HIGHER THAN THE ONE IN FRONT OF IT BY 2.

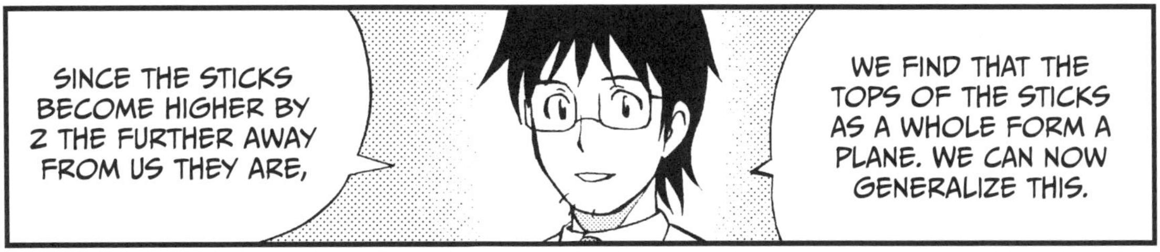
SINCE THE STICKS BECOME HIGHER BY 2 THE FURTHER AWAY FROM US THEY ARE,
WE FIND THAT THE TOPS OF THE STICKS AS A WHOLE FORM A PLANE. WE CAN NOW GENERALIZE THIS.

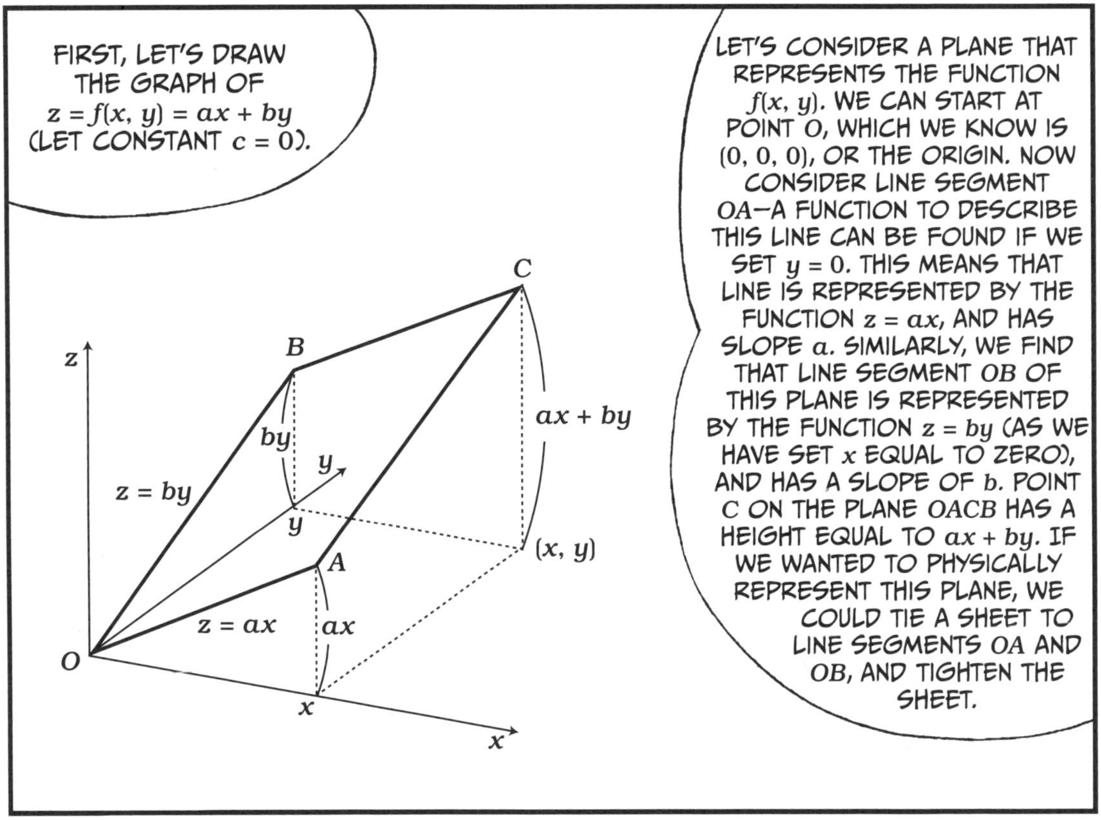
FIRST, LET'S DRAW THE GRAPH OF $z = f(x, y) = ax + by$ (LET CONSTANT $c = 0$).
C
B
z
$ax + by$
by
y
$z = by$
y
(x, y)
A
$z = ax$
ax
O
x
x
LET'S CONSIDER A PLANE THAT REPRESENTS THE FUNCTION $f(x, y)$. WE CAN START AT POINT O, WHICH WE KNOW IS $(0, 0, 0)$, OR THE ORIGIN. NOW CONSIDER LINE SEGMENT OA—A FUNCTION TO DESCRIBE THIS LINE CAN BE FOUND IF WE SET $y = 0$. THIS MEANS THAT LINE IS REPRESENTED BY THE FUNCTION $z = ax$, AND HAS SLOPE a. SIMILARLY, WE FIND THAT LINE SEGMENT OB OF THIS PLANE IS REPRESENTED BY THE FUNCTION $z = by$ (AS WE HAVE SET x EQUAL TO ZERO), AND HAS A SLOPE OF b. POINT C ON THE PLANE $OACB$ HAS A HEIGHT EQUAL TO $ax + by$. IF WE WANTED TO PHYSICALLY REPRESENT THIS PLANE, WE COULD TIE A SHEET TO LINE SEGMENTS OA AND OB, AND TIGHTEN THE SHEET.

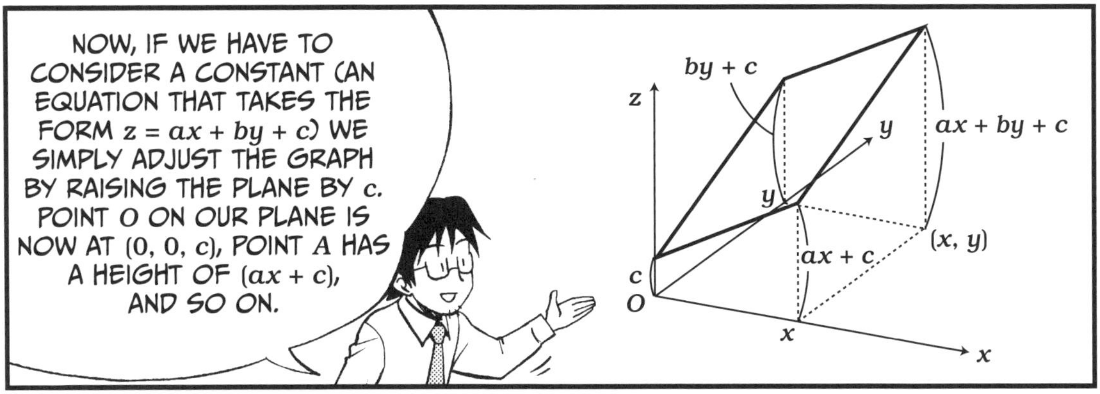
NOW, IF WE HAVE TO CONSIDER A CONSTANT (AN EQUATION THAT TAKES THE FORM $z = ax + by + c$) WE SIMPLY ADJUST THE GRAPH BY RAISING THE PLANE BY c. POINT O ON OUR PLANE IS NOW AT $(0, 0, c)$, POINT A HAS A HEIGHT OF $(ax + c)$, AND SO ON.
$by + c$
z
y
$ax + by + c$
y
(x, y)
$ax + c$
c
O
x
x

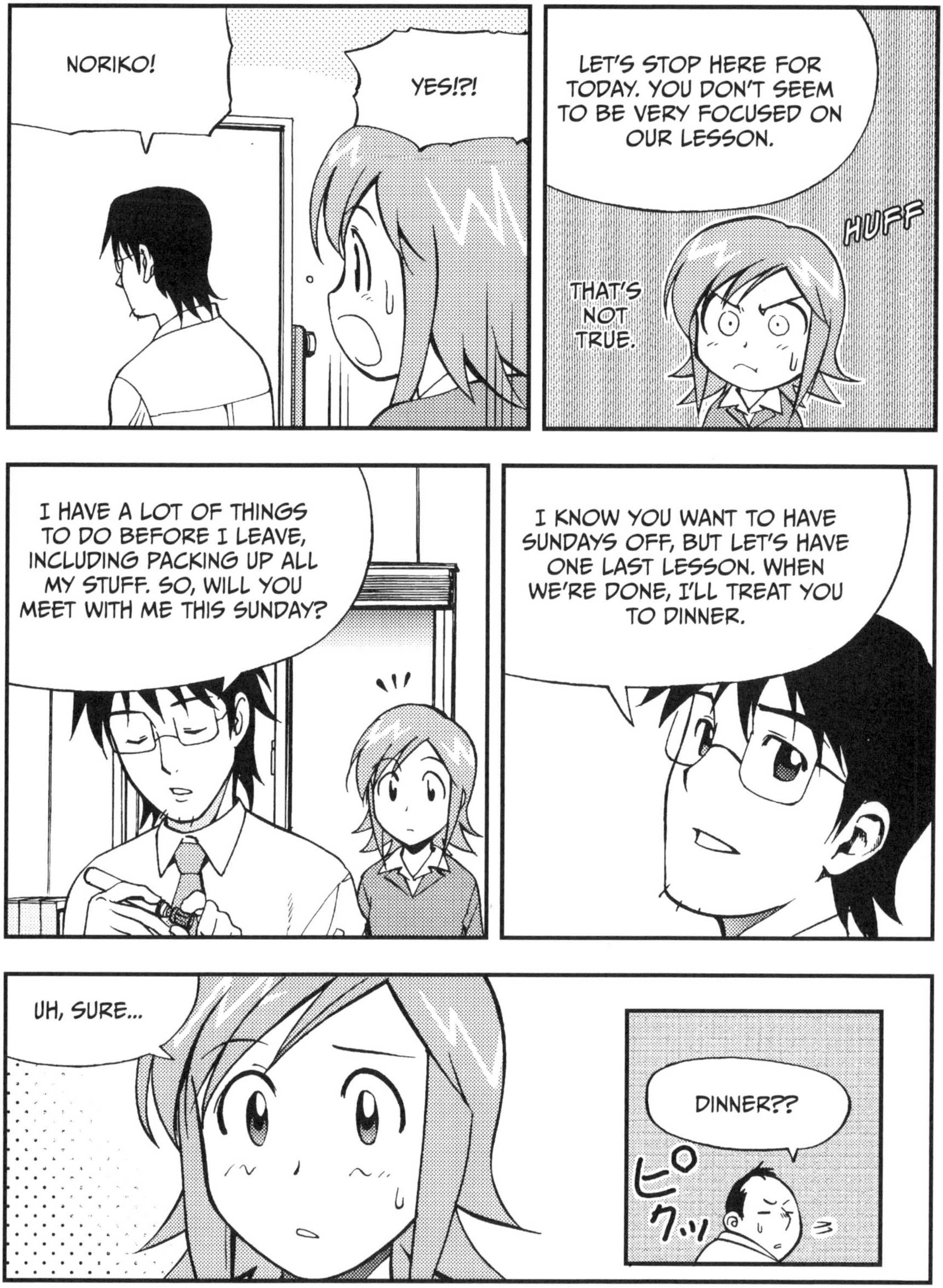
NORIKO!
YES!?!
LET'S STOP HERE FOR TODAY. YOU DON'T SEEM TO BE VERY FOCUSED ON OUR LESSON.
HUFF
THAT'S NOT TRUE.
I HAVE A LOT OF THINGS TO DO BEFORE I LEAVE, INCLUDING PACKING UP ALL MY STUFF. SO, WILL YOU MEET WITH ME THIS SUNDAY?
I KNOW YOU WANT TO HAVE SUNDAYS OFF, BUT LET'S HAVE ONE LAST LESSON. WHEN WE'RE DONE, I'LL TREAT YOU TO DINNER.
UH, SURE...
DINNER??
ピクッ

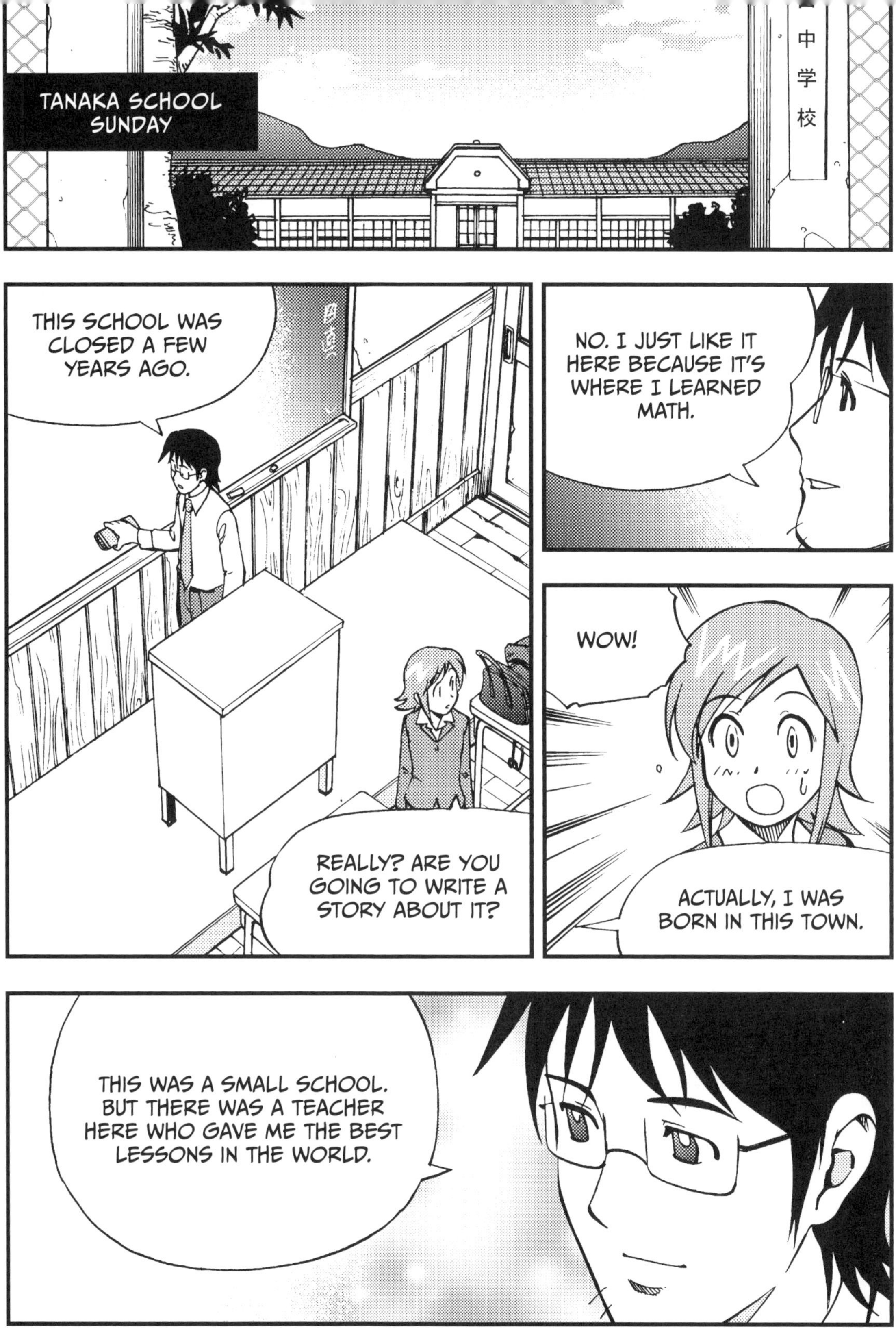
TANAKA SCHOOL SUNDAY
中学校
THIS SCHOOL WAS CLOSED A FEW YEARS AGO.
REALLY? ARE YOU GOING TO WRITE A STORY ABOUT IT?
NO. I JUST LIKE IT HERE BECAUSE IT'S WHERE I LEARNED MATH.
WOW!
ACTUALLY, I WAS BORN IN THIS TOWN.
THIS WAS A SMALL SCHOOL. BUT THERE WAS A TEACHER HERE WHO GAVE ME THE BEST LESSONS IN THE WORLD.

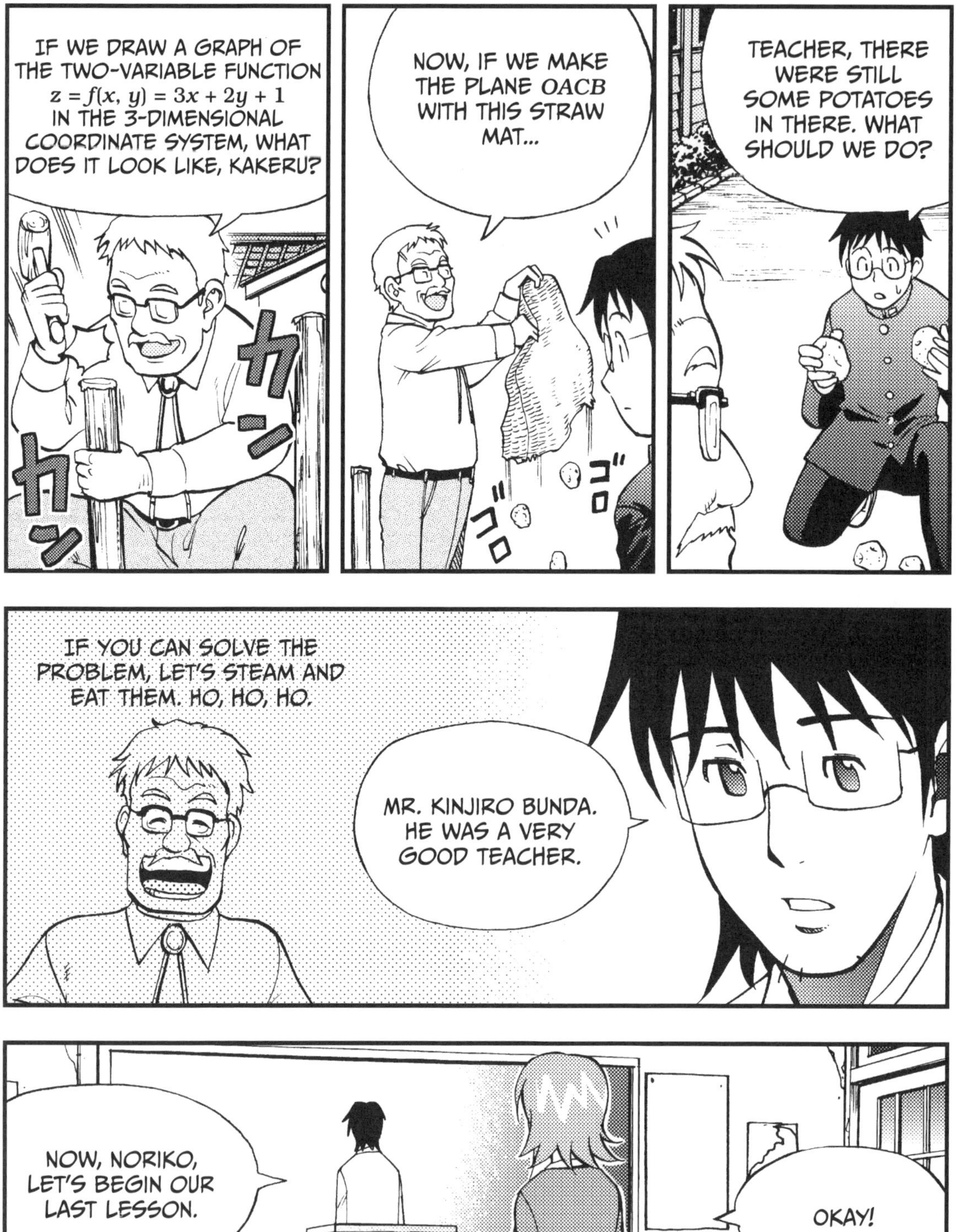
IF WE DRAW A GRAPH OF THE TWO-VARIABLE FUNCTION $z = f(x, y) = 3x + 2y + 1$ IN THE 3-DIMENSIONAL COORDINATE SYSTEM, WHAT DOES IT LOOK LIKE, KAKERU?
カン
カン
NOW, IF WE MAKE THE PLANE *OACB* WITH THIS STRAW MAT...
ゴロ
ゴロ
TEACHER, THERE WERE STILL SOME POTATOES IN THERE. WHAT SHOULD WE DO?
IF YOU CAN SOLVE THE PROBLEM, LET'S STEAM AND EAT THEM. HO, HO, HO.
MR. KINJIRO BUNDA. HE WAS A VERY GOOD TEACHER.
NOW, NORIKO, LET'S BEGIN OUR LAST LESSON.
OKAY!

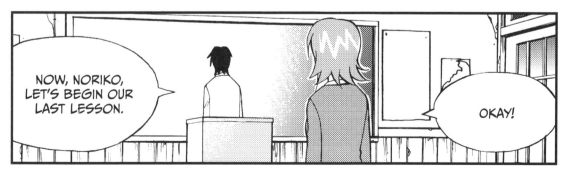

PARTIAL DIFFERENTIATION

OH, THERE'S THE FIRST PERIOD BELL! LET'S EXPLORE THE DIFFERENTIATION OF TWO-VARIABLE FUNCTIONS.

CLASS SCHEDULE	
1	PARTIAL DIFFERENTIATION
2	
3	

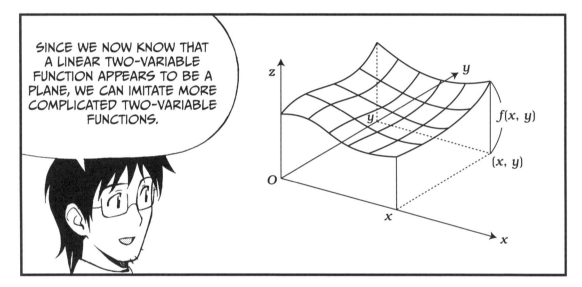

SINCE WE NOW KNOW THAT A LINEAR TWO-VARIABLE FUNCTION APPEARS TO BE A PLANE, WE CAN IMITATE MORE COMPLICATED TWO-VARIABLE FUNCTIONS.

OUR ORIGINAL FUNCTION LOOKS LIKE A FLAT-TOP TENT, DOESN'T IT?

IT LOOKS MORE LIKE A PIE TO ME.

WELL, THAT'S NOT AN IMPORTANT DISAGREEMENT. NOW, LET'S MAKE AN IMITATING TWO-VARIABLE LINEAR FUNCTION OF $f(x, y)$ NEAR A POINT (a, b) ($x = a$ AND $y = b$).

We make a two-variable linear function that has the same height as $f(a, b)$ at the point (a, b). The formula is $L(x, y) = p(x - a) + q(y - b) + f(a, b)$. Substituting a for x and b for y, we get $L(a, b) = f(a, b)$.

z = f(x, y)
z = L(x, y)
(Imitating
two-variable
linear function)
F
H
G
f(a + ε, b + δ)
C
D
E
δ
P = (a + ε, b + δ)
f(a, b)
A = (a, b)
ε
B
√(ε² + δ²)

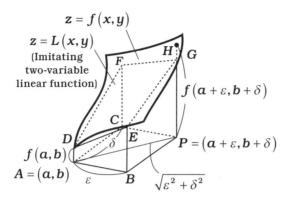

While the graph of $z = f(x, y)$ and that of $z = L(x, y)$ pass through the same point above the point $A = (a, b)$, they differ in height at the point $P = (a + \varepsilon, b + \delta)$. The error in this case is $f(a + \varepsilon, b + \delta) - L(a + \varepsilon, b + \delta) = f(a + \varepsilon, b + \delta) - f(a, b) - (p\varepsilon + q\delta)$, and *the relative error* expresses the ratio of the error to the distance AP.

$$\text{Relative error} = \frac{\text{difference between } f \text{ and } L}{\text{distance } AP}$$

$$❶ \quad = \frac{f(a + \varepsilon, b + \delta) - f(a, b) - (p\varepsilon + q\delta)}{\sqrt{\varepsilon^2 + \delta^2}}$$

We consider $L(x, y)$ as the difference between it and f becomes infinitely close to zero (when P is infinitely close to A) as the imitating linear function. For that case, we obtain p and q. p is the slope of DE and q that of DF in the figure. Since ε and δ are arbitrary, we first let $\delta = 0$ and analyze ❶. ❶ becomes

$$\text{Relative error} = \frac{f(a + \varepsilon, b + 0) - f(a, b) - (p\varepsilon + q \times 0)}{\sqrt{\varepsilon^2 + 0^2}}$$

$$= \frac{f(a + \varepsilon, b) - f(a, b)}{\varepsilon} - p$$

Thus, the statement "the relative error → 0 when $\varepsilon \to 0$" means the following:

❷ $\displaystyle \lim_{\varepsilon \to 0} \frac{f(a+\varepsilon, b) - f(a, b)}{\varepsilon} = p$

This is the slope of DE.

Here, we should realize that the left side of this expression is the same as single-variable differentiation. In other words, if we substitute b for y and keep it constant, we obtain $f(x, b)$, which is a function of x only. The left side of ❷ is then the calculation of finding the derivative of this function at $x = a$.

Although we are very much tempted to write the left side as $f'(a, b)$ since it is a derivative, it would then be impossible to tell with respect to which, x or y, we differentiated it.

So, we write "the derivative of f obtained at $x = a$ while y is fixed at b" as $f_x(a, b)$.

This f_x is called "the partial derivative of f in the direction of x". This is the notation corresponding to the "prime" in single-variable differentiation.

The notation $\dfrac{df}{dx}(a, b)$, that corresponds to $\dfrac{\partial f}{\partial x}$, is also used. In short, we have the following:

"The derivative of f in the direction of x obtained at $x = a$ while y is fixed at b"

$$f_x(a, b) = \frac{\partial f}{\partial x}(a, b) \quad \text{also written as} \quad \left(\left[\frac{\partial f}{\partial x} \right]_{x=a, y=b} \right)$$

$$= \text{Slope of } DE$$

> ∂ IS READ AS "PARTIAL DERIVATIVE."

In exactly the same way, we can obtain the following.

"The derivative of f in the direction of y obtained at $y = b$ while x is fixed at a"

$$f_y(a, b) = \frac{\partial f}{\partial y}(a, b)$$

$$= \text{Slope of } DF$$

We have now found the following.

If $z = f(x, y)$ has an imitating linear function near $(x, y) = (a, b)$, it is given by

❸ $\quad z = f_x(a,b)(x-a) + f_y(a,b)(y-b) + f(a,b)$

or* $\quad z = \dfrac{\partial f}{\partial x}(a,b)(x-a) + \dfrac{\partial f}{\partial y}(a,b)(y-b) + f(a,b)$

Consider a point (α, β) on a circle with radius 1 centered at the origin of the $x - y$ plane (the floor). We have $\alpha^2 + \beta^2 = 1$ (or $\alpha = \cos\theta$ and $\beta = \sin\theta$). We now calculate the derivative in the direction from $(0, 0)$ to (α, β). A displacement of distance t in this direction is expressed as $(a, b) \to (a + \alpha t, b + \beta t)$. If we set $\varepsilon = \alpha t$ and $\delta = \beta t$ in ❶, we get

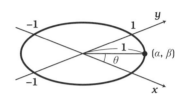

$$\text{Relative error} = \frac{f(a+\alpha t, b+\beta t) - f(a,b) - (p\alpha t + q\beta t)}{\sqrt{\alpha^2 t^2 + \beta^2 t^2}}$$

$$= \frac{f(a+\alpha t, b+\beta t) - f(a,b)}{t\sqrt{\alpha^2 + \beta^2}} - p\alpha - q\beta$$

$$= \frac{f(a+\alpha t, b+\beta t) - f(a,b)}{t} - p\alpha - q\beta$$

❹ $\quad$ Since $\quad \sqrt{\alpha^2 + \beta^2} = 1$

Assuming $p = f_x(a, b)$ and $q = f_y(a, b)$, we modify ❹ as follows:

❺ $\quad \dfrac{f(a+\alpha t, b+\beta t) - f(a, b+\beta t)}{t} + \dfrac{f(a, b+\beta t) - f(a,b)}{t} - f_x(a,b)\alpha - f_y(a,b)\beta$

Since the derivative of $f(x, b + \beta t)$, a function of x only, at $x = a$ is

$f_x(a, b+\beta t)$

we get, from the imitating single-variable linear function,

$f(a+\alpha t, b+\beta t) - f(a, b+\beta t) \approx f_x(a, b+\beta t)\alpha t$

---

* We have calculated the imitating linear function in such a way that its relative error approaches 0 when $AP \to 0$ in the x or y direction. It is not apparent, however, if the relative error $\to 0$ when $AP \to 0$ in any direction for the linear function that is made up of the derivatives $f_x(a, b)$ and $f_y(a, b)$. We'll now look into this in detail, although the discussion here will not be so strict.

Similarly, for y we get

$$f(a, b+\beta t) - f(a, b) \approx f_y(a, b)\beta t$$

Substituting this in ❺,

$$❺ \quad \approx f_x(a, b+\beta t)\alpha + f_y(a, b)\beta t - f_x(a, b)\alpha - f_y(a, b)\beta$$
$$= \left(f_x(a, b+\beta t) - f_x(a, b)\right)\alpha$$

Since $f_x(a, b + \beta t) - f_x(a, b) \approx 0$ if t is close enough to 0, the relative error = ❺ ≈ 0. Thus, we have shown "the relative error $\to 0$ when $AP \to 0$ in any direction."

It should be noted that f_x must be continuous to say $f_x(a, b + \beta t) - f_x(a, b) \approx 0$ ($t \approx 0$). Unless it is continuous, we don't know whether the derivative exists in every direction, even though f_x and f_y exist. Since such functions are rather exceptional, however, we won't cover them in this book.

EXAMPLES (FUNCTION OF EXAMPLE 1 FROM PAGE 183)

Let's find the partial derivatives of $h(v, t) = vt - 4.9t^2$ at $(v, t) = (100, 5)$.

In the v direction, we differentiate $h(v, 5) = 5v - 122.5$ and get

$$\frac{\partial h}{\partial v}(v, 5) = 5$$

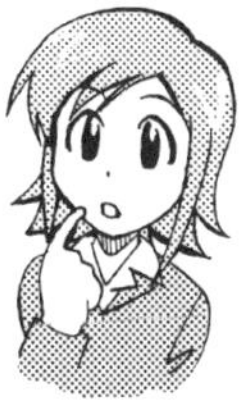

Thus,

$$\frac{\partial h}{\partial v}(100, 5) = h_v(100, 5) = 5$$

In the t direction, we differentiate $h(100, t) = 100t - 4.9t^2$ and get

$$\frac{\partial h}{\partial t}(100, t) = 100 - 9.8t$$

$$\frac{\partial h}{\partial t}(100, 5) = h_t(100, 5) = 100 - 9.8 \times 5 = 51$$

And the imitating linear function is

$$L(x, y) = 5(v - 100) + 51(t - 5) - 377.5$$

In general,

$$\frac{\partial h}{\partial v} = t, \frac{\partial h}{\partial v} = v - 9.8t$$

Therefore, from ❸ on page 194, near $(v, t) = (v_0, t_0)$,

$$h(v,t) \approx t_0(v - v_0) + (v_0 - 9.8t_0)(t - t_0) + h(v_0, t_0)$$

Next, we'll try imitating the concentration of sugar syrup given y grams of sugar in x grams of water.

$$f(x,y) = \frac{100y}{x+y}$$

$$\frac{\partial f}{\partial y} = f_y = -\frac{100y}{(x+y)^2}$$

$$\frac{\partial f}{\partial y} = f_y = \frac{100(x+y) - 100y \times 1}{(x+y)^2} = \frac{100x}{(x+y)^2}$$

Thus, near $(x, y) = (a, b)$, we have

$$f(x,y) \approx -\frac{100b}{(a+b)^2}(x-a) + \frac{100a}{(a+b)^2}(y-b) + \frac{100b}{a+b}$$

DEFINITION OF PARTIAL DIFFERENTIATION

When $z = f(x, y)$ is partially differentiable with respect to x for every point (x, y) in a region, the function $(x, y) \to f_x(x, y)$, which relates (x, y) to $f_x(x, y)$, the partial derivative at that point with respect to x, is called the partial differential function of $z = f(x, y)$ with respect to x and can be expressed by any of the following:

$$f_x, f_x(x,y), \frac{\partial f}{\partial x}, \frac{\partial z}{\partial x}$$

Similarly, when $z = f(x, y)$ is partially differentiable with respect to y for every point (x, y) in the region, the function

$$(x,y) \to f_y(x,y)$$

is called the partial differential function of $z = f(x, y)$ with respect to y and is expressed by any of the following:

$$f_y, f_y(x,y), \frac{\partial f}{\partial y}, \frac{\partial z}{\partial y}$$

Obtaining the partial derivatives of a function is called *partially differentiating* it.

TOTAL DIFFERENTIALS

カラーン
コローン
2ND PERIOD
TOTAL DIFFERENTIALS

From the imitating linear function of $z = f(x, y)$ at $(x, y) = (a, b)$, we have found

$$f(x,y) \approx f_x(a,b)(x-a) + f_y(a,b)(x-b) + f(a,b)$$

We now modify this as

$$❻ \quad f(x,y) - f(a,b) \approx \frac{\partial f}{\partial x}(a,b)(x-a) + \frac{\partial f}{\partial y}(a,b)(y-b)$$

Since $f(x, y) - f(a, b)$ means the increment of $z = f(x, y)$ when a point moves from (a, b) to (x, y), we write this as Δz, as we did for the single-variable functions.

Also, $(x - a)$ is Δx and $(y - b)$ is Δy.

Then, expression ❻ can be written as

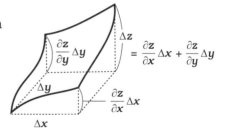

$\frac{\partial z}{\partial y}\Delta y$
Δz
$= \frac{\partial z}{\partial x}\Delta x + \frac{\partial z}{\partial y}\Delta y$
Δy
$\frac{\partial z}{\partial x}\Delta x$
Δx

$$❼ \quad \Delta z \approx \frac{\partial z}{\partial x}\Delta x + \frac{\partial z}{\partial y}\Delta y$$

This expression means, "If x increases from a by Δx and y from b by Δy in $z = f(x, y)$, z increases by

$$\frac{\partial z}{\partial x}\Delta x + \frac{\partial z}{\partial y}\Delta y$$

Since $\dfrac{\partial z}{\partial x}\,\Delta x$ is "the increment of z in the x direction when y is fixed at b" and $\dfrac{\partial z}{\partial y}\,\Delta y$ is "the increment in the y direction when x is fixed at a," expression ❼ also means "the increment of $z = f(x, y)$ is the sum of the increment in the x direction and that in the y direction."

When expression ❼ is idealized (made instantaneous), we have

❽ $\quad dz = \dfrac{\partial z}{\partial x}\,dx + \dfrac{\partial z}{\partial y}\,dy$

or

❾ $\quad df = f_x\,dx + f_y\,dy$

EXPRESSION ❽ OR ❾ IS CALLED THE FORMULA OF THE *TOTAL DIFFERENTIAL.*

(Δ has been changed to d.)

The meaning of the formula is as follows.

Increment of height of a curved surface =

| Partial derivative in the x direction | $\times$ | Increment in the x direction | $+$ | Partial derivative in the y direction | $\times$ | Increment in the y direction |

Now, let's look at the expression of a total differential from Example 4 (page 183).

By converting the unit properly, we rewrite the equation of temperature as $T = PV$.

$$\dfrac{\partial T}{\partial P} = \dfrac{\partial (PV)}{\partial P} = V \quad \text{and} \quad \dfrac{\partial T}{\partial V} = \dfrac{\partial (PV)}{\partial P} = P$$

Thus, the total differential can be written as $dT = VdP + PdV$.
In the form of an approximate expression, this is $\Delta T \approx V\Delta P + P\Delta V$.

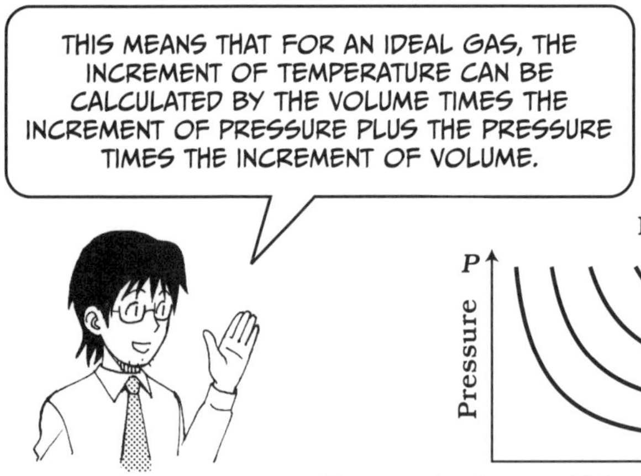

THIS MEANS THAT FOR AN IDEAL GAS, THE INCREMENT OF TEMPERATURE CAN BE CALCULATED BY THE VOLUME TIMES THE INCREMENT OF PRESSURE PLUS THE PRESSURE TIMES THE INCREMENT OF VOLUME.

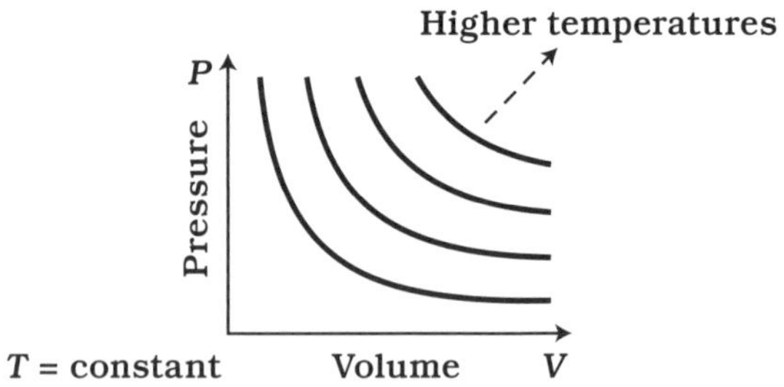

Higher temperatures

P

Pressure

T = constant   Volume V

CONDITIONS FOR EXTREMA

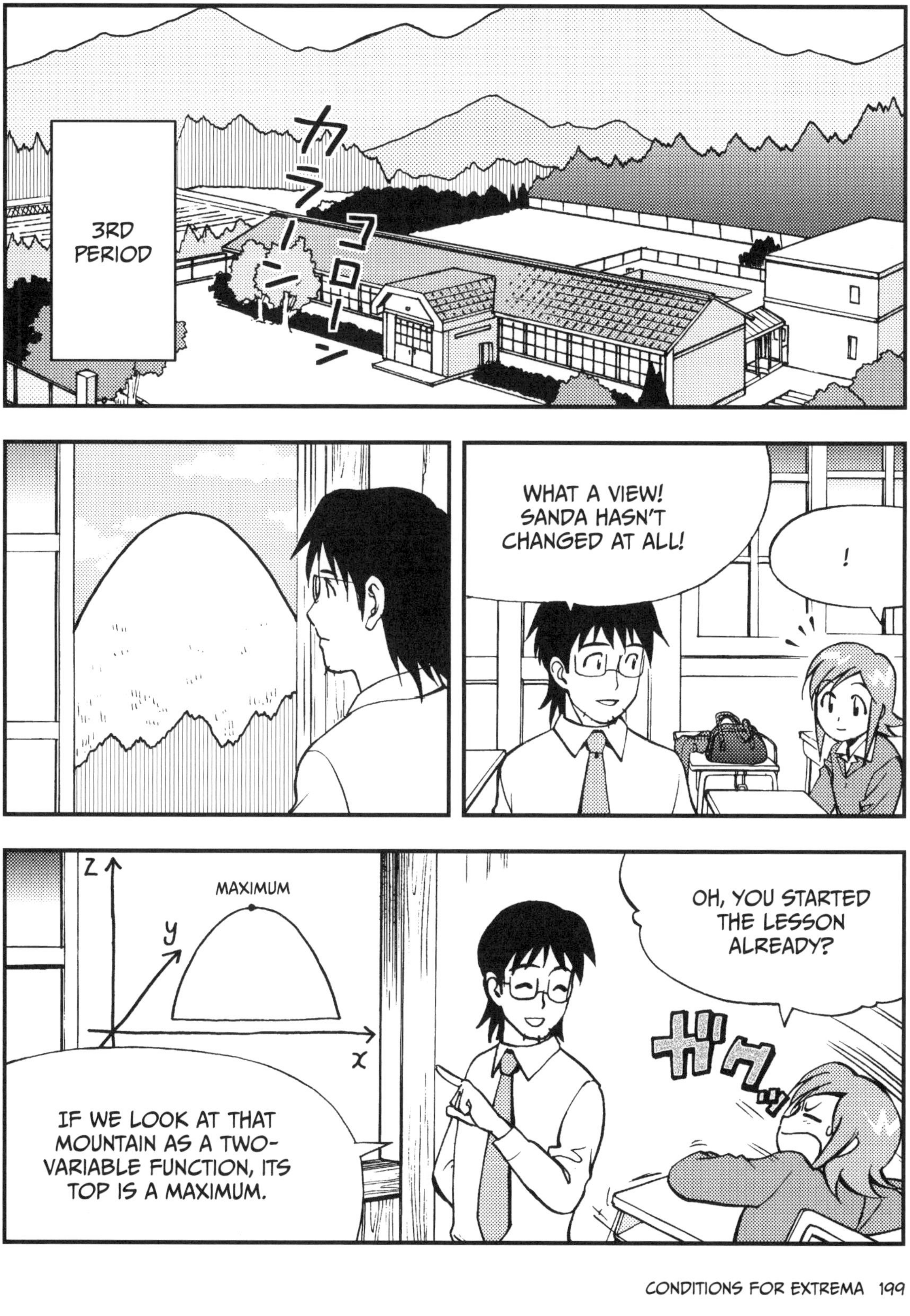
3RD PERIOD
カラーン
コローン
WHAT A VIEW! SANDA HASN'T CHANGED AT ALL!
!
z
MAXIMUM
y
x
OH, YOU STARTED THE LESSON ALREADY?
IF WE LOOK AT THAT MOUNTAIN AS A TWO-VARIABLE FUNCTION, ITS TOP IS A MAXIMUM.
ガクッ

The *extrema* of a two-variable function $f(x, y)$ are where its graph is at the top of a mountain or the bottom of a valley.

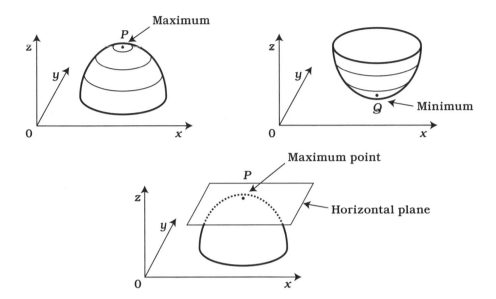

Since the plane tangent to the graph at point P or Q is parallel to the x-y plane, we should have

$$f(x,y) \approx p(x-a) + q(y-b) + f(a,b)$$

with $p = q = 0$ in the imitating linear function.
Since

$$p = \frac{\partial f}{\partial x}\left(= f_x\right) \quad q = \frac{\partial f}{\partial y}\left(= f_y\right)$$

the condition for extrema* is, if $f(x, y)$ has an extremum at $(x, y) = (a, b)$,

$$f_x\left(a,b\right) = f_y\left(a,b\right) = 0$$

or

$$\frac{\partial f}{\partial x}\left(a,b\right) = \frac{\partial f}{\partial y}\left(a,b\right) = 0$$

---

* The opposite of this is not true. In other words, even if $f_x(a, b) = f_y(a, b) = 0$, f will not always have an extremum at $(x, y) = (a, b)$. Thus, this condition only picks up the candidates for extrema.

AT THE EXTREMA OF A TWO-VARIABLE FUNCTION, THE PARTIAL DERIVATIVES IN BOTH THE x AND y DIRECTIONS ARE ZERO.

EXAMPLE

Let's find the minimum of $f(x, y) = (x - y)^2 + (y - 2)^2$. First, we'll find it algebraically.

Since

$$\left(x - y\right)^2 \geq 0 \quad \left(y - 2\right)^2 \geq 0$$

$$f\left(x, y\right) = \left(x - y\right)^2 + \left(y - 2\right)^2 \geq 0$$

If we substitute $x = y = 2$ here,

$$f\left(2, 2\right) = \left(2 - 2\right)^2 + \left(2 - 2\right)^2 = 0$$

From this, $f(x, y) \geq f(2, 2)$ for all (x, y). In other words, $f(x, y)$ has a minimum of zero at $(x, y) = (2, 2)$.

On the other hand, $\dfrac{\partial f}{\partial x} = 2\left(x - y\right)$ and $\dfrac{\partial f}{\partial y} = 2\left(x - y\right)\left(-1\right) + 2\left(y - 2\right) = -2x + 4y - 4$. If we set

$$\frac{\partial f}{\partial x} = \frac{\partial f}{\partial y} = 0$$

and solve these simultaneous equations,

$$\begin{cases} 2x - 2y = 0 \\ -2x + 4y - 4 = 0 \end{cases}$$

we find that $(x, y) = (2, 2)$, just as we found above.

THE SOLUTIONS ARE THE SAME!

APPLYING PARTIAL DIFFERENTIATION TO ECONOMICS

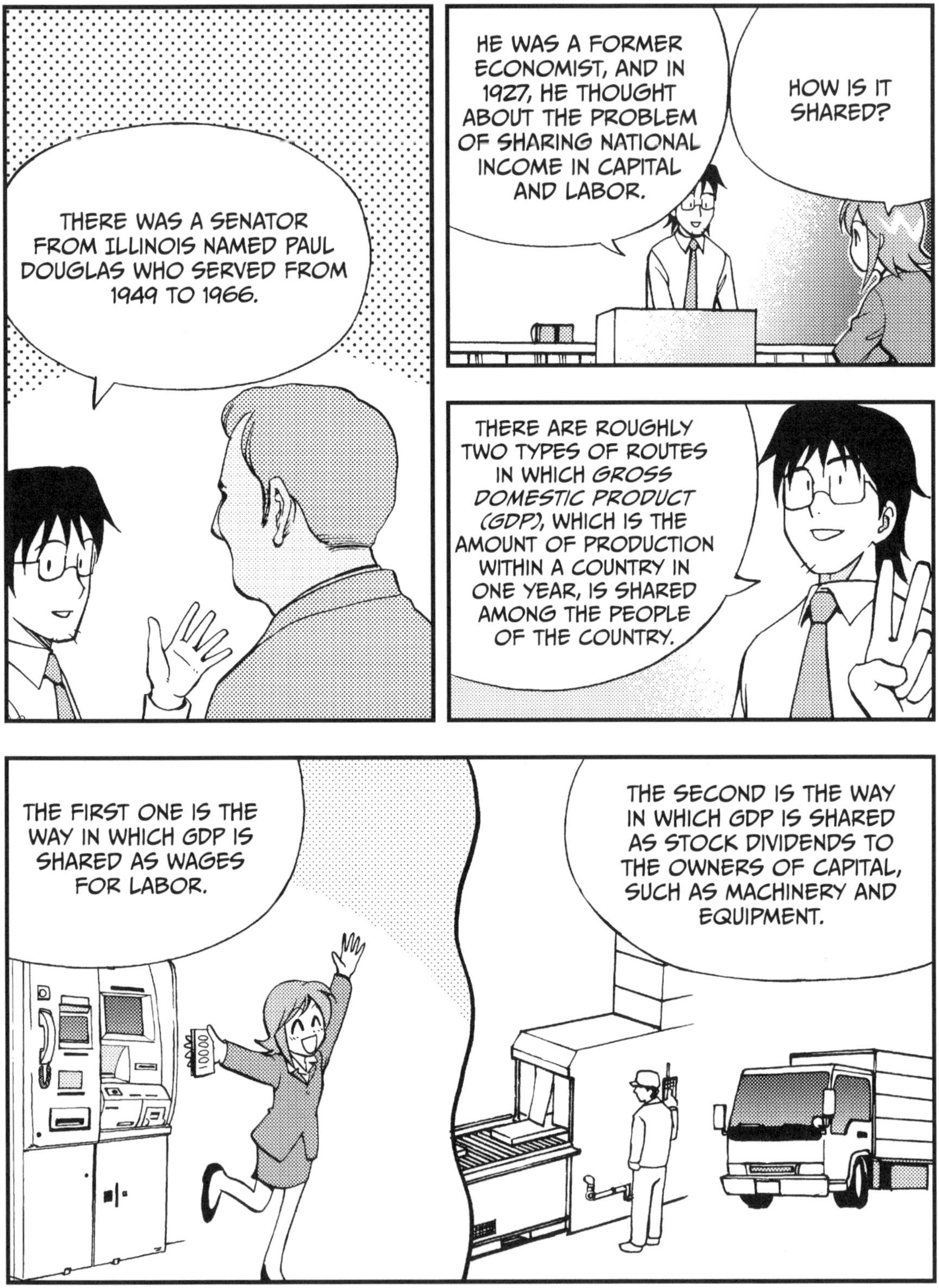
THERE WAS A SENATOR FROM ILLINOIS NAMED PAUL DOUGLAS WHO SERVED FROM 1949 TO 1966.
HE WAS A FORMER ECONOMIST, AND IN 1927, HE THOUGHT ABOUT THE PROBLEM OF SHARING NATIONAL INCOME IN CAPITAL AND LABOR.
HOW IS IT SHARED?
THERE ARE ROUGHLY TWO TYPES OF ROUTES IN WHICH *GROSS DOMESTIC PRODUCT (GDP)*, WHICH IS THE AMOUNT OF PRODUCTION WITHIN A COUNTRY IN ONE YEAR, IS SHARED AMONG THE PEOPLE OF THE COUNTRY.
THE FIRST ONE IS THE WAY IN WHICH GDP IS SHARED AS WAGES FOR LABOR.
THE SECOND IS THE WAY IN WHICH GDP IS SHARED AS STOCK DIVIDENDS TO THE OWNERS OF CAPITAL, SUCH AS MACHINERY AND EQUIPMENT.

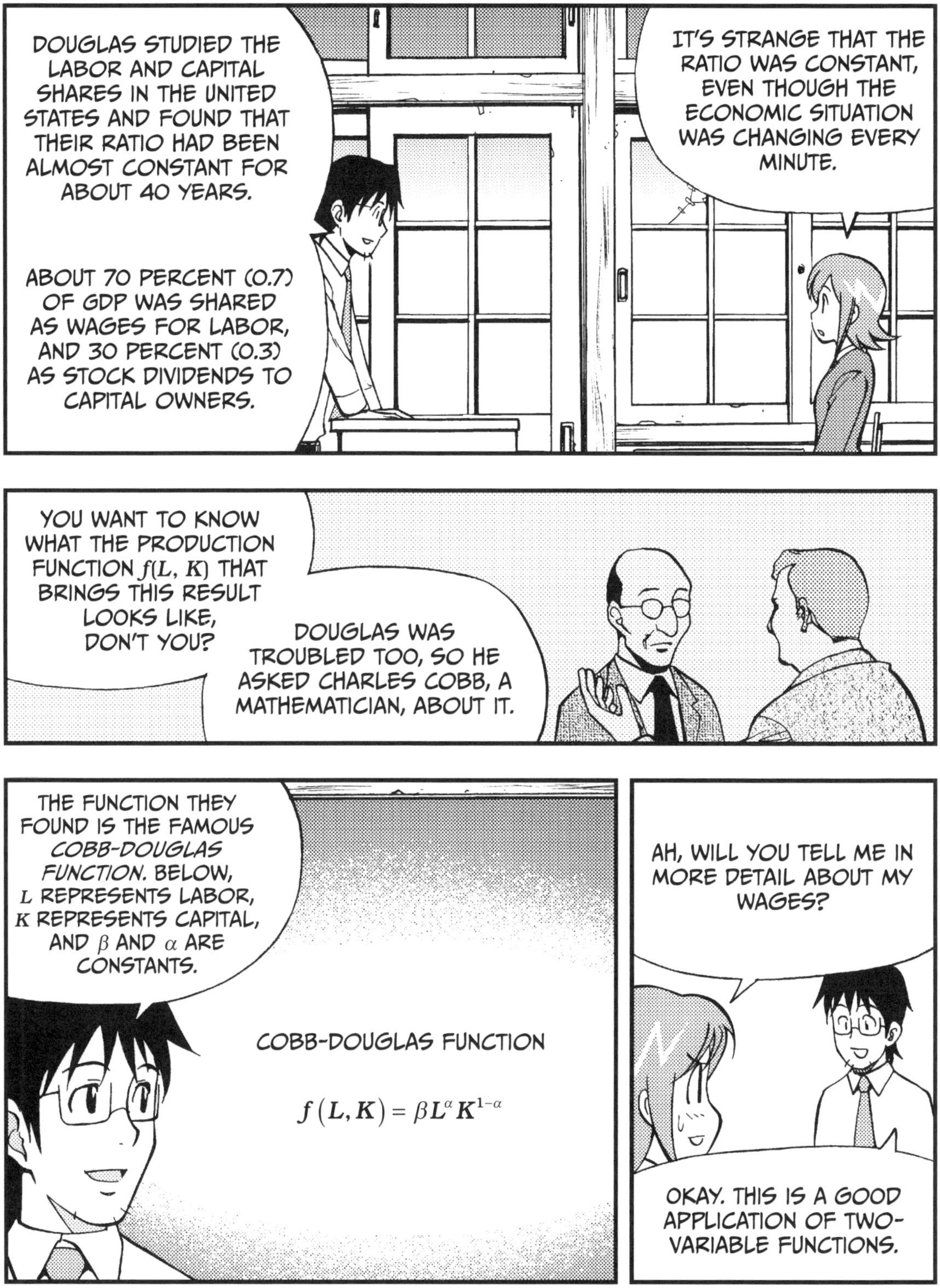
DOUGLAS STUDIED THE LABOR AND CAPITAL SHARES IN THE UNITED STATES AND FOUND THAT THEIR RATIO HAD BEEN ALMOST CONSTANT FOR ABOUT 40 YEARS.
ABOUT 70 PERCENT (0.7) OF GDP WAS SHARED AS WAGES FOR LABOR, AND 30 PERCENT (0.3) AS STOCK DIVIDENDS TO CAPITAL OWNERS.
IT'S STRANGE THAT THE RATIO WAS CONSTANT, EVEN THOUGH THE ECONOMIC SITUATION WAS CHANGING EVERY MINUTE.
YOU WANT TO KNOW WHAT THE PRODUCTION FUNCTION $f(L, K)$ THAT BRINGS THIS RESULT LOOKS LIKE, DON'T YOU?
DOUGLAS WAS TROUBLED TOO, SO HE ASKED CHARLES COBB, A MATHEMATICIAN, ABOUT IT.
THE FUNCTION THEY FOUND IS THE FAMOUS *COBB-DOUGLAS FUNCTION*. BELOW, L REPRESENTS LABOR, K REPRESENTS CAPITAL, AND β AND α ARE CONSTANTS.
COBB-DOUGLAS FUNCTION
$f(L, K) = \beta L^{\alpha} K^{1-\alpha}$
AH, WILL YOU TELL ME IN MORE DETAIL ABOUT MY WAGES?
OKAY. THIS IS A GOOD APPLICATION OF TWO-VARIABLE FUNCTIONS.

First, let's suppose that wages are measured in units of w, and capital is measured in units of r. We'll consider the production of the entire country to be given by the function $f(L, K)$ and assume the country is acting as a profit-maximizing business. The profit P is given by the equation:

$$P = f(L, K) - wL - rK$$

Because we know that a business chooses values of L and K to maximize profit (P), we get the following condition for extrema:

$$\frac{\partial P}{\partial L} = \frac{\partial P}{\partial K} = 0$$

❶ $\quad 0 = \dfrac{\partial P}{\partial L} = \dfrac{\partial f}{\partial L} - \dfrac{\partial(wL)}{\partial L} - \dfrac{\partial(rK)}{\partial L} = \dfrac{\partial f}{\partial L} - w \Rightarrow w = \dfrac{\partial f}{\partial L}$

❷ $\quad 0 = \dfrac{\partial P}{\partial K} = \dfrac{\partial f}{\partial K} - \dfrac{\partial(wL)}{\partial K} - \dfrac{\partial(rK)}{\partial K} = \dfrac{\partial f}{\partial K} - r \Rightarrow r = \dfrac{\partial f}{\partial K}$

The relations far to the right mean the following.

Wages = Partial derivative of the production function with respect to L

Capital share = Partial derivative of the production function with respect to K

Now, the reward the people of the country receive for labor is Wage × Labor = wL. When this is 70 percent of GDP, we have

❸ $\quad wL = 0.7f(L, K)$

Similarly, the reward the capital owners receive is

❹ $\quad rK = 0.3f(L, K)$

From ❶ and ❸,

❺ $\quad \dfrac{\partial f}{\partial L} \times L = 0.7f(L, K)$

From ❷ and ❹,

❻ $\quad \dfrac{\partial f}{\partial K} \times K = 0.3f(L, K)$

Cobb found $f(L, K)$ that satisfies these equations. It is

$$f\left(L,K\right)=\beta L^{0.7}K^{0.3}$$

where β is a positive parameter meaning the level of technology.

Let's check if this satisfies the above conditions.

$$\begin{aligned}\frac{\partial f}{\partial L}\times L&=\frac{\partial\left(\beta L^{0.7}K^{0.3}\right)}{\partial L}\times L=0.7\beta L^{(-0.3)}K^{0.3}\times L^{1}\\&=0.7\beta L^{0.7}K^{0.3}\\&=0.7f\left(L,K\right)\end{aligned}$$

$$\begin{aligned}\frac{\partial f}{\partial K}\times K&=\frac{\partial\left(\beta L^{0.7}K^{0.3}\right)}{\partial K}\times K=0.3\beta L^{0.7}K^{(-0.7)}\times K^{1}\\&=0.3\beta L^{0.7}K^{0.3}\\&=0.3f\left(L,K\right)\end{aligned}$$

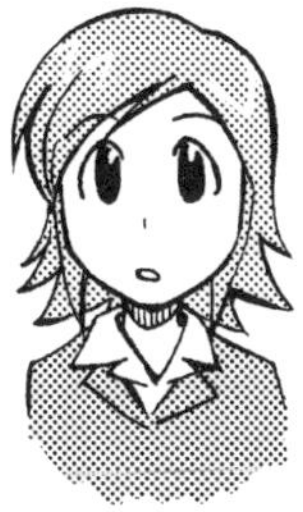

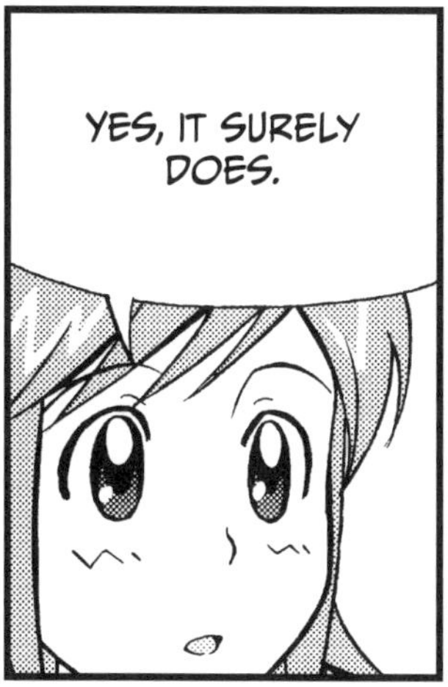
YES, IT SURELY DOES.

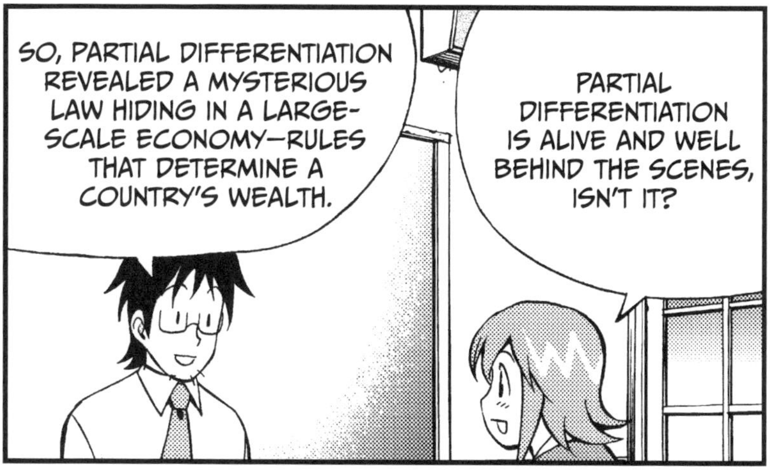
SO, PARTIAL DIFFERENTIATION REVEALED A MYSTERIOUS LAW HIDING IN A LARGE-SCALE ECONOMY—RULES THAT DETERMINE A COUNTRY'S WEALTH.
PARTIAL DIFFERENTIATION IS ALIVE AND WELL BEHIND THE SCENES, ISN'T IT?

THE CHAIN RULE

We have seen single-variable composite functions before (page 14).

$$y = f(x), \ z = g(y), \ z = g(f(x)),$$

$$g(f(x))' = g'(f(x))f'(x)$$

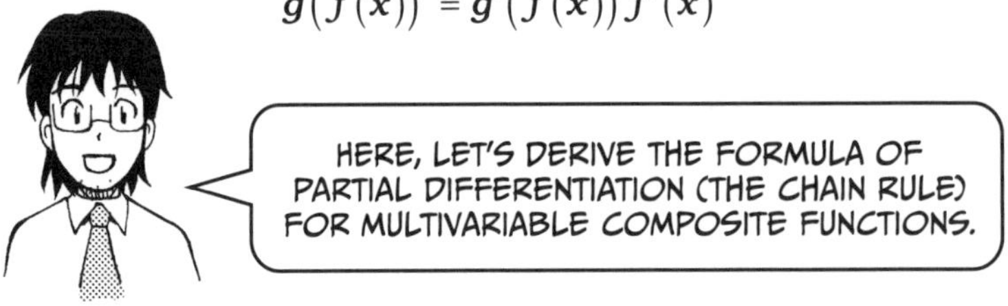

HERE, LET'S DERIVE THE FORMULA OF PARTIAL DIFFERENTIATION (THE CHAIN RULE) FOR MULTIVARIABLE COMPOSITE FUNCTIONS.

We assume that z is a two-variable function of x and y, expressed as $z = f(x, y)$, and that x and y are both single-variable functions of t, expressed as $x = a(t)$ and $y = b(t)$, respectively. Then, z can be expressed as a function of t only, as shown below.

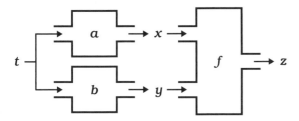

This relationship can be written as

$$z = f(x,y) = f(a(t), b(t))$$

What is the form of $\dfrac{dz}{dt}$ then?

We assume $a(t_0) = x_0$, $b(t_0) = y_0$ and $f(x_0, y_0) = f(a(t_0), b(t_0)) = z_0$ when $t = t_0$, and consider only the vicinities of t_0, x_0, y_0, and z_0.

If we obtain an α that satisfies

❶ $\quad z - z_0 \approx \alpha \times (t - t_0)$

it is $\dfrac{dz}{dt}(t_0)$.

First, from the approximation of $x = a(t)$,

❷ $$x - x_0 \approx \frac{da}{dt}(t_0)(t - t_0)$$

Similarly, from the approximation of $y = b(t)$,

❸ $$y - y_0 \approx \frac{db}{dt}(t_0)(t - t_0)$$

Next, from the formula of total differential for a two-variable function $f(x, y)$,

❹ $$z - z_0 \approx \frac{\partial f}{\partial x}(x_0, y_0)(x - x_0) + \frac{\partial f}{\partial y}(x_0, y_0)(y - y_0)$$

Substituting ❷ and ❸ in ❹,

❺ $$z - z_0 \approx \frac{\partial f}{\partial x}(x_0, y_0)\frac{da}{dt}(t_0)(t - t_0) + \frac{\partial f}{\partial y}(x_0, y_0)\frac{db}{dt}(t_0)(t - t_0)$$

$$= \left(\frac{\partial f}{\partial x}(x_0, y_0)\frac{da}{dt}(t_0) + \frac{\partial f}{\partial y}(x_0, y_0)\frac{db}{dt}(t_0) \right)(t - t_0)$$

Comparing ❶ and ❺, we get

$$\alpha = \frac{\partial f}{\partial x}(x_0, y_0)\frac{da}{dt}(t_0) + \frac{\partial f}{\partial y}(x_0, y_0)\frac{db}{dt}(t_0)$$

This is what we wanted, and we now have the following formula!

FORMULA 6-1: THE CHAIN RULE

When $z = f(x, y), x = a(t), y = b(t)$

$$\frac{dz}{dt} = \frac{\partial f}{\partial x}\frac{da}{dt} + \frac{\partial f}{\partial y}\frac{db}{dt}$$

MR. SEKI, WHY DON'T I GIVE YOU A LESSON NOW?

UH, OKAY. IT'LL BE FUN TO BE A STUDENT AGAIN.

OKAY! LET'S USE A MULTIVARIABLE FUNCTION TO THINK ABOUT...

AN ENVIRONMENTAL PROBLEM!

HERE, WE HAVE A FACTORY FROM WHICH WASTE IS RELEASED AS A RESULT OF PRODUCTION OF COMMODITIES. THE WASTE SUBSEQUENTLY POLLUTES THE SEA, CAUSING A REDUCTION IN THE LOCAL FISHERMAN'S CATCH.

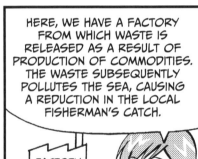

FACTORY

THE EFFECT THAT PRODUCTION ACTIVITIES OF A BUSINESS HAVE ON OTHER FIELDS WITHOUT GOING THROUGH THE MARKET, AS IS THIS CASE, IS CALLED AN *EXTERNALITY*. IN PARTICULAR, HARMFUL EXTERNALITIES, SUCH AS POLLUTION, ARE CALLED *NEGATIVE EXTERNALITIES*.

FACTORY

SUPPOSE THAT x WORKERS PRODUCE AN AMOUNT OF GOODS GIVEN BY $f(x)$. THE FACTORY ALSO RELEASES WASTE AS GOODS ARE MADE, WHICH AFFECTS THE CATCH OF FISH.

LET'S CALL THE QUANTITY OF WASTE $b = b(f(x))$. NOW...

We assume that the catch of fish can be expressed as a two-variable function $g(y, b)$ of the amount of labor y and the amount of waste b.

(The catch $g(y, b)$ decreases as b increases. Thus, $\dfrac{\partial g}{\partial b}$ is negative.)

Since the variable x is contained in $g(y, b) = g(y, b(f(x)))$, production at the factory influences fisheries without going through the market. This is an externality.

First, let's see what happens if the factory and the fishery each act (selfishly) only for their own benefit. If the wage is w for both of them, the price of a commodity produced at the factory p and the price of a fish q, the profit for the factory is given by

① $\quad P_1(x) = pf(x) - wx$

Thus, the factory wants to maximize this, and the condition for extrema is

② $\quad \dfrac{dP_1}{dx} = pf'(x) - w = 0 \Leftrightarrow pf'(x) = w$

Let s be such x that satisfies this condition. Thus, we have

③ $\quad pf'(s) = w$

This s is the amount of labor employed by the factory, the amount of production is $f(s)$, and the amount of waste is given by

$$b^* = b\big(f(s)\big)$$

Next, the profit P_2 for the fishery is given by

$$P_2 = qg(y, b) - wy$$

Since the amount of waste from the factory is given by $b^* = b(f(s))$,

④ $\quad P_2 = qg(y, b^*) - wy$

which is practically a single-variable function of y. To maximize P_2, we use only the condition about y for extrema of a two-variable function.

⑤ $\quad \dfrac{\partial P_2}{\partial y} = q\dfrac{\partial g}{\partial y}(y, b^*) - w = 0 \Leftrightarrow q\dfrac{\partial g}{\partial y}(y, b^*) = w$

Therefore, the optimum amount of labor t to be input satisfies

⑥ $\quad q\dfrac{\partial g}{\partial y}(t, b^*) = w$

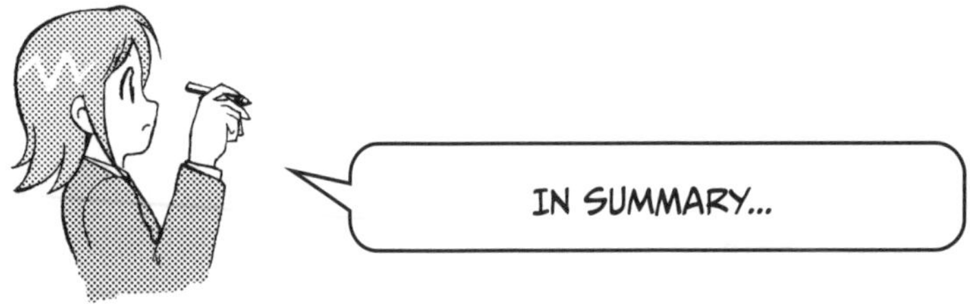
IN SUMMARY...

The production at the factory and the catch in the fishery when they act freely in this model are given by $f(s)$ and $g(t, b^*)$, respectively, where s and t satisfy the following.

③ $pf'(s) = w$

⑥ $b^* = b(f(s)), q\dfrac{\partial g}{\partial y}(t, b^*) = w$

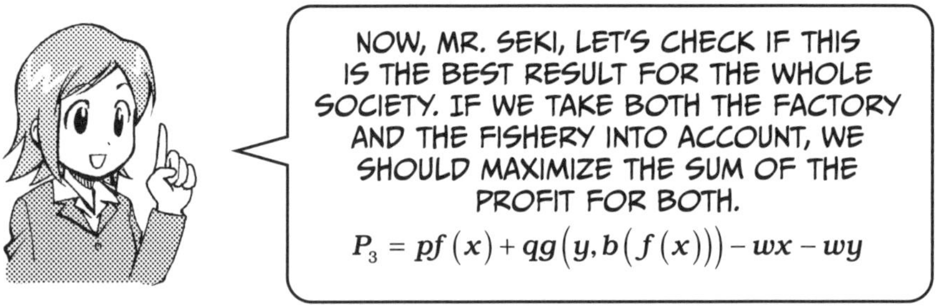
NOW, MR. SEKI, LET'S CHECK IF THIS IS THE BEST RESULT FOR THE WHOLE SOCIETY. IF WE TAKE BOTH THE FACTORY AND THE FISHERY INTO ACCOUNT, WE SHOULD MAXIMIZE THE SUM OF THE PROFIT FOR BOTH.
$P_3 = pf(x) + qg(y, b(f(x))) - wx - wy$

Since P_3 is a two-variable function of x and y, the condition for extrema is given by

$$\frac{\partial P_3}{\partial x} = \frac{\partial P_3}{\partial y} = 0$$

The first partial derivative is obtained as follows.

$$\begin{aligned}\frac{\partial P_3}{\partial x} &= pf'(x) + q\frac{\partial g(y, b(f(x)))}{\partial x} - w \\ &= pf'(x) + q\frac{\partial g}{\partial b}(y, b(f(x)))b'(f(x))f'(x) - w\end{aligned}$$

(Here, we used the chain rule.)

Thus,

$$\frac{\partial P_3}{\partial x} = 0 \Leftrightarrow \left(p + q\frac{\partial g}{\partial b}\left(y, b\left(f\left(x\right)\right)\right) b'\left(f\left(x\right)\right) \right) f'\left(x\right) = w$$

Similarly,

$$⑧ \quad \frac{\partial P_3}{\partial y} = 0 \Leftrightarrow q\frac{\partial g}{\partial y}\left(y, b\left(f\left(x\right)\right)\right) = w$$

Thus, if the optimum amount of labor is S for the factory and *T* for the fishery, they satisfy

$$⑨ \quad \left(p + q\frac{\partial g}{\partial b}\left(T, b\left(f\left(S\right)\right)\right) b'\left(f\left(S\right)\right) \right) f'\left(S\right) = w$$

$$⑩ \quad q\frac{\partial g}{\partial y}\left(T, b\left(f\left(S\right)\right)\right) = w$$

Although these equations look complicated, they are really just two-variable simultaneous equations.

If we compare these equations with equations ③ and ⑥, we find that ③ and ⑨ are different while ⑥ and ⑩ are the same. Then, how do they differ?

$$③ \quad p \times f'\left(s\right) = w$$

$$⑪ \quad \left(p + \heartsuit\right) \times f'\left(S\right) = w$$

As you see here, ♥ has appeared in the expression.

Since $\left(\heartsuit = q\frac{\partial g}{\partial b} b'\left(f\left(S\right)\right) \right)$ is negative, *p* + ♥ is smaller than *p*.

Since $f'(S)$ or $f'(s)$ is multiplied to the first part to give the same value *w*, $f'(S)$ must be larger than $f'(s)$.

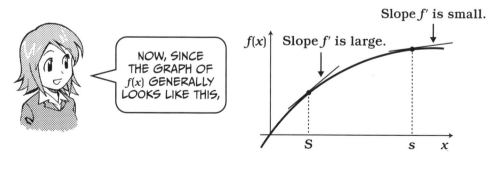

NOW, SINCE THE GRAPH OF f(x) GENERALLY LOOKS LIKE THIS,

Slope f′ is small.
f(x)
Slope f′ is large.
S
s
x

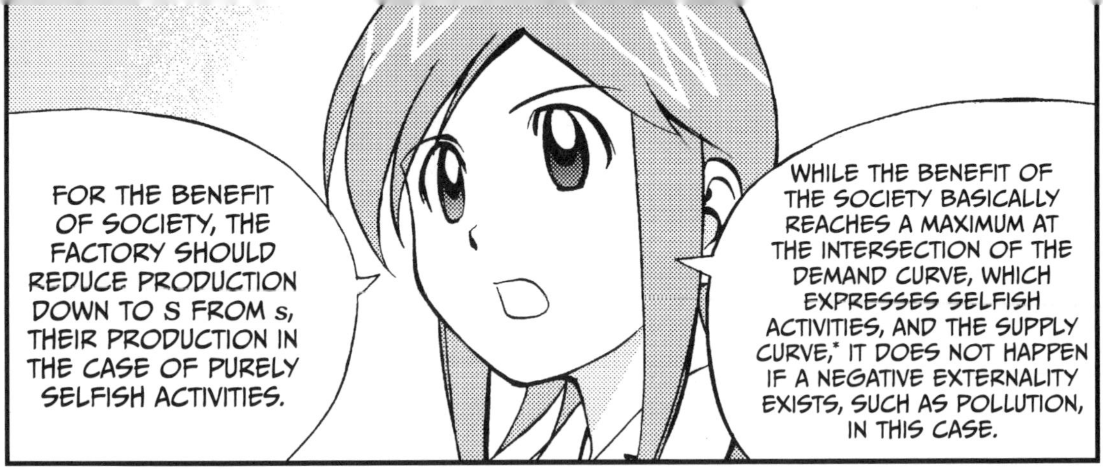
FOR THE BENEFIT OF SOCIETY, THE FACTORY SHOULD REDUCE PRODUCTION DOWN TO S FROM s, THEIR PRODUCTION IN THE CASE OF PURELY SELFISH ACTIVITIES.
WHILE THE BENEFIT OF THE SOCIETY BASICALLY REACHES A MAXIMUM AT THE INTERSECTION OF THE DEMAND CURVE, WHICH EXPRESSES SELFISH ACTIVITIES, AND THE SUPPLY CURVE,* IT DOES NOT HAPPEN IF A NEGATIVE EXTERNALITY EXISTS, SUCH AS POLLUTION, IN THIS CASE.

* SEE PAGE 105.

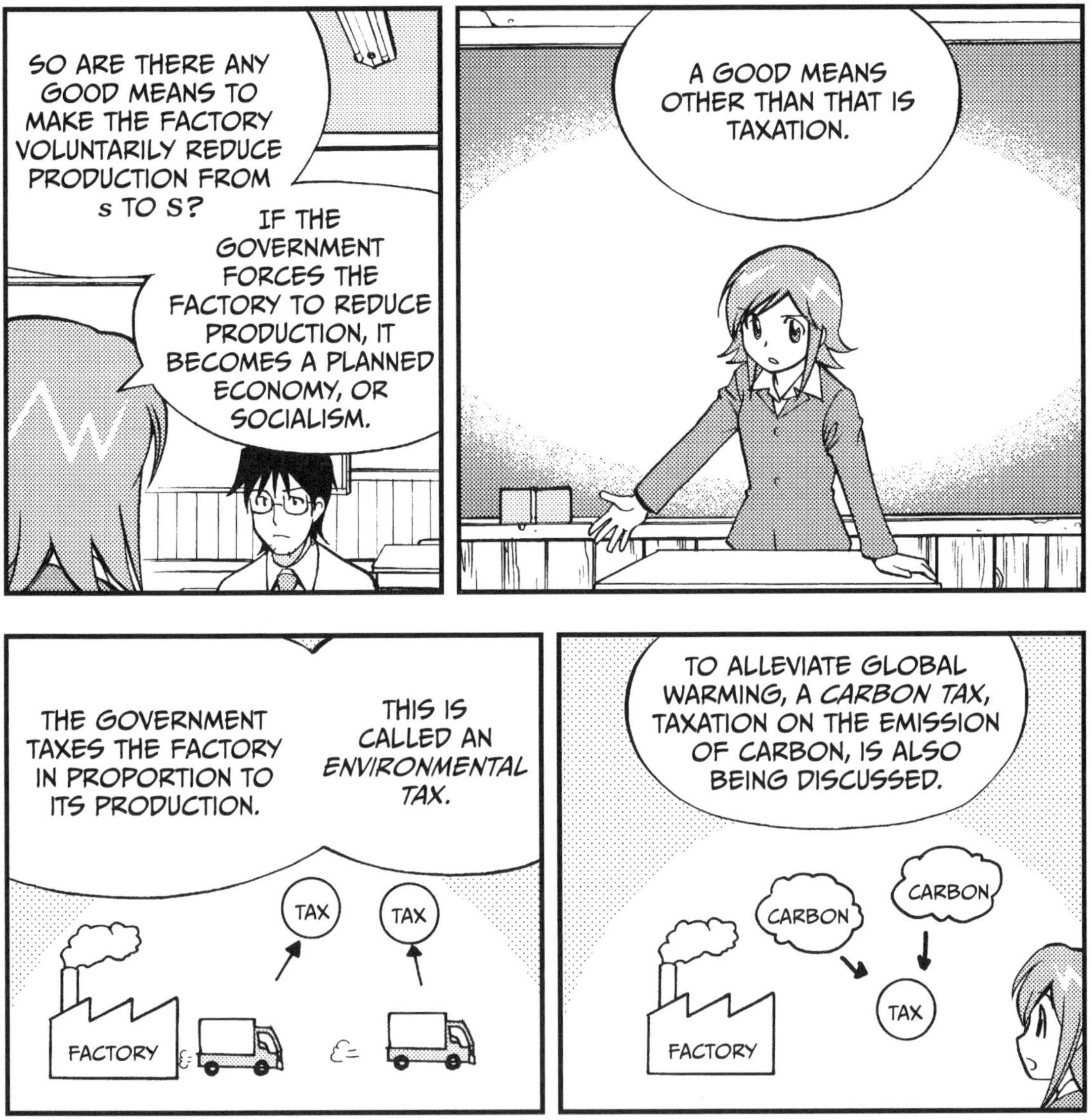
SO ARE THERE ANY GOOD MEANS TO MAKE THE FACTORY VOLUNTARILY REDUCE PRODUCTION FROM s TO S?
IF THE GOVERNMENT FORCES THE FACTORY TO REDUCE PRODUCTION, IT BECOMES A PLANNED ECONOMY, OR SOCIALISM.
A GOOD MEANS OTHER THAN THAT IS TAXATION.
THE GOVERNMENT TAXES THE FACTORY IN PROPORTION TO ITS PRODUCTION.
THIS IS CALLED AN *ENVIRONMENTAL TAX*.
TAX
TAX
FACTORY
TO ALLEVIATE GLOBAL WARMING, A *CARBON TAX*, TAXATION ON THE EMISSION OF CARBON, IS ALSO BEING DISCUSSED.
CARBON
CARBON
TAX
FACTORY

LET'S ASSUME THAT THE TAX ON A UNIT COMMODITY PRODUCED AT THE FACTORY IS $-\heartsuit$.

$$-\heartsuit = -q\frac{\partial g}{\partial b}b'\left(f\left(S\right)\right)$$

THIS IS A POSITIVE CONSTANT.

THEN, THE PROFIT ① IN THE CASE OF SELFISH ACTIVITIES BECOMES LIKE THIS.

$$⑫ \quad P_1\left(x\right) = pf\left(x\right) - wx - \left(-\heartsuit f\left(x\right)\right)$$

THE CONDITION FOR EXTREMA THAT MAXIMIZE THIS IS...

$$⑬ \quad \frac{\partial P_1}{\partial x} = pf'\left(x\right) - w + \heartsuit f'\left(x\right) = 0 \Leftrightarrow \left(p + \heartsuit\right)f'\left(x\right) = w$$

SINCE ⑬ IS THE SAME EQUATION AS ⑨, THE PRODUCTION AT THE FACTORY NOW MAXIMIZES THE BENEFIT FOR SOCIETY.

ORDINARY TAXES (INCOME TAX, CONSUMPTION TAX) ARE FOR PUBLIC INVESTMENT...

AN ENVIRONMENTAL TAX IS FOR MAINTAINING A HEALTHY ENVIRONMENT BY CONTROLLING THE ECONOMY.

HAVE YOU GOT IT, MR. SEKI?

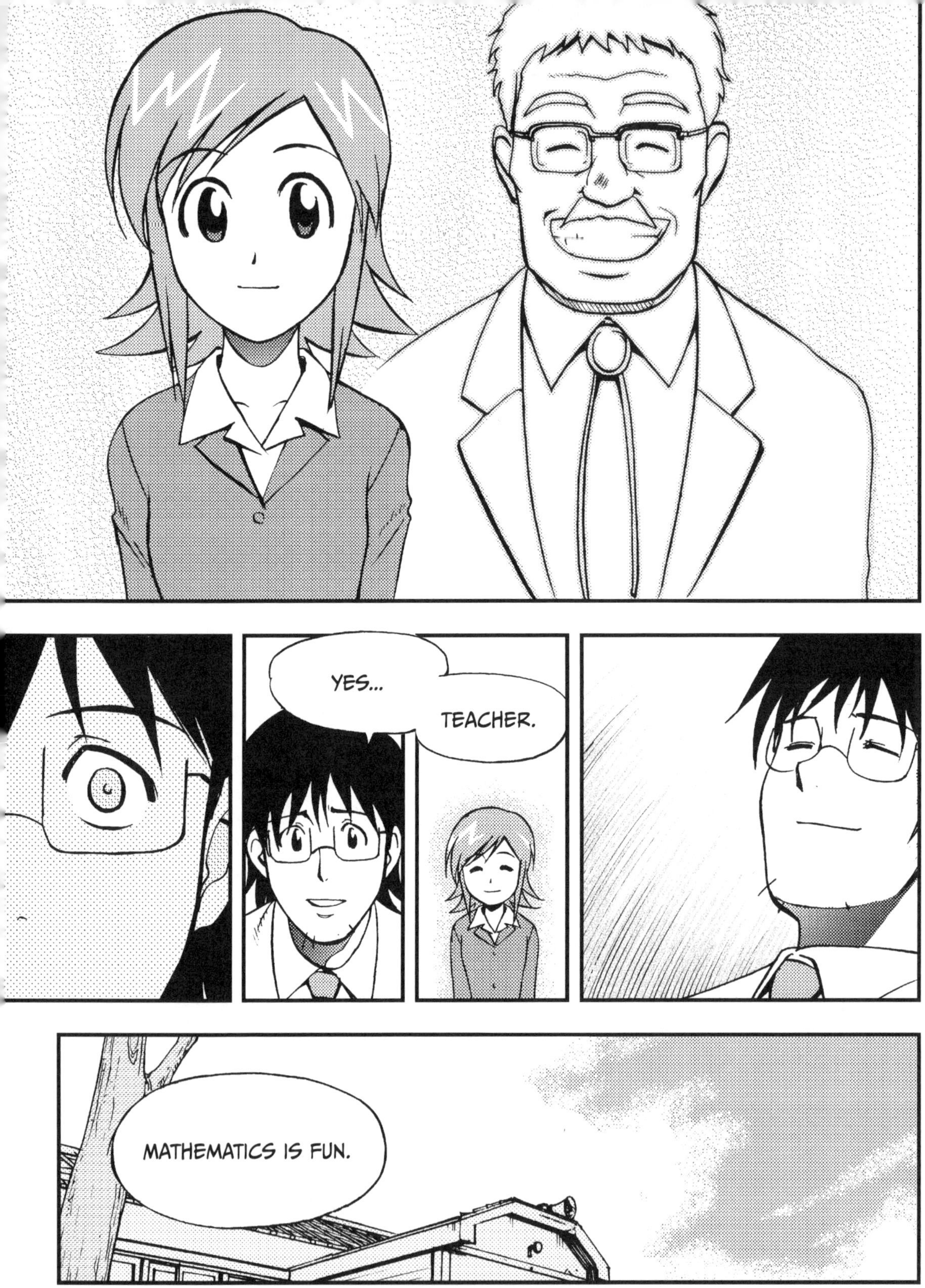
YES...
TEACHER.
MATHEMATICS IS FUN.

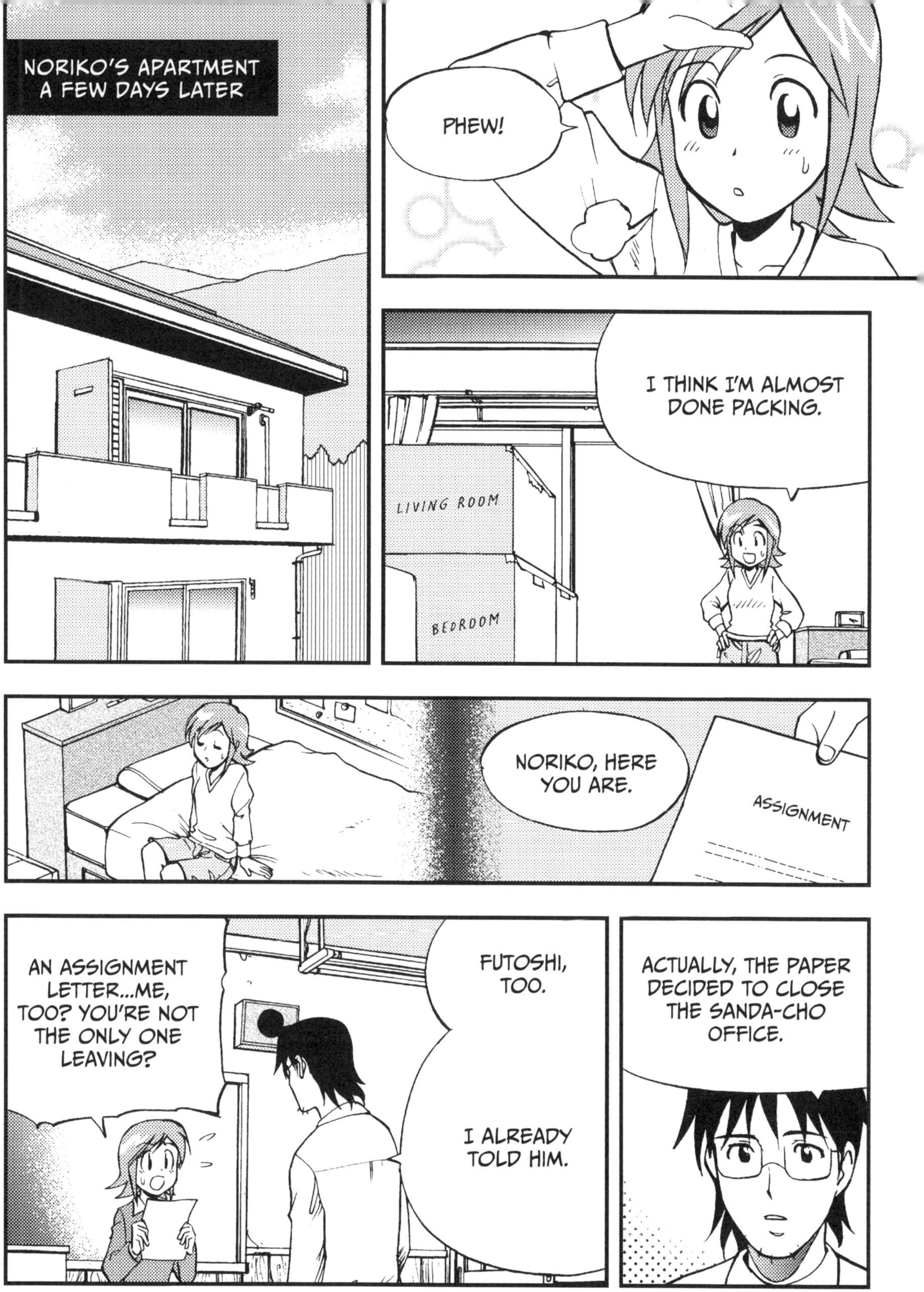
NORIKO'S APARTMENT A FEW DAYS LATER
PHEW!
LIVING ROOM
BEDROOM
I THINK I'M ALMOST DONE PACKING.
NORIKO, HERE YOU ARE.
ASSIGNMENT
AN ASSIGNMENT LETTER...ME, TOO? YOU'RE NOT THE ONLY ONE LEAVING?
FUTOSHI, TOO.
I ALREADY TOLD HIM.
ACTUALLY, THE PAPER DECIDED TO CLOSE THE SANDA-CHO OFFICE.

YOU WILL BE NOTIFIED SOON WHERE YOU ARE GOING.

I NEVER IMAGINED I'D BE GOING TO WORK IN OKINAWA.

TRANSFER TO OKINAWA

I DIDN'T EVEN KNOW OUR COMPANY HAD AN OKINAWA OFFICE.

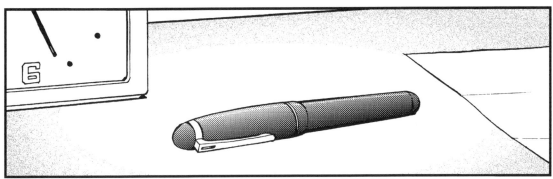

THIS IS A FAREWELL PRESENT FOR YOU. WRITE GOOD ARTICLES WITH THIS.

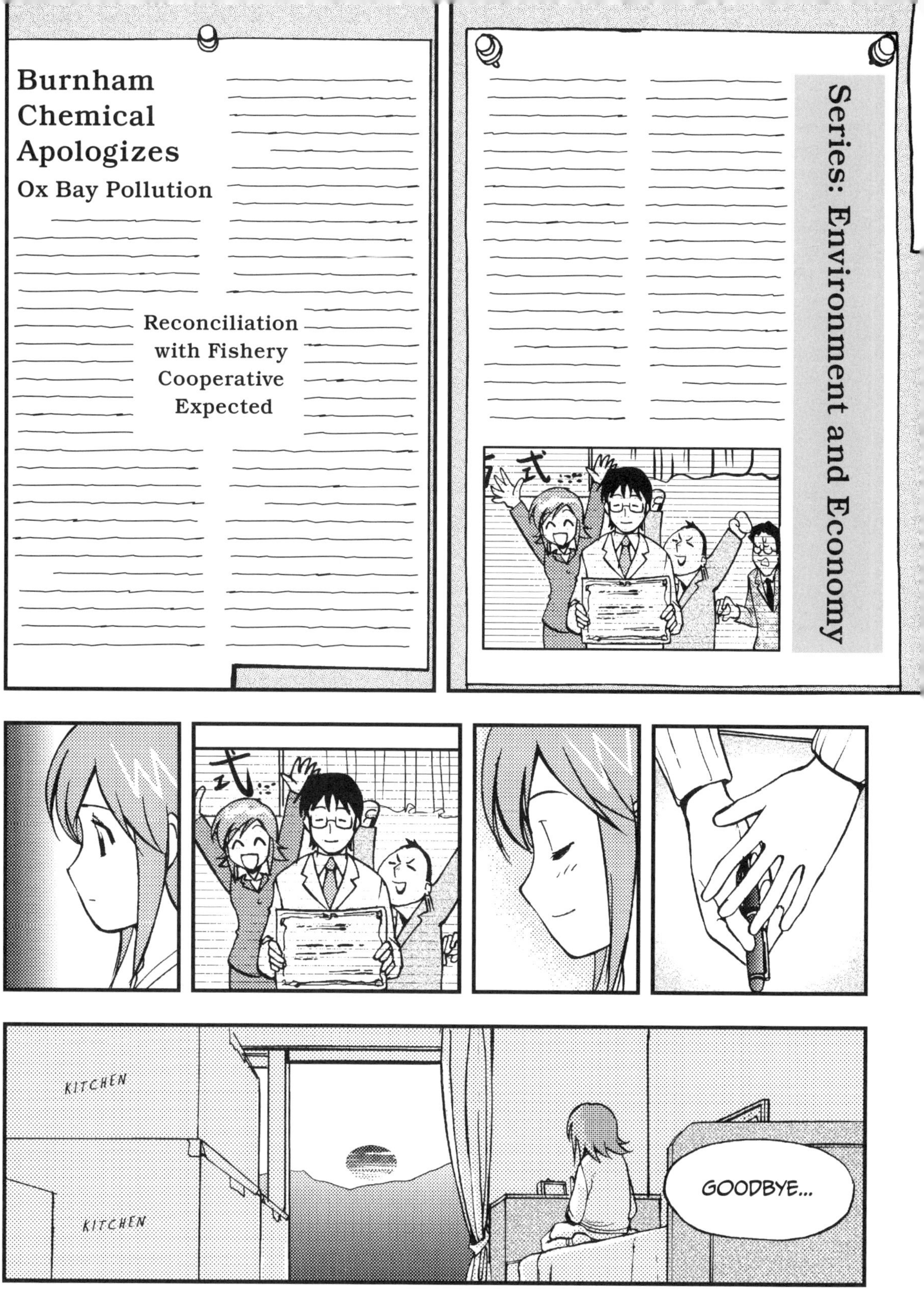

Burnham Chemical Apologizes
Ox Bay Pollution
Reconciliation with Fishery Cooperative Expected
Series: Environment and Economy
KITCHEN
KITCHEN
GOODBYE...

DERIVATIVES OF IMPLICIT FUNCTIONS

A point (x, y) for which a two-variable function $f(x, y)$ is equal to constant c describes a graph given by $f(x, y) = c$. When a part of the graph is viewed as a single-variable function $y = h(x)$, it is called an *implicit function*. An implicit function $h(x)$ satisfies $f(x, h(x)) = c$ for all x defined. We are going to obtain $h(x)$ here.

When $z = f(x, y)$, the formula of total differentials is written as $dz = f_x dx + f_y dy$. If (x, y) moves on the graph of $f(x, y) = c$, the value of the function $f(x, y)$ does not change, and the increment of z is 0, that is, $dz = 0$. Then, we get $0 = f_x dx + f_y dy$. Assuming $f_y \neq 0$ and modifying this, we get

$$\frac{dy}{dx} = -\frac{f_x}{f_y}$$

The left side of this equation is the ideal expression of the increment of y divided by the increment of x at a point on the graph. It is exactly the derivative of $h(x)$. Thus,

$$h'(x) = -\frac{f_x}{f_y}$$

EXAMPLE

$f(x, y) = r^2$, where $f(x, y) = x^2 + y^2$, describes a circle of radius r centered at the origin. Near a point that satisfies $x^2 \neq r^2$, we can solve $f(x, y) = x^2 + y^2 = r^2$ to find the implicit function $y = h(x) = r^2 - x^2$ or $y = h(x) = -\sqrt{r^2 - x^2}$. Then, from the formula, the derivative of these functions is given by

$$h'(x) = -\frac{f_x}{f_y} = -\frac{x}{y}$$

EXERCISES

1. Obtain f_x and f_y for $f(x, y) = x^2 + 2xy + 3y^2$.

2. Under the gravitational acceleration g, the period T of a pendulum having length L is given by

$$T = 2\pi\sqrt{\frac{L}{g}}$$

   (the gravitational acceleration g is known to vary depending on the height from the ground).
   Obtain the expression for total differential of T.
   If L is elongated by 1 percent and g decreases by 2 percent, about what percentage does T increase?

3. Using the chain rule, calculate the differential formula of the implicit function $h(x)$ of $f(x, y) = c$ in a different way than above.

EPILOGUE: WHAT IS MATHEMATICS FOR?

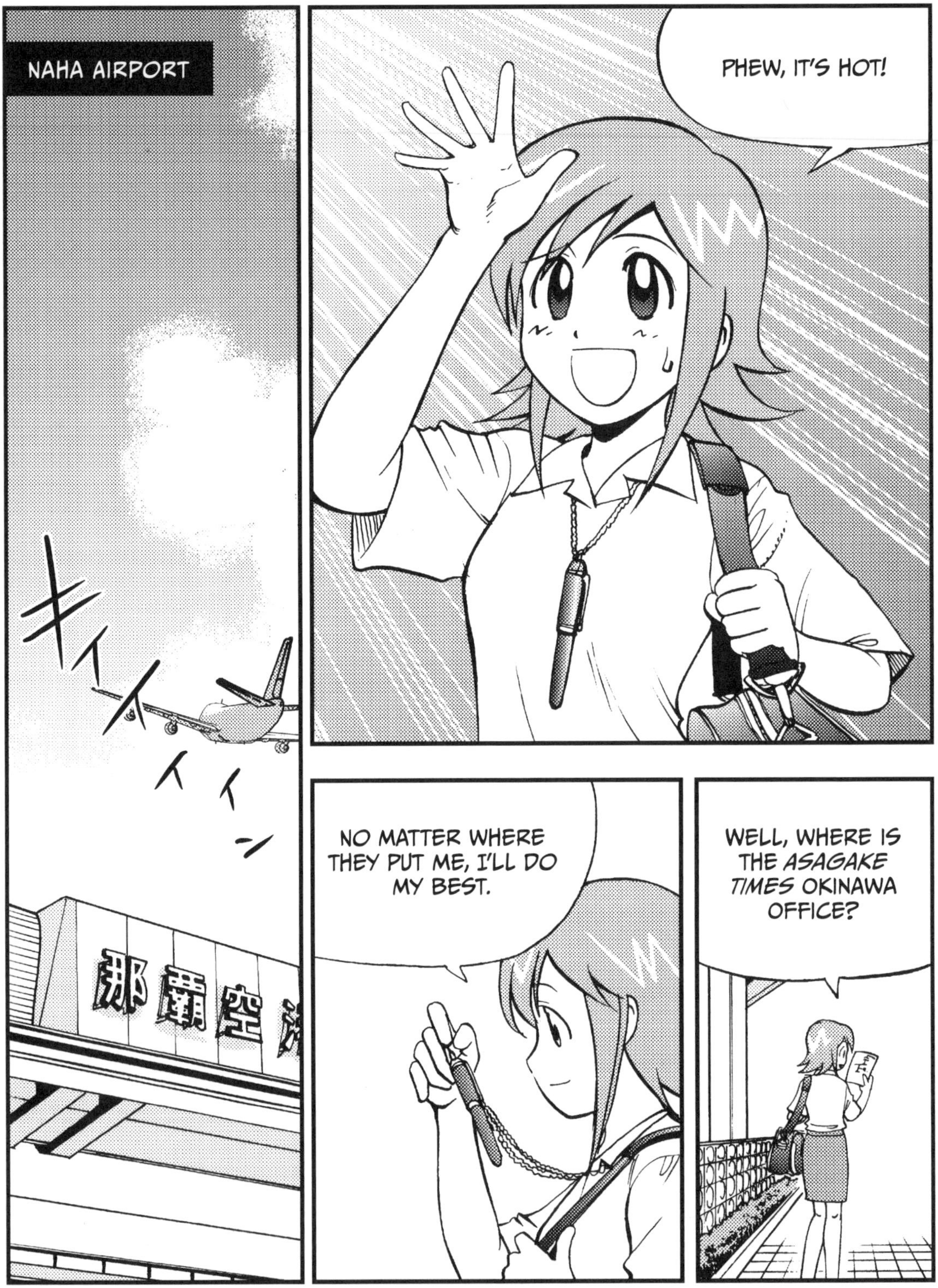
NAHA AIRPORT
キイイイン
那覇空
PHEW, IT'S HOT!
NO MATTER WHERE THEY PUT ME, I'LL DO MY BEST.
WELL, WHERE IS THE *ASAGAKE TIMES* OKINAWA OFFICE?

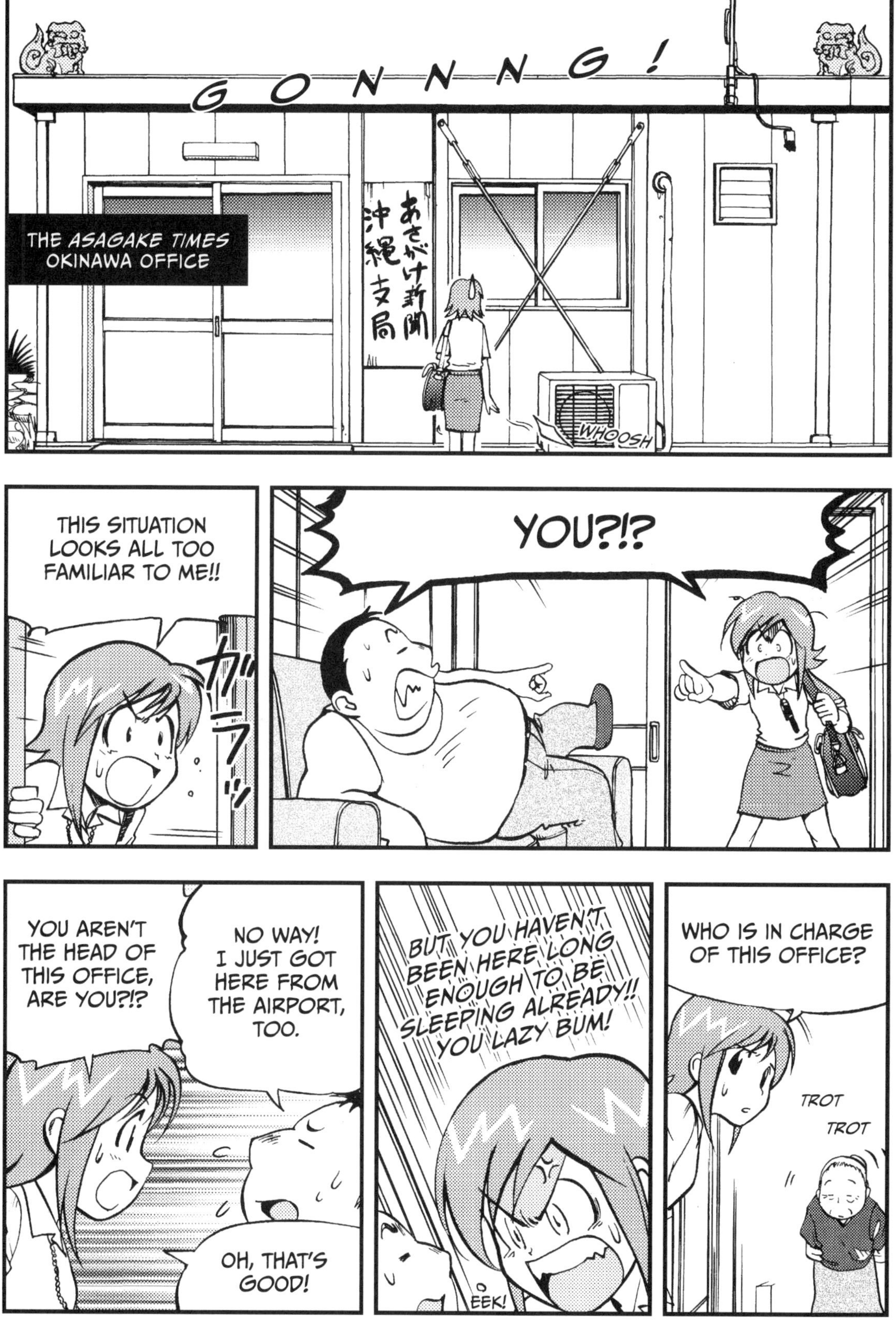
GONNNG!
THE ASAGAKE TIMES OKINAWA OFFICE
あさがけ新聞
沖縄支局
WHOOSH
THIS SITUATION LOOKS ALL TOO FAMILIAR TO ME!!
ガラッ
YOU?!?
YOU AREN'T THE HEAD OF THIS OFFICE, ARE YOU?!?
NO WAY! I JUST GOT HERE FROM THE AIRPORT, TOO.
OH, THAT'S GOOD!
BUT YOU HAVEN'T BEEN HERE LONG ENOUGH TO BE SLEEPING ALREADY!! YOU LAZY BUM!
EEK!
WHO IS IN CHARGE OF THIS OFFICE?
TROT
TROT

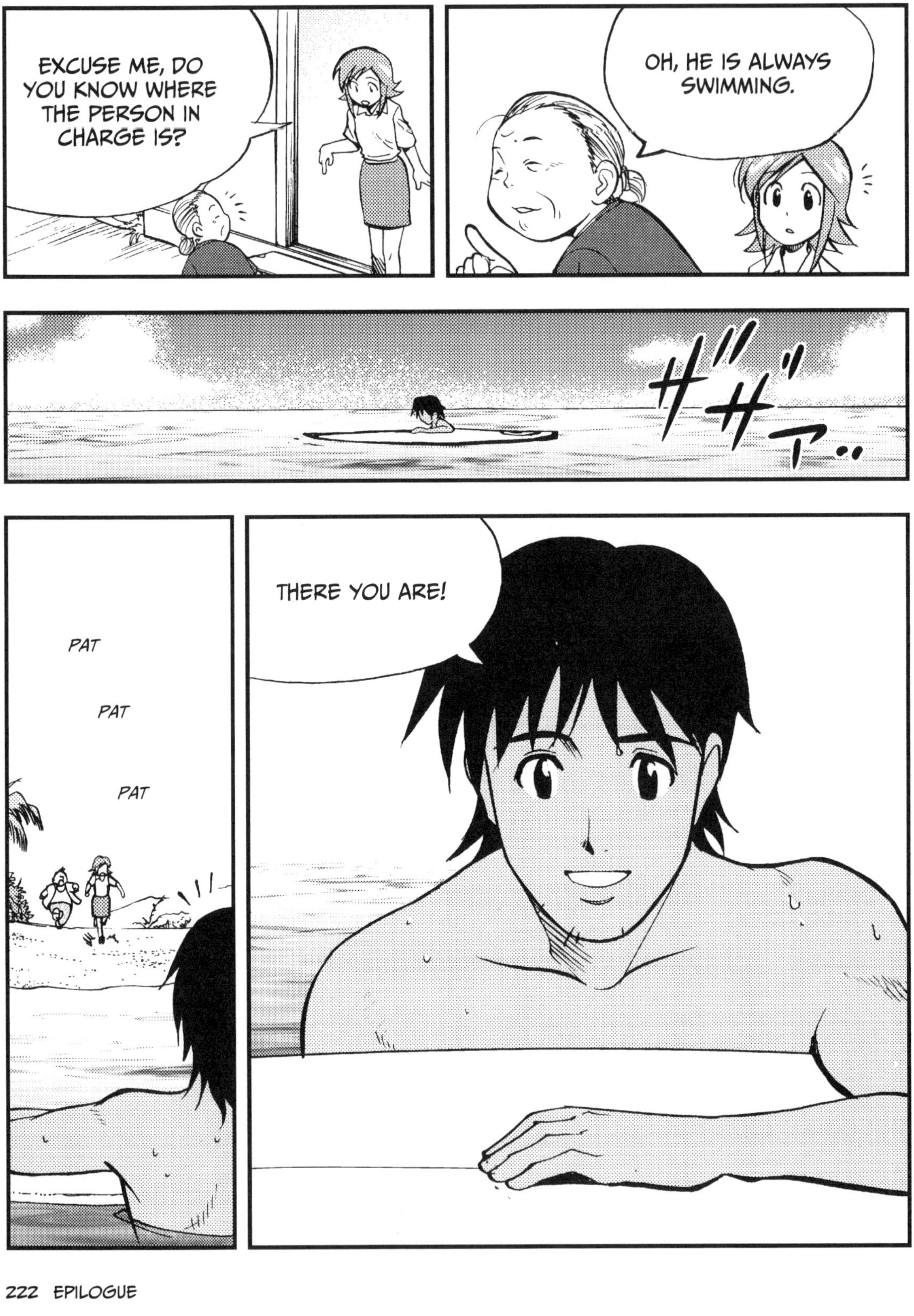
EXCUSE ME, DO YOU KNOW WHERE THE PERSON IN CHARGE IS?
OH, HE IS ALWAYS SWIMMING.
ザザァ..
PAT
PAT
PAT
THERE YOU ARE!

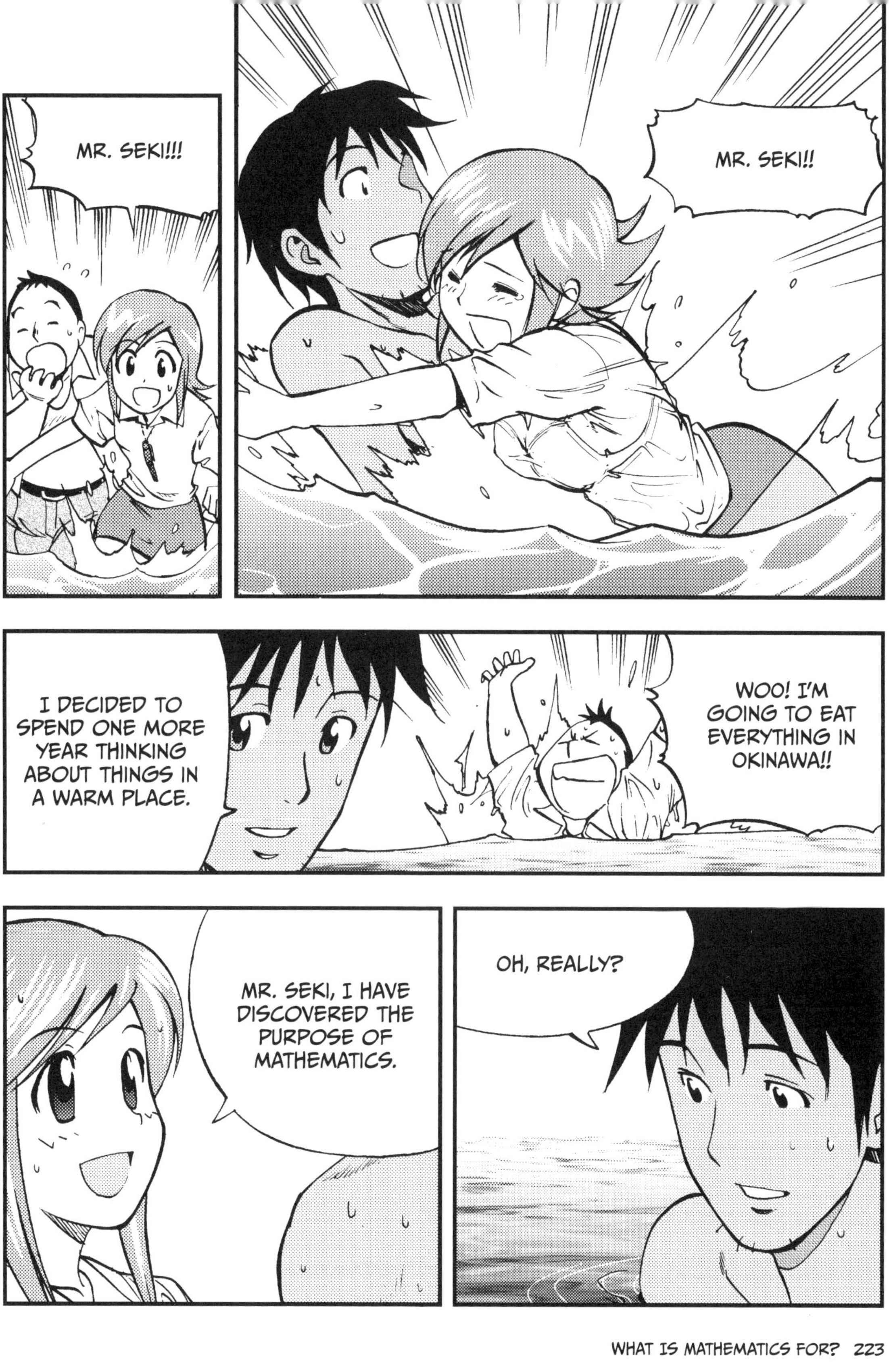

MR. SEKI!!!
MR. SEKI!!
I DECIDED TO SPEND ONE MORE YEAR THINKING ABOUT THINGS IN A WARM PLACE.
WOO! I'M GOING TO EAT EVERYTHING IN OKINAWA!!
MR. SEKI, I HAVE DISCOVERED THE PURPOSE OF MATHEMATICS.
OH, REALLY?

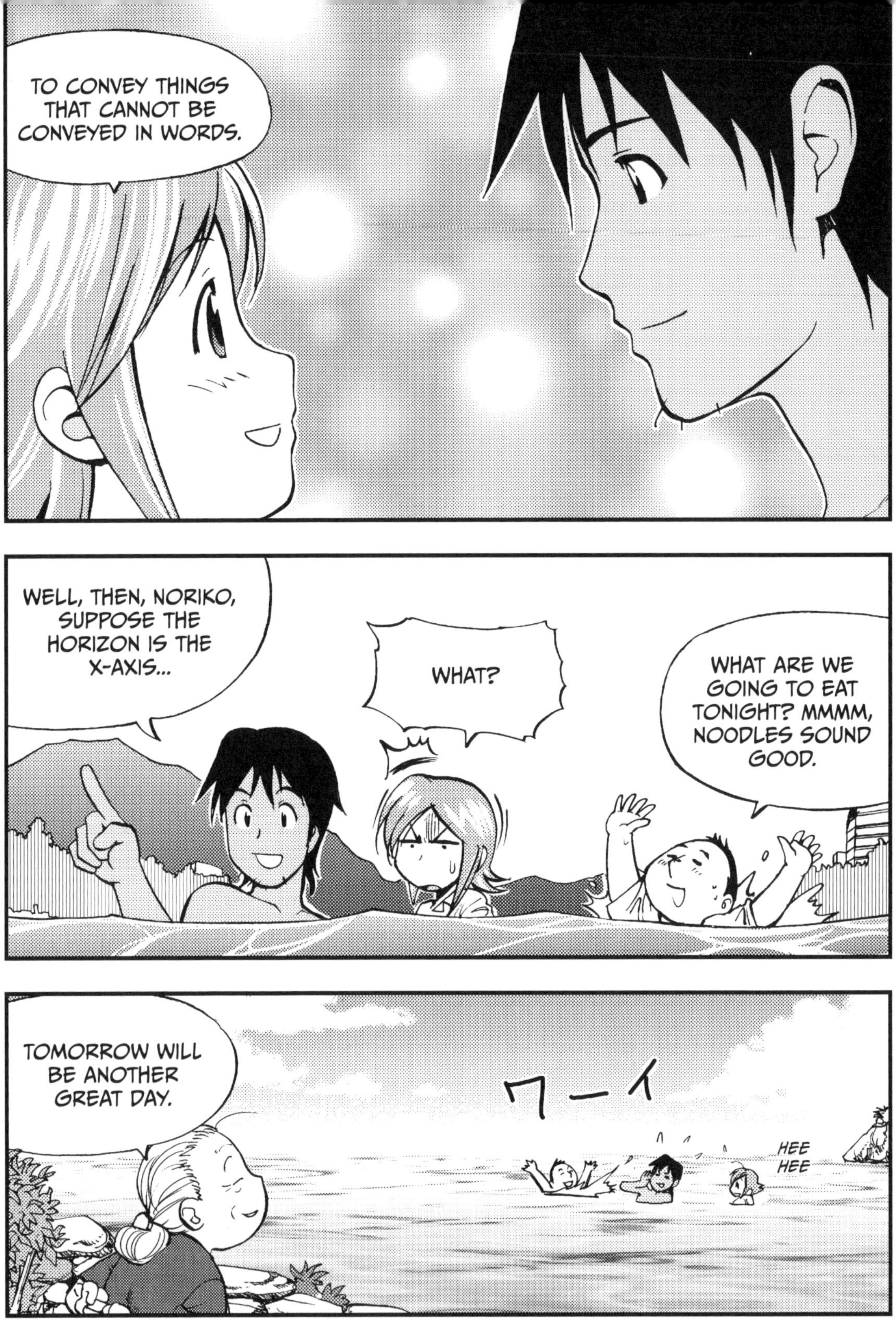
TO CONVEY THINGS THAT CANNOT BE CONVEYED IN WORDS.
WELL, THEN, NORIKO, SUPPOSE THE HORIZON IS THE X-AXIS...
WHAT?
WHAT ARE WE GOING TO EAT TONIGHT? MMMM, NOODLES SOUND GOOD.
TOMORROW WILL BE ANOTHER GREAT DAY.
ワーイ
HEE HEE

A
SOLUTIONS TO EXERCISES

PROLOGUE

1. Substituting

$$y = \frac{5}{9}(x - 32) \text{ in } z = 7y - 30, z = \frac{35}{9}(x - 32) - 30$$

CHAPTER 1

1. A. $f(5) = g(5) = 50$
   B. $f'(5) = 8$

2. $$\lim_{\varepsilon \to 0} \frac{f(a + \varepsilon) - f(a)}{\varepsilon} = \lim_{\varepsilon \to 0} \frac{(a + \varepsilon)^3 - a^3}{\varepsilon} = \lim_{\varepsilon \to 0} \frac{3a^2\varepsilon + 3a\varepsilon^2 + \varepsilon^3}{\varepsilon}$$
   $$= \lim_{\varepsilon \to 0}\left(3a^2 + 3a\varepsilon + \varepsilon^2\right) = 3a^2$$

   Thus, the derivative of $f(x)$ is $f'(x) = 3x^2$.

CHAPTER 2

1. The solution is

   $$f'(x) = -\frac{\left(x^n\right)'}{\left(x^n\right)^2} = -\frac{nx^{n-1}}{x^{2n}} = -\frac{n}{x^{n+1}}$$

2. $f'(x) = 3x^2 - 12 = 3(x-2)(x+2)$

    When $x < -2$, $f'(x) > 0$, when $-2 < x < 2$, $f'(x) < 0$, and when $x > 2$, $f'(x) > 0$. Thus at $x = -2$, we have a maximum with $f(-2) = 16$, and at $x = 2$, we have a maximum with $f(2) = -16$.

3. Since $f(x) = (1-x)^3$ is a function $g(h(x))$ combining $g(x) = x^3$ and $h(x) = 1 - x$.

$$f'(x) = g'(h(x))h'(x) = 3(1-x)^2(-1) = -3(1-x)^2$$

4. Differentiating $g(x) = x^2(1-x)^3$ gives

$$g'(x) = (x^2)'(1-x)^3 + x^2\left((1-x)^3\right)'$$
$$= 2x(1-x)^3 + x^2\left(-3(1-x)^2\right)$$
$$= x(1-x)^2(2(1-x) - 3x)$$
$$= x(1-x)^2(2 - 5x)$$
$$g'(x) = 0 \text{ when } x = \frac{2}{5} \text{ or } x = 1, \text{ and } g(1) = 0.$$

    Thus it has the maximum $g\left(\dfrac{2}{5}\right) = \dfrac{108}{3125}$ at $x = \dfrac{2}{5}$

CHAPTER 3

1. The solutions are

❶ $\displaystyle\int_1^3 3x^2 dx = x^3\Big|_1^3 = 3^3 - 1^3 = 26$

❷ $\displaystyle\int_2^4 \frac{x^3+1}{x^2}\,dx = \int_2^4\left(x + \frac{1}{x^2}\right)dx = \int_2^4 x\,dx + \int_2^4 \frac{1}{x^2}\,dx$
$$= \frac{1}{2}(4^2 - 2^2) - \left(\frac{1}{4} - \frac{2}{4}\right) = \frac{25}{4}$$

❸ $\displaystyle\int_0^5 x + (1+x^2)^7 dx + \int_0^5 x - (1+x^2)^7 dx = \int_0^5 2x\,dx = 5^2 - 0^2 = 25$

2. A. The area between the graph of $y = f(x) = x^2 - 3x$ and the x-axis equals

$$-\int_0^3 x^2 - 3x\,dx$$

B. $-\int_0^3 x^2 - 3x\,dx = -\left(\frac{1}{3}x^3 - \frac{3}{2}x^2\right)\Big|_0^3 = -\frac{1}{3}\left(3^3 - 0^3\right) + \frac{3}{2}\left(3^2 - 0^2\right) = \frac{9}{2}$

CHAPTER 4

1. The solution is

$$\left(\tan x\right)' = \left(\frac{\sin x}{\cos x}\right)' = \frac{\left(\sin x\right)' \cos x - \sin x \left(\cos x\right)'}{\cos^2 x}$$
$$= \frac{\cos^2 x + \sin^2 x}{\cos^2 x} = \frac{1}{\cos^2 x}$$

2. Since

$$\left(\tan x\right)' = \frac{1}{\cos^2 x}$$

$$\int_0^{\frac{\pi}{4}} \frac{1}{\cos^2 x}\,dx = \tan\frac{\pi}{4} - \tan 0 = 1$$

3. From

$$f'(x) = (x)' e^x + x\left(e^x\right)' = e^x + xe^x = (1 + x)e^x$$

the minimum is

$$f(-1) = -\frac{1}{e}$$

4. Setting $f(x) = x^2$ and $g(x) = \ln x$, integrate by parts.

$$\int_1^e \left(x^2\right)' \ln x\,dx + \int_1^e x^2 \left(\ln x\right)'\,dx = e^2 \ln e - \ln 1$$

Thus,

$$\int_1^e 2x \ln x \, dx + \int_1^e x^2 \frac{1}{x} dx = e^2$$

$$\int_1^e 2x \ln x \, dx = -\int_1^e x \, dx + e^2 = -\frac{1}{2}\left(e^2 - 1\right)^2 + e^2$$

$$= \frac{1}{2}e^2 + \frac{1}{2}$$

CHAPTER 5

1. For

$$f(x) = e^{-x}, \; f'(x) = -e^{-x}, \; f''(x) = e^{-x}, \; f'''(x) = -e^{-x}$$
$$f(0) = 1, \; f'(0) = -1, \; f''(0) = 1, \; f'''(0) = -1 \ldots$$
$$f(x) = 1 - x + \frac{1}{2!}x^2 - \frac{1}{3!}x^3 + \ldots$$

2. Differentiate

$$f(x) = (\cos x)^{-1}, \; f'(x) = (\cos x)^{-2} \sin x$$
$$f''(x) = 2(\cos x)^{-3}(\sin x)^2 + (\cos x)^{-2} \cos x$$
$$= 2(\cos x)^{-3}(\sin x)^2 + (\cos x)^{-1}$$

from $f(0) = 1, \; f'(0) = 0, \; f''(0) = 1$

3. Proceed in exactly the same way as on page 155 by differentiating $f(x)$ repeatedly. Since you are centering the expansion around $x = a$, plugging in a will let you work out the c_ns. You should get $c_n = 1/n! \, f^{(n)}(a)$, as shown in the formula on page 159.

CHAPTER 6

1. For $f(x, y) = x^2 + 2xy + 3y^2$, $f_x = 2x + 2y$, and $f_y = 2x + 6y$.

2. The total differential of

$$T = 2\pi\sqrt{\frac{L}{g}} = 2\pi g^{-\frac{1}{2}}L^{\frac{1}{2}}$$

is given by

$$dT = \frac{\partial T}{\partial g}dg + \frac{\partial T}{\partial L}dL = -\pi g^{-\frac{3}{2}}L^{\frac{1}{2}}dg + \pi g^{-\frac{1}{2}}L^{-\frac{1}{2}}dL$$

Thus,

$$\Delta T \approx -\pi g^{-\frac{3}{2}}L^{\frac{1}{2}}\Delta g + \pi g^{-\frac{1}{2}}L^{-\frac{1}{2}}\Delta L$$

Substituting $\Delta g = -0.02g$, $\Delta L = 0.01L$, we get

$$\Delta T \approx 0.02\pi g^{-\frac{3}{2}}L^{\frac{1}{2}}g + 0.01\pi g^{-\frac{1}{2}}L^{-\frac{1}{2}}L$$

$$= 0.03\pi g^{-\frac{1}{2}}L^{\frac{1}{2}} = 0.03\frac{T}{2} = 0.015T$$

So T increases by 1.5%.

3. If we suppose $y = h(x)$ is the implicit function of $f(x, y) = c$.
   Thus, since the left side is a constant in this region, $f(x, h(x)) = c$ near x.
   From the chain rule formula

$$\frac{df}{dx} = 0, \frac{df}{dx} = f_x + f_y h'(x) = 0$$

Therefore

$$h'(x) = -\frac{f_x}{f_y}$$

B

MAIN FORMULAS, THEOREMS, AND FUNCTIONS COVERED IN THIS BOOK

LINEAR EQUATIONS (LINEAR FUNCTIONS)

The equation of a line that has slope m and passes through a point (a, b):

$$y = m(x - a) + b$$

DIFFERENTIATION

DIFFERENTIAL COEFFICIENTS

$$f'(a) = \lim_{h \to 0} \frac{f(a + h) - f(a)}{h}$$

DERIVATIVES

$$f'(x) = \lim_{h \to 0} \frac{f(x + h) - f(x)}{h}$$

Other notations for derivatives

$$\frac{dy}{dx}, \frac{df}{dx}, \frac{d}{dx} f(x)$$

CONSTANT MULTIPLES

$$\{\alpha f(x)\}' = \alpha f'(x)$$

DERIVATIVES OF NTH-DEGREE FUNCTIONS

$$\{x^n\}' = nx^{n-1}$$

SUM RULE OF DIFFERENTIATION

$$\{f(x) + g(x)\}' = f'(x) + g'(x)$$

PRODUCT RULE OF DIFFERENTIATION

$$\{f(x) g(x)\}' = f'(x) g(x) + f(x) g'(x)$$

QUOTIENT RULE OF DIFFERENTIATION

$$\left\{\frac{g(x)}{f(x)}\right\}' = \frac{g'(x) f(x) - g(x) f'(x)}{\{f(x)\}^2}$$

DERIVATIVES OF COMPOSITE FUNCTIONS

$$\{g(f(x))\}' = g'(f(x)) f'(x)$$

DERIVATIVES OF INVERSE FUNCTIONS
When $y = f(x)$ and $x = g(y)$

$$g'(y) = \frac{1}{f'(x)}$$

EXTREMA
If $y = f(x)$ has a maximum or a minimum at $x = a$, $f'(a) = 0$.
$y = f(x)$ is increasing around $x = a$, if $f'(a) > 0$.
$y = f(x)$ is decreasing around $x = a$, if $f'(a) < 0$.

THE MEAN VALUE THEOREM
For a, b $(a < b)$, there is a c with $a < c < b$, so that

$$f(b) = f'(c)(b - a) + f(a)$$

DERIVATIVES OF POPULAR FUNCTIONS

TRIGONOMETRIC FUNCTIONS

$$\{\cos\theta\}' = -\sin\theta, \{\sin\theta\}' = \cos\theta$$

EXPONENTIAL FUNCTIONS

$$\{e^x\}' = e^x$$

LOGARITHMIC FUNCTIONS

$$\{\log x\}' = \frac{1}{x}$$

INTEGRALS

DEFINITE INTEGRALS

When $F'(x) = f(x)$

$$\int_a^b f(x)\,dx = F(b) - F(a)$$

CONNECTION OF INTERVALS OF DEFINITE INTEGRALS

$$\int_a^b f(x)\,dx + \int_b^c f(x)\,dx = \int_a^c f(x)\,dx$$

SUM OF DEFINITE INTEGRALS

$$\int_a^b \{f(x) + g(x)\}\,dx = \int_a^b f(x)\,dx + \int_a^b g(x)\,dx$$

CONSTANT MULTIPLES OF DEFINITE INTEGRALS

$$\int_a^b \alpha f(x)\,dx = \alpha \int_a^b f(x)\,dx$$

SUBSTITUTION OF INTEGRALS

When $x = g(y)$, $b = g(\beta)$, $a = g(\alpha)$

$$\int_a^b f(x)\,dx = \int_\alpha^\beta f(g(y))\,g'(y)\,dy$$

INTEGRATION BY PARTS

$$\int_a^b f'(x)\,g(x)\,dx + \int_a^b f(x)\,g'(x)\,dx = f(b)\,g(b) - f(a)\,g(a)$$

TAYLOR EXPANSION

When $f(x)$ has a Taylor expansion near $x = a$,

$$f(x) = f(a) + \frac{1}{1!}f'(a)(x-a) + \frac{1}{2!}f''(a)(x-a)^2$$
$$+ \frac{1}{3!}f'''(a)(x-a)^3 + \ldots + \frac{1}{n!}f^{(n)}(a)(x-a)^{(n)} + \ldots$$

TAYLOR EXPANSIONS OF VARIOUS FUNCTIONS

$$\cos x = 1 - \frac{1}{2!}x^2 + \frac{1}{4!}x^4 + \ldots + (-1)^n \frac{1}{(2n)!}x^{2n} + \ldots$$

$$\sin x = x - \frac{1}{3!}x^3 + \frac{1}{5!}x^5 + \ldots + (-1)^{n-1} \frac{1}{(2n-1)!}x^{2n-1} + \ldots$$

$$e^x = 1 + \frac{1}{1!}x + \frac{1}{2!}x^2 + \frac{1}{3!}x^3 + \frac{1}{4!}x^4 + \ldots + \frac{1}{n!}x^n + \ldots$$

$$\ln(1+x) = x - \frac{1}{2}x^2 + \frac{1}{3}x^3 - \frac{1}{4}x^4 + \ldots + (-1)^{n+1}\frac{1}{n}x^n + \ldots$$

PARTIAL DERIVATIVES

PARTIAL DERIVATIVES

$$\frac{\partial f}{\partial x} = \lim_{h \to 0} \frac{f(x+h,y) - f(x,y)}{h}$$

$$\frac{\partial f}{\partial y} = \lim_{k \to 0} \frac{f(x,y+k) - f(x,y)}{k}$$

TOTAL DIFFERENTIALS

$$dz = \frac{\partial z}{\partial x}dx + \frac{\partial z}{\partial y}dy$$

FORMULA OF CHAIN RULE
When $z = f(x, y)$, $x = a(t)$, $y = b(t)$

$$\frac{dz}{dt} = \frac{\partial f}{\partial x}\frac{da}{dt} + \frac{\partial f}{\partial y}\frac{db}{dt}$$

INDEX

Q

R

S

T

U

V

X

DIFFERENTIAL AND INTEGRAL CALCULUS DANCE
SONG FOR TRIGONOMETRIC FUNCTIONS

IT SOUNDS QUITE BOOKISH, INDEED. BUT THIS DANCE SONG MAKES IT SO EASY!

DIFFERENTIATE SINE AND YOU GET COSINE!

IN THE LOOP OF SINE AND COSINE, IT'S SO NATURAL—SOLUTIONS TAKE TURNS FOR THE DIFFERENTIAL AND INTEGRAL.

IN THE LOOP OF SINE AND COSINE, DIFFERENTIAL AND INTEGRAL CALCULUS SKILLS ARE MINE.

INTEGRATE COSINE AND YOU GET SINE!

ABOUT THE AUTHOR

Hiroyuki Kojima was born in 1958. He received his PhD in Economics from the Graduate School of Economics, Faculty of Economics, at the University of Tokyo. He has worked as a lecturer and is now an associate professor in the Faculty of Economics at Teikyo University in Tokyo, Japan. While highly praised as an economist, he is also active as an essayist and has published a wide range of books on mathematics and economics at the fundamental, practical, and academic levels.

PRODUCTION TEAM FOR THE JAPANESE EDITION

PRODUCTION: BECOM CO., LTD.

Since its founding in 1998 as an editorial and design production studio, Becom has produced many books and magazines in the fields of medicine, education, and communication. In 2001, Becom established a team of comic designers, and the company has been actively involved in many projects using manga, such as corporate guides, product manuals, as well as manga books. More information about the company is available at its website, *http://www.becom.jp/.*

Yurin Bldg 5F, 2-40-7 Kanda-Jinbocho, Chiyoda-ku, Tokyo, Japan 101-0051
Tel: 03-3262-1161; Fax: 03-3262-1162

SCENARIO WRITERS: SHINJIRO NISHIDA AND EIJI SHIMADA

ILLUSTRATOR: SHIN TOGAMI

COVER DESIGN: YUUKI ITAMI

COLOPHON

The Manga Guide to Calculus was laid out in Adobe InDesign. The fonts are CCMeanwhile and Bookman.

The book was printed and bound at Malloy Incorporated in Ann Arbor, Michigan. The paper is Glatfelter Spring Forge 60# Eggshell, which is certified by the Sustainable Forestry Initiative (SFI).